T0336901

The Logic in Philosophy of Science

Major figures of twentieth-century philosophy were enthralled by the revolution in formal logic, and many of their arguments are based on novel mathematical discoveries. Hilary Putnam claimed that the Löwenheim–Skølem theorem refutes the existence of an objective, observer-independent world; Bas van Fraassen claimed that arguments against empiricism in philosophy of science are ineffective against a semantic approach to scientific theories; W. v. O. Quine claimed that the distinction between analytic and synthetic truths is trivialized by the fact that any theory can be reduced to one in which all truths are analytic. This book dissects these and other arguments through in-depth investigation of the mathematical facts undergirding them. It presents a systematic, mathematically rigorous account of the key notions arising from such debates, including theory, equivalence, translation, reduction, and model. The result is a far-reaching reconceptualization of the role of formal methods in answering philosophical questions.

Hans Halvorson is Stuart Professor of Philosophy at Princeton University, New Jersey. He has written extensively on philosophical issues in physics and the other sciences, on mathematical logic, and on the relationship between science and religion.

The Logic in Philosophy of Science

HANS HALVORSON
Princeton University

CAMBRIDGE
UNIVERSITY PRESS

University Printing House, Cambridge CB2 8BS, United Kingdom

One Liberty Plaza, 20th Floor, New York, NY 10006, USA

477 Williamstown Road, Port Melbourne, VIC 3207, Australia

314-321, 3rd Floor, Plot 3, Splendor Forum, Jasola District Centre, New Delhi - 110025, India

79 Anson Road, #06-04/06, Singapore 079906

Cambridge University Press is part of the University of Cambridge.

It furthers the University's mission by disseminating knowledge in the pursuit of education, learning and research at the highest international levels of excellence.

www.cambridge.org
Information on this title: www.cambridge.org/9781107110991
DOI: 10.1017/9781316275603

First published 2019

A catalogue record for this publication is available from the British Library

Library of Congress Cataloging in Publication data
Names: Halvorson, Hans, author.
Title: The logic in philosophy of science / Hans Halvorson (Princeton University, New Jersey).
Description: Cambridge ; New York, NY : Cambridge University Press, 2019. |
Includes bibliographical references and index.
Identifiers: LCCN 2018061724 | ISBN 9781107110991 (hardback : alk. paper) |
ISBN 9781107527744 (pbk. : alk. paper)
Subjects: LCSH: Science–Philosophy.
Classification: LCC Q175 .H2475 2019 | DDC 501–dc23
LC record available at https://lccn.loc.gov/2018061724

ISBN 978-1-107-11099-1 Hardback
ISBN 978-1-107-52774-4 Paperback

Contents

Preface

The twentieth century's most interesting philosophers were enthralled by the revolution in mathematical logic, and they accordingly clothed many of their arguments in a formal garb. For example, Hilary Putnam claimed that the Löwenheim-Skølem theorem reduces metaphysical realism to absurdity; Bas van Fraassen claimed that arguments against empiricism presuppose the syntactic view of theories; and W. v. O. Quine claimed that Carnap's notion of an "external question" falls apart because every many-sorted theory is equivalent to a single-sorted theory. These are only a few of the many arguments of twentieth-century philosophy that hinge upon some or other metalogical theorem.

Lack of understanding of the logical theorems can be a huge obstacle to assessing these philosophers' arguments, and this book is my attempt to help remove that obstacle. However, my ideal reader is not the casual tourist of twentieth-century philosophy who wants the minimal amount of logic needed to get the big picture. My ideal reader is the (aspiring) logician-philosopher who wants to strip these arguments down to their logical nuts and bolts.

Although my motivation for writing this book wasn't to get across some particular philosophical point, a few such points emerged along the way. First, the distinction between realism and antirealism really boils down to one's attitude toward theoretical equivalence. Realists are people with a conservative notion of equivalence, and antirealists are people with a liberal notion of equivalence. Second, and relatedly, to give a philosophical account of a relation between theories (e.g., equivalence and reducibility) is tantamount to recommending certain norms of inquiry. For example, if you say that two theories T and T' are equivalent, then you mean (among other things) that any reason for accepting T is also a reason for accepting T'. Hence, you won't bother trying to design an experiment that would test T against T'. Similarly, if you think that T and T' are equivalent, then you'll consider as confused anyone who argues about which of the two is better. In short, to adopt a view on relations between theories is to adopt certain rules about how to use those theories.

I should explain one glaring omission from this book: modal logic. I didn't leave out modal logic because I'm a Quinean extensionalist. To the contrary, I've come to think that the metatheory of first-order logic, and of scientific theories more generally, is chock full of intensional concepts. For example, the models of a scientific theory represent the nomologically possible worlds according to the theory. Furthermore, a scientific theory comes equipped with a notion of "natural property" (in the sense of David Lewis), and these natural properties determine a notion of similarity between possible worlds, which in turn licenses certain counterfactual inferences. So, while my goal is to theorize about the extensional logic that forms the backbone of the sciences, I believe that doing so calls for the use of intensional concepts.

A final note on how to read this book: Chapters 1–3 are introductory but are not strictly prerequisite for the subsequent chapters. Chapters 1 and 3 treat the metatheory of propositional logic, teaching some Boolean algebra and topology along the way. In Chapter 3, we go through the proof of the Stone duality theorem, because it exemplifies the duality between syntax and semantics that informs the remaining chapters. Chapter 2 covers the basics of both category and set theory in one go, and it's the most technically demanding (and least philosophical) chapter of the book. You don't have to know categorical set theory in order to benefit from the other chapters, it would be enough to know some set theory (e.g., Halmos' *Naive Set Theory*) and to flip back occasionally to look up category-theoretic concepts.

Acknowledgements: Thanks to Bas van Fraassen for the inspiration to pursue philosophy of science both as a science and as an art.

The idea behind this book arose during a year I spent in Utrecht studying category theory. I thank the Mellon New Directions Fellowship for financing that year. Thanks to my Dutch hosts (Klaas Landsman, Ieke Moerdijk, and Jaap van Oosten) for their warm hospitality.

When I returned home, I rediscovered that it's difficult to do two (or fifty) things at once. The project might have foundered, had it not been for the theorem-proving wizardry of Thomas Barrett, Neil Dewar, Dimitris Tsementzis, and Evan Washington. I also found my philosophical views shaped and sharpened by conversations with several students and colleagues, especially John Burgess, Ellie Cohen, Robbie Hirsch, Laurenz Hudetz, Michaela McSweeney, Alex Meehan, Gideon Rosen, Elliot Salinger, David Schroeren, and Jim Weatherall. I probably left somebody out, and I'm sorry about that. For comments and corrections on earlier versions of the manuscript, I thank Thomas Barrett, Gordon Belot, Neil Dewar, Harvey Lederman, Dimitris Tsementzis, Jim Weatherall and Isaac Wilhelm.

Finally, thank you to Hilary and Sophie at CUP for their initial belief in the project and for persevering with me to the end.

Introduction

A New Kind of Philosophy

Some people think that philosophy never makes progress. In fact, professional philosophers might think that more frequently – and feel it more acutely – than anyone else. At the beginning of the twentieth century, some philosophers were so deeply troubled that they decided to cast all previous philosophy on the scrap heap and to rebuild from scratch. "Why shouldn't philosophy be like science?" they asked. "Why can't it also make genuine progress?"

Now, you might guess that these philosophers would have located philosophy's problems in its lack of empirical data and experiments. One advantage of the empirical sciences is that bad ideas (such as "leeches cure disease") can be falsified through experiments. However, this wasn't the diagnosis of the first philosophers of science; they didn't see empirical testability as the sine qua non of a progressive science. Their guiding light was not the empirical sciences, but mathematics, and mathematical physics.

The nineteenth century had been a time of enormous progress in mathematics, not only in answering old questions and extending applications, but but also in clarifying and strengthening the foundations of the discipline. For example, George Boole had clarified the structure of logical relations between propositions, and Georg Cantor had given a precise account of the concept of "infinity," thereby setting the stage for the development of the new mathematical theory of sets. The logician Gottlob Frege had proposed a new kind of symbolic logic that gave a precise account of all the valid argument forms in mathematics. And the great German mathematician David Hilbert, building on a rich tradition of analytic geometry, proposed an overarching axiomatic method in which all mathematical terminology is "de-interpreted" so that the correctness of proofs is judged on the basis of purely formal criteria.

For a younger generation of thinkers, there was a stark contrast between the ever more murky terminology of speculative philosophy and the rising standards of clarity and rigor in mathematics. "What is the magic that these mathematicians have found?" asked some philosophically inclined scientists at the beginning of the twentieth century. "How is it that mathematicians have a firm grip on concepts such as 'infinity' and 'continuous function,' while speculative philosophers continue talking in circles?" It was time, according to this new generation, to rethink the methods of philosophy as an academic discipline.

The first person to propose that philosophy be recreated in the image of nineteenth-century mathematics was Bertrand Russell. And Russell was not at all modest in what he thought this new philosophical method could accomplish. Indeed, Russell cast himself as a direct competitor with the great speculative philosophers, most notably with Hegel. That is, Russell thought that, with the aid of the new symbolic logic, he could describe the fundamental structure of reality more clearly and accurately than Hegel himself did. Indeed, Russell's "logical atomism" was intended as a replacement for Hegel's monistic idealism.

Russell's grand metaphysical ambitions were cast upon the rocks by his student Ludwig Wittgenstein. In essence, Wittgenstein's *Tractatus Logico Philosophicus* was intended to serve as a *reductio ad absurdum* of the idea that the language of mathematical logic is suited to mirror the structure of reality in itself. To the extent that Russell himself accepted Wittgenstein's rebuke, this first engagement of philosophy and mathematical logic came to an unhappy end. In order for philosophy to become wedded to mathematical logic, it took a distinct second movement, this time involving a renunciation of the ambitions of traditional speculative metaphysics. This second movement proposed not only a new method of philosophical inquiry but also a fundamental reconstrual of its aims.

As mentioned before, the nineteenth century was a golden age for mathematics in the broad sense, and that included mathematical physics. Throughout the century, Newtonian physics has been successfully extended to describe systems that had not originally been thought to lie within its scope. For example, prior to the late nineteenth century, changes in temperature had been described by the science of thermodynamics, which describes heat as a sort of continuous substance that flows from one body to another. But then it was shown that the predictions of thermodynamics could be reproduced by assuming that these bodies are made of numerous tiny particles obeying the laws of Newtonian mechanics. This reduction of thermodynamics to statistical mechanics led to much philosophical debate over the existence of unobservable entities, e.g., tiny particles (atoms) whose movement is supposed to explain macroscopic phenomena such as heat. Leading scientists such as Boltzmann, Mach, Planck, and Poincaré sometimes took opposing stances on these questions, and it led to more general reflection on the nature and scope of scientific knowledge.

These scientists couldn't have predicted what would happen to physics at the beginning of the twentieth century. The years 1905–1915 saw no fewer than three major upheavals in physics. These upheavals began with Einstein's publication of his special theory of relativity, and continued with Bohr's quantum model of the hydrogen atom, and then Einstein's general theory of relativity. If anything became obvious through these revolutions, it was that we didn't understand the nature of science as well as we thought we did. We had believed we understood how science worked, but people like Einstein and Bohr were changing the rules of the game. It was high time to reflect on the nature of the scientific enterprise as a whole.

The new theories in physics also raised further questions, specifically about the role of mathematics in physical science. All three of the new theories – special and general relativity, along with quantum theory – used highly abstract mathematical notions, the likes of which physicists had not used before. Even special relativity, the most intuitive of the three theories, uses four-dimensional geometry and a notion of "distance" that takes both positive and negative values. Things only got worse when, in the 1920s, Heisenberg proposed that the new quantum theory make use of non-commutative algebras that had no intuitive connection whatsoever to things happening in the physical world.

The scientists of the early twentieth century were decidedly philosophical in outlook. Indeed, reading the reflections of the young Einstein or Bohr, one realizes that the distinction between "scientist" and "philosopher" had not yet been drawn as sharply as it is today. Nonetheless, despite their philosophical proclivities, Einstein, Bohr, and the other scientific greats were not philosophical system builders, if only because they were too busy publicizing their theories and then working for world peace. Thus, the job of "making sense of how science works" was left to some people who we now consider to be philosophers of science.

If we were to call anybody the first "philosopher of science" in the modern sense of the term, then it should probably be **Moritz Schlick** (1882–1936). Schlick earned his PhD in physics at Berlin under the supervision of Max Planck and thereafter began studying philosophy. During the 1910s, Schlick became one of the first philosophical interpreters of Einstein's new theories, and in doing so, he developed a distinctive view in opposition to Marburg neo-Kantianism. In 1922, Schlick was appointed chair of *Naturphilosophie* in Vienna, a post that had earlier been held by Boltzmann and then by Mach.

When Schlick formulated his epistemological theories, he did so in a conscious attempt to accommodate the newest discoveries in mathematics and physics. With particular reference to mathematical knowledge, Schlick followed nineteenth-century mathematicians – most notably Pasch and Hilbert – in saying that mathematical claims are true by definition and that the words that occur in the axioms are thereby implicitly defined. In short, those words have no meaning beyond that which accrues to them by their role in the axioms.

While Schlick was planting the roots of philosophy of science in Vienna, the young **Hans Reichenbach** (1891–1953) had found a way to combine the study of philosophy, physics, and mathematics by moving around between Berlin, Göttingen, and Munich – where he studied philosophy with Cassirer, physics with Einstein, Planck, and Sommerfeld; and mathematics with Hilbert and Noether. He struggled at first to find a suitable academic post, but eventually Reichenbach was appointed at Berlin in 1926. It was in Berlin that Reichenbach took on a student named Carl Hempel (1905–1997), who would later bring this new philosophical approach to the elite universities in the United States. Hempel's students include several of the major players in twentieth-century philosophy of science, such as Adolf Grünbaum, John Earman, and Larry Sklar. Reichenbach himself eventually relocated to UCLA, where he had two additional students of no little renown: Wesley Salmon and Hilary Putnam.

However, back in the 1920s, shortly before he took the post at Berlin, Reichenbach had another auspicious meeting at a philosophy conference in Erlangen. Here he met a young man named Rudolf Carnap who, like Reichenbach, found himself poised at the intersection of philosophy, physics, and mathematics. Reichenbach introduced Carnap to his friend Schlick, the latter of whom took an eager interest in Carnap's ambition to develop a "scientific philosophy." A couple of short years later, Carnap was appointed assistant professor of philosophy in Vienna – and so began the marriage between mathematical logic and philosophy of science.

Carnap

Having been a student of Frege's in Jena, Rudolf Carnap (1891–1970) was an early adopter of the new logical methods. He set to work immediately trying to employ these methods in the service of a new style of philosophical inquiry. His first major work – *Der Logische Aufbau der Welt* (1928) – attempted the ultra-ambitious project of constructing all scientific concepts out of primitive (fundamental) concepts. What is especially notable for our purposes was the notion of *construction* that Carnap employed, for it was a nearby relative to the notion of *logical construction* that Russell had employed, and which descends from the mathematician's idea that one kind of mathematical object (e.g., real numbers) can be constructed from another kind of mathematical object (e.g., natural numbers). What's also interesting is that Carnap takes over the idea of *explication*, which arose in mathematical contexts – e.g., when one says that a function f is "continuous" just in case for each $\epsilon > 0$, there is a $\delta > 0$ such that ...

When assessing philosophical developments such as these, which are so closely tied to developments in the exact sciences, we should keep in mind that ideas that are now clear to us might have been quite opaque to our philosophical forebears. For example, these days we know quite clearly what it means to say that a theory T is complete. But to someone like Carnap in the 1920s, the notion of completeness was vague and hazy, and he struggled to integrate it into his philosophical thoughts. We should keep this point in mind as we look toward the next stage of Carnap's development, where he attempted a purely "syntactic" analysis of the concepts of science.

In the late 1920s, the student Kurt Gödel (1906–1978) joined in the discussions of the Vienna circle, and Carnap later credited Gödel's influence for turning his interest to questions about the language of science. Gödel gave the first proof of the completeness of the predicate calculus in his doctoral dissertation (1929), and two years later, he obtained his famous incompleteness theorem, which shows that there is some truth of arithmetic that cannot be derived from the first-order Peano axioms.

In proving incompleteness, Gödel's technique was "metamathematical" – i.e., he employed a theory M *about* the first-order theory T of arithmetic. Moreover, this metatheory M employed purely syntactic concepts – e.g., the length of a string of symbols, or the number of left parentheses in a string, or being the last formula in a valid proof that begins from the axioms of arithmetic. This sort of approach proved to be fascinating for Carnap, in particular, because it transformed questions that seemed hopelessly vague and "philosophical" into questions that were tractable – and indeed tractable by means of the very methods that scientists themselves employed. In short, Gödel's approach indicated the possibility of an exact science of the exact sciences.

And yet, Gödel's inquiry was restricted to one little corner of the exact sciences: arithmetic. Carnap's ambitions went far beyond elementary mathematics; he aspired to apply these new methods to the entire range of scientific theories, and especially the new theories of physics. Nonetheless, Carnap quickly realized that he faced additional problems beyond those faced by the metamathematician, for scientific theories – unlike their mathematical cousins – purport to say something *contingently true* – i.e., something that could have been otherwise. Hence, the logical approach to philosophy of science isn't finished when one has analyzed a theory T qua mathematical object; one must also say something about how T latches on to empirical reality.

Carnap's first attempts in this direction were a bit clumsy, as he himself recognized. In the 1920s and 1930s, philosophers of science were just learning the basics of formal logic. It would take another forty years until "model theory" was a well-established discipline, and the development of mathematical logic continues today (as we hope to make clear in this book). However, when mathematical logic was still in its infancy, philosophers often tried the "most obvious" solution to their problems – not realizing that it couldn't stand up to scrutiny. Consider, for example, Carnap's attempt to specify the empirical content of a theory T. Carnap proposes that the vocabulary Σ in which a theory T is formulated must include an empirical subvocabulary $O \subseteq \Sigma$, in which case the empirical content of T can be identified with the set $T|_O$ of consequences of T restricted to the vocabulary O. Similarly, in attempting to cash out the notion of "reduction" of one theory to another, Carnap initially said that the concepts of the reduced theory needed to be explicitly defined in terms of the concepts of the reducing theory – not realizing that he was thereby committing to a far more narrow notion of reduction than was being used in the sciences.

In Carnap's various works, however, we do find the beginnings of an approach that is still relevant today. Carnap takes a "language" and a "theory" to be objects of his inquiries, and he notes explicitly that there are choices to be made along the way. So, for example, the classical mathematician chooses a certain language and then adopts certain transformation rules. In contrast, the intuitionistic mathematician chooses a different language and adopts different transformation rules. Thus, Carnap allows himself to ascend semantically – to look at scientific theories from the outside, as it were. From this vantage point, he is no longer asking the "internal questions" that the theorist herself is asking. He is not asking, for example, whether there is a greatest prime number. Instead, the philosopher of science is raising "external questions" – i.e., questions about the theory T, and especially those questions that have precise syntactic formulations. For example, Carnap proposes that the notion of a sentence's being "analytic relative to T" is an external notion that we metatheorists use to describe the structure of T.

The twentieth-century concern with analytic truth didn't arise in the seminar rooms of philosophy departments – or at least not in philosophy departments like the ones of today. In fact, this concern began rather with nineteenth-century geometers, faced with two parallel developments: (1) the discovery of non-Euclidean geometries, and (2) the need to raise the level of rigor in mathematical arguments. Together, these two developments led mathematical language to be disconnected from the physical world. In other words, one key outcome of the development of modern mathematics was the *de-interpretation* of mathematical terms such as "number" or "line." These terms were replaced by symbols that bore no intuitive connection to external reality.

It was this de-interpretation of mathematical terms that gave rise to the idea that analytic truth is *truth by postulation*, the very idea that was so troubling to Russell, and then to Quine. But in the middle of the nineteenth century, the move that Russell called "theft" enabled mathematicians to proceed with their investigations in absence of the fear that they lacked insight into the meanings of words such as "line" or "continuous function." In their view, it didn't matter what words you used, so long as you clearly explained the rules that governed their use. Accordingly, for leading mathematicians such as Hilbert, mathematical terms such as "line" mean nothing more nor less than what axioms say of them, and it's simply impossible to write down false mathematical postulates. There is no external standard against which to measure the truth of these postulates.

It's against this backdrop that Carnap developed his notion of analytic truth in a framework; and that Quine later developed his powerful critique of the analytic–synthetic distinction. However, Carnap and Quine were men of their time, and their thoughts operated at the level of abstraction that science had reached in the 1930s. The notion of logical metatheory was still in its infancy, and it had hardly dawned on logicians that "frameworks" or "theories" could themselves be treated as objects of investigation.

Quine

If one was a philosophy student in the late twentieth century, then one learned that Quine "demolished" logical positivism. In fact, the errors of positivism were used as classroom lessons in how not to commit the most obvious philosophical blunders. How silly to state a view that, if true, entails that one cannot justifiably believe it!

During his years as an undergraduate student at Oberlin, **Willard van Orman Quine** (1908–2000) had become entranced with Russell's mathematical logic. After getting his PhD from Harvard in 1932, Quine made a beeline for Vienna just at the time that Carnap was setting his "logic of science" program into motion. Quine quickly became Carnap's strongest critic. As the story is often told, Quine was single-handedly responsible for the demise of Carnap's program, and of logical positivism more generally.

Of course, Quine was massively influential in twentieth-century philosophy – not only for the views he held, but also via the methods he used for arriving at those views. In short, the Quinean methodology looks something like this:

1. One cites some theorem ϕ in logical metatheory.
2. One argues that ϕ has certain philosophical consequences, e.g., makes a certain view untenable.

Several of Quine's arguments follow this pattern, even if he doesn't always explicitly mention the relevant theorem from logical metatheory. One case where he is explicit is in his 1940 paper with Nelson Goodman, where he "proves" that every synthetic truth can be converted to an analytic truth. Whatever one may think of Quine's later arguments against analyticity, there is no doubt, historically speaking, that this metatheoretical result played a role in Quine's arriving at the conclusion that there is no analytic–synthetic distinction. And it would only be reasonable to think that *our* stance on the analytic–synthetic distinction should be responsive to what this mathematical result can be supposed to show.

As the story is typically told, Quine's "Two Dogmas of Empiricism" dealt the death blow to logical positivism. However, Carnap presented Quine with a moving target, as his views continued to develop. In "Empiricism, Semantics, and Ontology" (1950), Carnap further developed the notion of a *framework*, which bears striking resemblances both to the notion of a *scientific theory* and, hence, to the notion of a theory T in first-order logic. Here Carnap distinguishes two types of questions – the questions that are *internal* to the framework and the questions that are *external* to the framework. The internal questions are those that can be posed in the language of the framework and for which the framework can (in theory) provide an answer. In contrast, the external questions are those that we ask *about* a framework.

Carnap's abstract idea can be illustrated by simple examples from first-order logic. If we write down a vocabulary Σ for a first-order language, and a theory T in this vocabulary, then a typical internal question might be something like, "Does anything satisfy the predicate $P(x)$?" In contrast, a typical external question might be, "How many predicate symbols are there in Σ?" Thus, the internal–external distinction corresponds roughly to the older distinction between object language and metalanguage that frames Carnap's discussion in *Logische Syntax der Sprache* (1934).

The philosophical point of the internal–external distinction was supposed to be that one's answers to external questions are not held to the same standards as one's answers to internal questions. A framework includes rules, and an internal question should be answered in accordance with these rules. So, to take one of Carnap's favorite examples, "Are there numbers?" can naturally construed as an external question, since no mathematician is actively investigating that question. This question is *not* up for grabs in mathematical science – instead, it's a presupposition of mathematical science. In contrast, "Is there a greatest prime number?" is internal to mathematical practice; i.e., it is a question to which mathematics aspires to give an answer.

Surely most of us can grasp the intuition that Carnap is trying to develop here. The external questions must be answered in order to set up the game of science; the internal questions are answered in the process of playing the game of science. But Carnap wants to push this idea beyond the intuitive level – he wants to make it a cornerstone of his theory of knowledge. Thus, Carnap says that we may single out a certain special class of predicates – the so-called *Allwörter* – to label a domain of inquiry. For example, the number theorist uses the word "number" to pick out her domain of inquiry – she doesn't investigate whether something falls under the predicate "x is a number." In contrast, a number theorist might investigate whether there are numbers x, y, z such that $x^3 + y^3 = z^3$; and she simply doesn't consider whether some other things, which are not themselves numbers, satisfy this relation.

Quine (1951a, 1960) takes up the attack against Carnap's internal–external distinction. While Quine's attack has several distinct maneuvers, his invocation of hard logical facts typically goes unquestioned. In particular, Quine appeals to the supposedly hard logical fact that every theory in a language that has several distinct quantifiers (i.e., many-sorted logic) is equivalent to a theory in a language with a single unrestricted quantifier.

> It is evident that the question whether there are numbers will be a category question only with respect to languages which appropriate a separate style of variables for the exclusive purpose of referring to numbers. If our language refers to numbers through variables that also take classes other than numbers as values, then the question whether there are numbers becomes a subclass question ... Even the question whether there are classes, or whether there are physical objects becomes a subclass question if our language uses a single style of variables to range over both sorts of entities. Whether the statement that there are physical objects and the statement that there are black swans should be put on the same side of the dichotomy, or on opposite sides, comes to depend upon the rather trivial consideration of whether we use one style of variables or two for physical objects and classes. (Quine, 1976, p. 208)

Thus, suggests Quine, there is a metatheoretical result – that a many-sorted theory is equivalent to a single-sorted theory – that destroys Carnap's attempt to distinguish between *Allwörter* and other predicates in our theories.

We won't weigh in on this issue here, in our introduction. It would be premature to do so, because the entire point of this book is to lay out the mathematical facts in a clear fashion so that the reader can judge the philosophical claims for herself.

In "Two Dogmas of Empiricism," Quine argues that it makes no sense to talk about a statement's admitting of confirming or infirming (i.e., disconfirming) instances, at least when taken in isolation. Just a decade later, **Hilary Putnam**, in his paper "What Theories Are Not" (Putnam, 1962) applied Quine's idea to entire scientific theories. Putnam, student of the ur-positivist Reichenbach, now turns the positivists' primary weapon against them, to undercut the very distinctions that were so central to their program. In this case, Putnam argues that the set $T|_O$ of "observation sentences" does not accurately represent a theory T's empirical content. Indeed, he argued that a scientific theory cannot properly be said to have empirical content and, hence, that the warrant for believing it cannot flow from the bottom (the empirical part) to the top (the theoretical part). The move here is paradigmatic Putnam: a little bit of mathematical logic deftly invoked to draw a radical philosophical conclusion. This isn't the last time that we will see Putnam wield mathematical logic in the service of a far-reaching philosophical claim.

The Semantic Turn

In the early 1930s, the Vienna circle made contact with the group of logicians working in Warsaw, and in particular with **Alfred Tarski** (1901–1983). As far as twentieth-century analytic philosophy is concerned, Tarski's greatest influence has been through his bequest of **logical semantics**, along with his explications of the notions of **structure** and **truth in a structure**. Indeed, in the second half of the twentieth century, analytic philosophy has been deeply intertwined with logical semantics, and ideas from model theory have played a central role in debates in metaphysics, epistemology, philosophy of science, and philosophy of mathematics.

The promise of a purely syntactic metatheory for mathematics fell into question already in the 1930s when Kurt Gödel proved the incompleteness of Peano arithmetic. At the time, a new generation of logicians realized that not all interesting questions about theories could be answered merely by looking at theories "in themselves", and without relation to other mathematical objects. Instead, they claimed, the interesting questions about theories include questions about how they might relate to antecedently understood mathematical objects, such as the universe of sets. Thus was born the discipline of logical semantics. The arrival of this new approach to metatheory was heralded by Alfred Tarski's famous definitions of "truth in a structure" and "model of a theory." Thus, after Tarski, to understand a theory T, we have more than the theory qua syntactic object, we also have a veritable universe $\mathrm{Mod}(T)$ of models of T.

Bas van Fraassen was one of the earliest adopters of logical semantics as a tool for philosophy of science, and he effectively marshaled it in developing an alternative to the dominant outlook of scientific realism. Van Fraassen ceded Putnam's argument that the empirical content of a theory cannot be isolated syntactically. And then, in good philosophical fashion, he transformed Putnam's modus ponens into a modus tollens: the problem is not with empirical content, per se, but with the attempt to explicate is syntactically. Indeed, van Fraassen claimed that one needs the tools of logical semantics in order to make sense of the notion of empirical content; and equipped with this new explication of empirical content, empiricism can be defended against scientific realism. Thus, both the joust and the parry were carried on within an explicitly metalogical framework.

Since the 1970s, philosophical discussions of science have been profoundly influenced by this little debate about the place of syntax and semantics. Prior to the criticisms – by Putnam, van Fraassen, et al. – of the "syntactic view of theories" philosophical discussions of science frequently drew upon new results in mathematical logic. As was pointed out by van Fraassen particularly, these discussions frequently degenerated, as philosophers found themselves hung up on seemingly trivial questions, e.g., whether the observable consequences of a recursively axiomatized theory are also recursively axiomatizable. Part of the shift from syntactic to semantic methods was supposed to be a shift toward a more faithful construal of science in practice. In other words, philosophers were supposed to start asking the questions that arise in the practice of science, rather than the questions that were suggested by an obsessive attachment to mathematical logic.

The move away from logical syntax has had some healthy consequences in terms of philosophers engaging more closely with actual scientific theories. It is probably not a coincidence that since the fall of the syntactic view of theories, philosophers of science have turned their attention to specific theories in physics, biology, chemistry, etc. As was correctly pointed out by van Fraassen, Suppes, and others, scientists themselves don't demand first-order axiomatizations of these theories – and so it would do violence to those theories to try to encode them in first-order logic. Thus, the demise of the syntactic view allowed philosophers to freely draw upon the resources of set-theoretic structures, such as topological spaces, Riemannian manifolds, Hilbert spaces, C^*-algebras, etc.

Nonetheless, the results of the semantic turn have not been uniformly positive. For one, philosophy of science has seen a decline in standards of rigor, with the unfortunate consequence that debating parties more often than not talk past each other. For example, two philosophers of science might take up a debate about whether isomorphic models represent the same or different possibilities. However, these two philosophers of science may not have a common notion of "model" or of "isomorphism." In fact, many philosophers of science couldn't even tell you a precise formal explication of the word "isomorphism" – even though they rely on the notion in many of their arguments. Instead, their arguments rely on some vague sense that isomorphisms preserve structure, and an even more vague sense of what structure is.

In this book, we'll see many cases in point, where a technical term from science (physics, math, or logic) has made its way into philosophical discussion but has then lost touch with its technical moorings. The result is almost always that philosophers add to the stock of confusion rather than reducing it. How unfortunate it is that philosophy of science has fallen into this state, given the role we could play as prophets of clarity and logical rigor. One notable instance where philosophers of science could help increase clarity is the notion of *theoretical equivalence*. Scientists, and especially physicists, frequently employ the notion of two theories being equivalent. Their judgments about equivalence are not merely important for their personal attitudes toward their theories, but also for determining their actions – e.g., will they search for a crucial experiment to determine whether T_1 or T_2 is true? For example, students of classical mechanics are frequently told that the Lagrangian and Hamiltonian frameworks are equivalent, and on that basis, they are discouraged from trying to choose between them.

Now, it's not that philosophers don't talk about such issues. However, in my experience, philosophers tend to bring to bear terminology that is alien to science, and which sheds no further light on the problems. For example, if an analytic philosopher is asked, "when do two sentences ϕ and ψ mean the same thing?" then he is likely to say something like, "if they pick out the same proposition." Here the word "proposition" is alien to the physicist; and what's more, it doesn't help to solve real-life problems of synonymy. Similarly, if an analytic philosopher is asked, "when do two theories T_1 and T_2 say the same thing?" then he might say something like, "if they are true in the same possible worlds." This answer may conjure a picture in the philosopher's head, but it won't conjure any such picture in a physicist's head – and even if it did, it wouldn't help decide controversial cases. We want to know whether Lagrangian mechanics is equivalent to Hamiltonian mechanics, and whether Heisenberg's matrix mechanics is equivalent to Schrödinger's wave mechanics. The problem here is that space of possible worlds (if it exists) cannot be surveyed easily, and the task of comparing the subset of worlds in which T_1 is true with the subset of worlds in which T_2 is true is hardly tractible. Thus, the analytic philosopher's talk about "being true in the same possible worlds" doesn't amount to an *explication* of the concept of equivalence. An explication, in the Carnapian sense, should supply clear guidelines for how to use a concept.

Now, don't get me wrong. I am not calling for a Quinean ban on propositions, possible worlds, or any of the other concepts that analytic philosophers have found so interesting. I only want to point out that these concepts are descendants, or cousins, of similar concepts that are used in the exact sciences. Thus, it's important that analytic philosophers – to the extent that they want to understand and/or clarify science – learn to tie their words back down into their scientific context. For example, philosophers' possible worlds are the descendant of the logician's "models of a theory," the mathematician's "solutions of a differential equation," and the physicist's "points in state space." Thus, it's fine to talk about possible worlds, but it would be advisable to align our usage of the concept with the way it's used in the sciences.

As we saw before, Carnap had self-imposed the constraint that a philosophical explication of a concept must be *syntactic*. So, for example, to talk about "observation sentences," one must construct a corresponding predicate in the language of syntactic metalogic – a language whose primitive concepts are things like "predicate symbol" and "binary connective." Carnap took a swing at defining such predicates, and Quine, Putnam, and friends found his explications to be inadequate. There are many directions that one could go from here – and one of these directions remains largely unexplored. First, one can do as Quine and Putnam themselves did: stick with logical syntax and change one's philosophical views. Second, one can do as van Fraassen did: move to logical semantics and stick with Carnap's philosophical views. (To be fair, van Fraassen's philosophical views are very different than Carnap's – I only mean to indicate that there are certain central respects in which van Fraassen's philosophical views are closer to Carnap's than to Quine's.) The third option is to say perhaps logical syntax had not yet reached a fully mature stage in 1950, and perhaps new developments will make it more feasible to carry out syntactic explications of philosophical concepts. That third option is one of the objectives of this book – i.e. to raise syntactic analysis to a higher level of nuance and sophistication.

Model Theoretic Madness

By the 1970s, scientific realism was firmly entrenched as the dominant view in philosophy of science. Most the main players in the field – Boyd, Churchland, Kitcher, Lewis, Salmon, Sellars, etc. – had taken up the realist cause. Then, with a radical about-face, Putnam again took up the tools of mathematical logic, this time to argue for the incoherence of realism. In his famous "model-theoretic argument," Putnam argued that logical semantics – in particular, the Löwenheim-Skølem theorem – implies that any consistent theory is true. In effect, then, Putnam proposed a return to a more liberal account of theoretical equivalence, indeed, something even more liberal than the logical positivists' notion of empirical equivalence. Indeed, in the most plausible interpretation of Putnam's conclusion, it entails that any two consistent theories are equivalent to each other.

Whatever you might think of Putnam's radical claim, there is no doubt that it stimulated some interesting responses. In particular, Putnam's claim prompted the arch-realist David Lewis to clarify the role that *natural properties* play in his metaphysical system. According to Lewis, the defect in Putnam's argument is the assumption that a predicate P can be assigned to any subset of objects in the actual world. This assumption is mistaken, says Lewis, because not every random collection of things corresponds to some natural class, and we should only consider interpretations in which predicates that occur in T are assigned to natural classes of objects in the actual world. Even if T is consistent, there may be no such interpretation relative to which T is actually true.

There are mixed views on whether Lewis' response to Putnam is effective. However, for our purposes, the important point is that the upshot of Lewis' response would be to move in the direction of a more conservative account of theoretical equivalence. And now the question is whether the notion of theoretical equivalence that Lewis is proposing goes too far in the other direction. On one interpretation of Lewis, his claim is that two theories T and T' are equivalent only if they share the same "primitive notions." If we apply that claim literally to first-order theories, then we might think that theories T and T' are equivalent only if they are written with the same symbols. However, this condition wouldn't even allow notationally variant theories to be equivalent.

While Lewis was articulating the realist stance, Putnam was digging up more arguments for a liberal and inclusive criterion of theoretical equivalence. Here he drew on his extensive mathematical knowledge to find examples of theories that mathematicians call equivalent, but which metaphysical realists would call inequivalent. One of Putnam's favorite examples here was axiomatic Euclidean geometry, which some mathematicians formulate with points as primitives, and other mathematicians formulate with lines as primitives — but they never argue that one formulation is more correct than the other. Thus, Putnam challenges the scientific credentials of realism by giving examples of theories that scientists declare to be equivalent, but which metaphysical realists would declare to be inequivalent.

At the time when Putnam put forward these examples, analytic philosophy was unfortunately growing more distant from its logical and mathematical origins. What this meant, in practice, is that while Putnam's examples were extensively discussed, the discussion never reached a high level of logical precision. For example, nobody clearly explained how the word "equivalence" was being used.

These exciting, and yet imprecise, discussions continued with reference to a second example that Putnam had given. In this second example, Putnam asks how many things are on the following line:

∗ ∗

There are two schools of metaphysicians who give different answers to this question. According to the mereological nihilists, there are exactly two things on the line, and both are asterisks. According to the mereological universalists, there are three things on the line: two individual asterisks, and one composite of two asterisks. Putnam, however, declares that the debate between these two schools of metaphysicians is a "purely verbal dispute", and neither party is more correct than the other.

Again, what's important for us here it that Putnam's claim amounts to a proposal to liberalize the standards of theoretical equivalence. By engaging in this dispute, metaphysicians have implicitly adopted a rather conservative standard of equivalence – where it matters whether you think that a pair of asterisks is something more beyond the individuals that constitute it. Putnam urges us to adopt a more liberal criterion of theoretical equivalence, according to which it simply doesn't matter whether we say that the pair "really exists", or whether we don't.

From Reduction to Supervenience

The logical positivists – Schlick, Carnap, Neurath, etc. – aspired to uphold the highest standards of scientific rationality. Most of them believed that commitment to scientific rationality demands a commitment to physicalism, i.e. the thesis that physical science is the final arbiter on claims of ontology. In short, they said that we ought to believe that something exists only if physics licenses that belief.

Of course, we don't much mind rejecting claims about angels, demons, witches, and fairies. But what are we supposed to do with the sorts of statements that people make in the ordinary course of life – about each other, and about themselves? For example, if I say, "Søren is in pain," then I seem to be committed to the existence of some object denoted by "Søren", that has some property "being in pain." How can physical science license such a claim, when it doesn't speak of an object Søren or the property of being in pain?

The general thesis of physicalism, and the particular thesis that a person is his body, were not 20th century novelties. However, it was a 20th century novelty to attempt to explicate these theses using the tools of symbolic logic. To successfully explicate this concept would transform it from a vague ideological stance to a sharp scientific hypothesis. (There is no suggestion here that the hypothesis would be empirically verifiable – merely that it would be clear enough to be vulnerable to counterargument.)

For example, suppose that $r(x)$ denotes the property of being in pain. Then it would be natural for the physicalist to propose either (1) that statements using $r(x)$ are actually erroneous, or (2) that there is some predicate $\phi(x)$ in the language of fundamental physics such that $\forall x(r(x) \leftrightarrow \phi(x))$. In other words, if statements using $r(x)$ are legitimate, then $r(x)$ must actually pick out some underlying physical property $\phi(x)$.

The physicalist will want to clarify what he means by saying that $\forall x(r(x) \leftrightarrow \phi(x))$, for even a Cartesian dualist could grant that this sentence is contingently true. That is, a Cartesian dualist might say that there is a physical description $\phi(x)$ which happens, as a matter of contingent fact, to pick out exactly those things that are in pain. The reductionist, in contrast, wants to say more. He wants to say that there is a more thick connection between pain experiences and happenings in the physical world. At the very least, a reductionist would say that

$$T \vdash r(x) \leftrightarrow \phi(x),$$

where T is our most fundamental theory of the physical world. That is, to the extent that ordinary language ascriptions are correct, they can be translated into true statements of fundamental physics.

This sort of linguistic reductionism seems to have been the favored view among early-twentieth-century analytic philosophers – or, at least among the more scientifically inclined of them. Certainly, reductionism had vocal proponents, such as U.T. Place and Herbert Feigl. Nonetheless, by the third quarter of the twentieth century, this view had fallen out of fashion. In fact, some of the leading lights in analytic philosophy – such as Putnam and Fodor – had arguments which were taken to demonstrate the utter implausibility of the reductionist point of view. Nonetheless, what had not fallen out of favor among analytic philosophers was the naturalist stance that had found its precise explication in the reductionist thesis. Thus, analytic philosophers found themselves on the hunt for a new, more plausible way to express their naturalistic sentiments.

There was another movement afoot in analytic philosophy – a movement away from the formal mode, back toward the material mode, i.e., from a syntactic point of view, to a semantic point of view. What this movement entailed in practice was a shift from syntactic explications of concepts to semantic explications of concepts. Thus, it is only natural that having discarded the syntactic explication of mind–body reduction, analytic philosophers would cast about for a semantic explication of the idea. Only, in this case, the very word "reduction" had so many negative associations that a new word was needed. To this end, analytic philosophers co-opted the word "supervenience." Thus Donald Davidson:

> Mental characteristics are in some sense dependent, or supervenient, on physical characteristics. Such supervenience might be taken to mean that there cannot be two events alike in all physical respects but differing in some mental respect, or that an object cannot alter in some mental respect without altering in some physical respect. (Davidson, 1970)

Davidson's prose definition of supervenience is so clear that it is begging for formalization. Indeed, as we'll later see, when the notion of supervenience is formalized, then it is none other than the model theorist's notion of implicit definability.

It must have seemed to the 1970s philosophers that significant progress had been made in moving from the thin syntactic concept of reduction to the thick semantic concept of supervenience. Indeed, by the 1980s, the concept of supervenience had begun to play a major role in several parts of analytic philosophy. However, with the benefit of hindsight, we ought to be suspicious if we are told that an implausible philosophical position can be converted into a plausible one merely by shifting from a syntactic to a semantic explication of the relevant notions. In this case, there is a worry that the concept of supervenience is nothing but a reformulation, in semantic terms, of the notion of reducibility. As we will discuss in Section 6.7, if supervenience is cashed out as the notion of implicit definability, then **Beth's theorem** shows that supervenience is equivalent to reducibility.

Why did philosophers decide that mind-brain reductionism was implausible? We won't stop here to review the arguments, as interesting as they are, since that has been done in many other places (see Bickle, 2013). We are interested rather in claims (see, e.g., Bickle (1998)) that the arguments against reduction are only effective against syntactic accounts thereof – and that semantics permits a superior account of reduction that is immune to these objections.

Throughout this book, we argue for a fundamental duality between logical syntax and semantics. To the extent that this duality holds, it is mistaken to think that semantic accounts of concepts are more intrinsic, or that they allow us to transcend the human reliance on representations, or that they provide a bridge to the "world" side of the mind-world divide.

To the contrary, logical semantics is ... wait for it ... just more mathematics. As such, while semantics can be used to represent things in the world, including people and their practice of making claims about the world, its means of representation are no different than those of any other part of mathematics. Hence, every problem and puzzle and confusion that arises in logical syntax – most notably, the problem of language dependence – will rear its ugly head again in logical semantics. Thus, for example, if scientific antirealism falls apart when examined under a syntactic microscope, then it will also fall apart when examined under a semantic microscope. Similarly, if mind-body reductionism isn't plausible when explicated syntactically, then it's not going to help to explicate it semantically.

What I am saying here should not be taken as a blanket criticism of attempts to explicate concepts semantically. In fact, I'll be the first to grant that model theory is not only a beautiful mathematical theory, but is also particularly useful for philosophical thinking. However, we should be suspicious of any claims that a philosophical thesis (e.g. physicalism, antirealism, etc.) is untenable when explicated syntactically, but becomes tenable when explicated semantically. We should also be suspicious of any claims that semantic methods are any less prone to creating pseudoproblems than syntactic methods.

Realism and Equivalence

As we have seen, many of these debates in twentieth-century philosophy ultimately turn on the question of how one theory is related to another. For example, the debate about the mind-body relation can be framed as a question about how our folk theory of mind is related to the theories of the brain sciences, and ultimately to the theories of physics.

If we step up a level of abstraction, then even the most general divisions in 20th century philosophy have to do with views on the relations of theories. Among the logical positivists, the predominant view was a sort of antirealism, certainly about metaphysical claims, but also about the theoretical claims of science. Not surprisingly, the preferred view of theoretical equivalence among the logical positivists was empirical equivalence: two theories are equivalent just in case they make the same predictions. That notion of equivalence is quite liberal in that it equates theories that intuitively seem to be inequivalent.

If we leap forward to the end of the twentieth century, then the outlook had changed radically. Here we find analytic metaphysicians engaged in debates about mereological nihilism versus universalism, or about presentism versus eternalism. We also find philosophers of physics engaged in debates about Bohmian mechanics versus Everettian interpretations of quantum mechanics, or about substantivalism versus relationalism about spacetime. The interesting point here is that there obviously had been a radical change in the regnant standard of theoretical equivalence in the philosophical community. Only seventy years prior, these debates would have been considered pseudo-debates, for they attempt to choose between theories that are empirically equivalent. In short, the philosophical community as a whole had shifted from a more liberal to a more conservative standard of theoretical equivalence.

There have been, however, various defections from the consensus view on theoretical equivalence. The most notable example here is the Hilary Putnam of the 1970s. At this time, almost all of Putnam's efforts were devoted to liberalizing standards of theoretical equivalence. We can see this not only in his model-theoretic argument, but also in the numerous examples that he gave of theories with "different ontologies," but which he claimed are equivalent. Putnam pointed to different formulations of Euclidean geometry, and also the famous example of "Carnap and the mereologist," which has since become a key example of the quantifier variance debate. We discuss the geometry example in Section 7.4, and the mereology example in Section 5.4.

One benefit of the formal methods developed in this book is a sort of taxonomy of views in twentieth-century philosophy. The realist tendency is characterized by the adoption of more conservative standards of theoretical equivalence; and the antirealist tendency is characterized by the adoption of more liberal standards of theoretical equivalence. Accordingly, we shouldn't think of "realism versus antirealism" on the model of American politics, with its binary division between Republicans and Democrats. Indeed, philosophical opinions on the realism–antirealism question lie on a continuum, corresponding to a continuum of views on theoretical equivalence. (In fact, views on theoretical equivalence really form a multidimensional continuum; I'm merely using the one-dimensional language for intuition's sake.) Most of us will find ourselves with a view of theoretical equivalence that is toward the middle of the extremes, and many of the philosophical questions we consider are questions about whether to move – if ever so slightly – in one direction or the other.

In this book, we will develop three moderate views of theoretical equivalence. The first two views say that theories are equivalent just in case they are intertranslatable – only they operate with slightly different notions of "translation." The first, and more conservative, view treats quantifier statements as an invariant, so that a good translation must preserve them intact. (We also show that this first notion of intertranslatability corresponds to "having a common definitional extension." See Theorems 4.6.17 and 6.6.21.) The second, and more liberal, view allows greater freedom in translating one language's quantifier statements into a complex of the other language's quantifier statements. (We also show that this second notion of intertranslatability corresponds to "having a common Morita extension." See Theorems 7.5.3 and 7.5.5.) The third view of equivalence we consider is the most liberal, and is motivated not by linguistic considerations, but by scientific practice. In particular, scientists seem to treat theories as equivalent if they can "do the same things with them." We will explicate this notion of what a scientific theory can do in terms of its "category of models." We then suggest that two theories are equivalent in this sense if their categories of models are equivalent in the precise, category-theoretic sense.

Summary and Prospectus

The following seven chapters try to accomplish two things at once: to introduce some formal techniques, and to use these techniques to gain philosophical insight. Most of the philosophical discussions are interspersed between technical results, but there is one concluding chapter that summarizes the major philosophical themes. We include here a chart of some of the philosophical issues that arise in the course of these chapters. The left column states a technical result, the middle column states the related philosophical issue, and the right column gives the location (section number) where the discussion can be found. To be fair, I don't mean to say that the philosophers mentioned on the right explicitly endorse the argument from the metalogical result to the philosophical conclusion. In some cases they do; but in other cases, the philosopher seems rather to presuppose the metalogical result.

Logic	Philosophy	Location
Translate into empty theory	Analytic–synthetic distinction (Quine)	**3.7.10**
Translate into empty theory	Implicit definition (Quine)	**3.7.10**
Eliminate sorts	Ontological monism (Quine)	**5.3**
Eliminate sorts	No external questions (Quine)	**5.4.17**
Eliminate sorts	Against quantifier variance	**5.4.4, 5.4.16**
Indivisible vocabulary	Against empiricism (Putnam, Boyd)	**4.4**
Beth's theorem	Supervenience implies reduction	**6.7**
Löwenheim–Skølem	Against realism (Putnam)	**8.3**
Equivalent geometries	Against realism (Putnam)	**7.4**
Ramsefication	Structural realism	**8.1**
Ramsefication	Functionalism	**8.1**

Notes

- In this chapter, our primary objective was to show the philosopher-in-training some of the payoffs for learning the metatheory of first-order logic: the better she understands the logic, the better she will understand twentieth-century philosophy, and the options going forward. Although we've tried to be reasonably faithful to the historical record, we've focused on just one part of this history. The curious reader should consult more detailed studies, such as Coffa (1993); Friedman (1999); Hylton (2007); Soames (2014).
- For Russell's program for rebuilding philosophy on the basis of formal logic, see Russell (1901, 1914a).
- Carnap's personal recollections can be found in Carnap and Schilpp (1963).
- Frege and Russell were early critics of Hilbert's view of implicit definition (see, e.g., Blanchette, 2012). In contrast, Schlick (1918, I.7) explicitly endorses Hilbert's view. For Carnap's view, see Park (2012). The discussion later got muddled up in discussions of Ramsey sentences (see, e.g., Winnie, 1967; Lewis,

1970), which we will discuss in Chapter 8. For an extended discussion of implicit definition and its relation to 20th century philosophical issues, see Ben-Menahem (2006).

- For more on the 19th century backdrop to analyticity, see Coffa (1986).
- For overviews of logical methods in philosophy of science, see van Benthem (1982); Winnie (1986); Van Fraassen (2011); Leitgeb (2011). The primary novelty of the present book is our use of category-theoretic methods. We have tried not to mention category theory more than necessary, but we use it frequently.

1 Invitation to Metatheory

This chapter is meant to serve as a preview, and for motivation to work through the chapters to come. In the next chapter, we'll move quickly into "categorical set theory" – which isn't all that difficult, but which is not yet well known among philosophers. For the past fifty years or so, it has almost been mandatory for analytic philosophers to know a little bit of set theory. However, it has most certainly not been mandatory for philosophers to know a little bit of category theory. Indeed, most analytic philosophers are familiar with the words "subset" and "powerset" but not the words "natural transformation" or "equivalence of categories." Why should philosophers bother learning these unfamiliar concepts?

The short answer is that is that category theory (unlike set theory) was designed to explicate *relations* between mathematical structures. Since philosophers want to think about relations between theories (e.g., equivalence, reducibility) and since theories can be modeled as mathematical objects, philosophers' aims will be facilitated by gaining some fluency in the language of category theory. At least that's one of the main premises of this book. So, in this chapter, we'll review some of the basics of the metatheory of propositional logic. We will approach the issues from a slightly different angle than usual, placing less emphasis on what individual theories say and more emphasis on the relations between these theories.

To repeat, the aim of **metatheory** is to theorize about theories. For simplicity, let's use M to denote this hypothetical theory about theories. Thus, M is not the *object* of our study; it is the *tool* we will use to study other theories and the relations between them. In this chapter, I will begin using this tool M to talk about theories – without explicitly telling you anything about M itself. In the next chapter, I'll give you the user's manual for M.

1.1 Logical Grammar

DEFINITION 1.1.1 A **propositional signature** Σ is a collection of items, which we call **propositional constants**. Sometimes these propositional constants are also called **elementary sentences**. (Sometimes people call them atomic sentences, but we will be using the word "atomic" for a different concept.)

These propositional constants are assumed to have no independent meaning. Nonetheless, we assume a primitive notion of identity between propositional constants; the fact that two propositional constants are equal or non-equal is not explained by any more fundamental fact. This assumption is tantamount to saying that Σ is a **bare set** (and it stands in gross violation of Leibniz's principle of the identity of indiscernibles).

ASSUMPTION 1.1.2 The **logical vocabulary** consists of the symbols $\neg, \wedge, \vee, \rightarrow$. We also use two further symbols for punctuation: a left and a right parenthesis.

DEFINITION 1.1.3 Given a propositional signature Σ, we define the set $\mathsf{Sent}(\Sigma)$ of Σ-sentences as follows:

1. If $\phi \in \Sigma$, then $\phi \in \mathsf{Sent}(\Sigma)$.
2. If $\phi \in \mathsf{Sent}(\Sigma)$, then $(\neg\phi) \in \mathsf{Sent}(\Sigma)$.
3. If $\phi \in \mathsf{Sent}(\Sigma)$ and $\psi \in \mathsf{Sent}(\Sigma)$, then $(\phi \wedge \psi) \in \mathsf{Sent}(\Sigma)$, $(\phi \vee \psi) \in \mathsf{Sent}(\Sigma)$, and $(\phi \rightarrow \psi) \in \mathsf{Sent}(\Sigma)$.
4. Nothing is in $\mathsf{Sent}(\Sigma)$ unless it enters via one of the previous clauses.

The symbol ϕ here is a variable that ranges over finite strings of symbols drawn from the alphabet that includes Σ; the connectives $\neg, \wedge, \vee, \rightarrow$; and (when necessary) left and right parentheses "(" and ")". We will subsequently play it fast and loose with parentheses, omitting them when no confusion can result. In particular, we take a negation symbol $\neg$ always to have binding precedence over the binary connectives.

Note that each sentence is, by definition, a finite string of symbols and, hence, contains finitely many propositional constants.

Since the set $\mathsf{Sent}(\Sigma)$ is defined inductively, we can prove things about it using "proof by induction." A proof by induction proceeds as follows:

1. Show that the property of interest, say P, holds of the elements of Σ.
2. Show that if P holds of ϕ, then P holds of $\neg\phi$.
3. Show that if P holds of ϕ and ψ, then P holds of $\phi \wedge \psi$, $\phi \vee \psi$, and $\phi \rightarrow \psi$.

When these three steps are complete, one may conclude that all things in $\mathsf{Sent}(\Sigma)$ have property P.

DEFINITION 1.1.4 A **context** is essentially a finite collection of sentences. However, we write contexts as sequences, for example $\phi_1, \ldots, \phi_n$ is a context. But ϕ_1, ϕ_2 is the same context as ϕ_2, ϕ_1, and is the same context as ϕ_1, ϕ_1, ϕ_2. If Δ and Γ are contexts, then we let Δ, Γ denote the union of the two contexts. We also allow an empty context.

1.2 Proof Theory

We now define the relation $\Delta \vdash \phi$ of derivability that holds between contexts and sentences. This relation is defined **recursively** (aka **inductively**), with base case $\phi \vdash \phi$ (Rule of Assumptions). Here we use a horizontal line to indicate that if $\vdash$ holds between the things above the line, then $\vdash$ also holds for the things below the line.

Rule of Assumptions
$$\frac{}{\phi \vdash \phi}$$

$\wedge$ **elimination**
$$\frac{\Gamma \vdash \phi \wedge \psi}{\Gamma \vdash \phi} \qquad \frac{\Gamma \vdash \phi \wedge \psi}{\Gamma \vdash \psi}$$

$\wedge$ **introduction**
$$\frac{\Gamma \vdash \phi \qquad \Delta \vdash \psi}{\Gamma, \Delta \vdash \phi \wedge \psi}$$

$\vee$ **introduction**
$$\frac{\Gamma \vdash \phi}{\Gamma \vdash \phi \vee \psi} \qquad \frac{\Gamma \vdash \psi}{\Gamma \vdash \phi \vee \psi}$$

$\vee$ **elimination**
$$\frac{\Gamma \vdash \phi \vee \psi \qquad \Delta, \phi \vdash \chi \qquad \Theta, \psi \vdash \chi}{\Gamma, \Delta, \Theta \vdash \chi}$$

$\rightarrow$ **elimination**
$$\frac{\Gamma \vdash \phi \rightarrow \psi \qquad \Delta \vdash \phi}{\Gamma, \Delta \vdash \psi}$$

$\rightarrow$ **introduction**
$$\frac{\Gamma, \phi \vdash \psi}{\Gamma \vdash \phi \rightarrow \psi}$$

RA
$$\frac{\Gamma, \phi \vdash \psi \wedge \neg\psi}{\Gamma \vdash \neg\phi}$$

DN
$$\frac{\Gamma \vdash \neg\neg\phi}{\Gamma \vdash \phi}$$

The definition of the turnstyle $\vdash$ is then completed by saying that $\vdash$ is the smallest relation (between sets of sentences and sentences) such that

1. $\vdash$ is closed under the previously given clauses, and
2. If $\Delta \vdash \phi$ and $\Delta \subseteq \Delta'$, then $\Delta' \vdash \phi$.

The second property here is called **monotonicity**.

There are a variety of ways that one can explicitly generate pairs Δ, ϕ such that $\Delta \vdash \phi$. A method for doing such is typically called a **proof system**. We will not explicitly introduce any proof system here, but we will adopt the following definitions.

DEFINITION 1.2.1 A pair Δ, ϕ is called a **sequent** or **proof** just in case $\Delta \vdash \phi$. A sentence ϕ is said to be **provable** just in case $\vdash \phi$. Here $\vdash \phi$ is shorthand for $\emptyset \vdash \phi$. We use $\top$ as shorthand for a sentence that is provable – for example, $p \rightarrow p$. We could then add as an inference rule "$\top$ introduction," which allowed us to write $\Delta \vdash \top$. It can be proven that the resulting definition of $\vdash$ would be the same as the original definition. We also sometimes use the symbol $\bot$ as shorthand for $\neg\top$. It might then be convenient to restate RA as a rule that allows us to infer $\Delta \vdash \neg\phi$ from $\Delta, \phi \vdash \bot$. Again, the resulting definition of $\vdash$ would be the same as the original.

DISCUSSION 1.2.2 The rules we have given for $\vdash$ are sometimes called the **classical propositional calculus** or just the **propositional calculus**. Calling it a "calculus" is meant to indicate that the rules are purely formal and don't require any understanding of the meaning of the symbols. If one deleted the DN rule and replaced it with Ex Falso Quodlibet, the resulting system would be the **intuitionistic propositional calculus**. However, we will not pursue that direction here.

1.3 Semantics

DEFINITION 1.3.1 An **interpretation** (sometimes called a **valuation**) of Σ is a function from Σ to the set {true, false}, i.e., an assignment of truth-values to propositional constants. We will usually use 1 as shorthand for "true" and 0 as shorthand for "false."

Clearly, an interpretation v of Σ extends naturally to a function $v : \mathsf{Sent}(\Sigma) \to \{0, 1\}$ by the following clauses:

1. $v(\neg\phi) = 1$ if and only if $v(\phi) = 0$.
2. $v(\phi \wedge \psi) = 1$ if and only if $v(\phi) = 1$ and $v(\psi) = 1$.
3. $v(\phi \vee \psi) = 1$ if and only if either $v(\phi) = 1$ or $v(\psi) = 1$.
4. $v(\phi \to \psi) = v(\neg\phi \vee \psi)$.

DISCUSSION 1.3.2 The word "interpretation" is highly suggestive, but it might lead to confusion. It is sometimes suggested that elements of $\mathsf{Sent}(\Sigma)$ are part of an uninterpreted calculus without intrinsic meaning, and that an intepretation $v : \Sigma \to \{0, 1\}$ endows these symbols with meaning. However, to be clear, $\mathsf{Sent}(\Sigma)$ and $\{0, 1\}$ are both mathematical objects; neither one of them is more linguistic than the other, and neither one of them is more "concrete" than the other.

This point becomes even more subtle in predicate logic, where we might be tempted to think that we can interpret the quantifiers so that they range over all actually existing things. To the contrary, the domain of a predicate logic interpretation must be a *set*, and a set is something whose existence can be demonstrated by ZF set theory. Since the existence of the world is not a consequence of ZF set theory, it follows that the world is not a set. (Put slightly differently: a set is an abstract object, and the world is a concrete object. Therefore, the world is not a set.)

DEFINITION 1.3.3 A **propositional theory** T consists of a signature Σ, and a set Δ of sentences in Σ. Sometimes we will simply write T in place of Δ, although it must be understood that the identity of a theory also depends on its signature. For example, the theory consisting of a single sentence p is different depending on whether it's formulated in the signature $\Sigma = \{p\}$ or in the signature $\Sigma' = \{p, q\}$.

DEFINITION 1.3.4 (Tarski truth) Given an interpretation v of Σ and a sentence ϕ of Σ, we say that ϕ is **true** in v just in case $v(\phi) = 1$.

DEFINITION 1.3.5 For a set Δ of Σ sentences, we say that v is a **model** of Δ just in case $v(\phi) = 1$, for all ϕ in Δ. We say that Δ is **consistent** if Δ has at least one model, and that Δ is **inconsistent** if it has no models.

Any time we define a concept for sets of sentences (e.g., consistency), we can also extend that concept to theories, as long as it's understood that a theory is technically a pair consisting of a signature and a set of sentences in that signature.

DISCUSSION 1.3.6 The use of the word "model" here has its origin in consistency proofs for non-Euclidean geometries. In that case, one shows that certain non-Euclidean geometries can be translated into models of Euclidean geometry. Thus, if Euclidean geometry is consistent, then non-Euclidean geometry is also consistent. This kind of maneuver is what we now call a **proof of relative consistency**.

In our case, it may not be immediately clear what sits on the "other side" of an interpretation, because it's certainly not Euclidean geometry. What kind of mathematical thing are we interpreting our logical symbols into? The answer here – as will become apparent in Chapter 3 – is either a Boolean algebra or a fragment of the universe of sets.

DEFINITION 1.3.7 Let Δ be a set of Σ sentences, and let ϕ be a Σ sentence. We say that Δ **semantically entails** ϕ, written $\Delta \vDash \phi$, just in case ϕ is true in all models of Δ. That is, if v is a model of Δ, then $v(\phi) = 1$.

EXERCISE 1.3.8 Show that if $\Delta, \phi \vDash \psi$, then $\Delta \vDash \phi \to \psi$.

EXERCISE 1.3.9 Show that $\Delta \vDash \phi$ if and only if $\Delta \cup \{\neg\phi\}$ is inconsistent. Here $\Delta \cup \{\neg\phi\}$ is the theory consisting of $\neg\phi$ and all sentences in Δ.

We now state three main theorems of the metatheory of propositional logic.

THEOREM 1.3.10 (Soundness) *If $\Delta \vdash \phi$, then $\Delta \vDash \phi$.*

The soundness theorem can be proven by an argument directly analogous to the substitution theorem in Section 1.4. We leave the details to the reader.

THEOREM 1.3.11 (Completeness) *If $\Delta \vDash \phi$, then $\Delta \vdash \phi$.*

The completeness theorem can be proven in various ways. In this book, we will give a topological proof via the Stone duality theorem (see Chapter 3).

THEOREM 1.3.12 (Compactness) *Let Δ be a set of sentences. If every finite subset Δ_F of Δ is consistent, then Δ is consistent.*

The compactness theorem can be proven in various ways. One way of proving it – although not the most illuminating – is as a corollary of the completeness theorem. Indeed, it's not hard to show that if $\Delta \vdash \phi$, then $\Delta_F \vdash \phi$ for some finite subset Δ_F of Δ. Thus, if Δ is inconsistent, then $\Delta \vdash \bot$, hence $\Delta_F \vdash \bot$ for a finite subset Δ_F of Δ. But then Δ_F is inconsistent.

DEFINITION 1.3.13 A theory T, consisting of axioms Δ in signature Σ, is said to be **complete** just in case Δ is consistent and for every sentence ϕ of Σ, either $\Delta \vDash \phi$ or $\Delta \vDash \neg\phi$.

Be careful to distinguish between the completeness of our proof system (which is independent of any theory) and completeness of some particular theory T. Famously, Kurt Gödel proved that the theory of Peano arithmetic is incomplete – i.e., there is a sentence ϕ of the language of arithmetic such that neither $T \vdash \phi$ nor $T \vdash \neg\phi$. However, there are much simpler examples of incomplete theories. For example, if $\Sigma = \{p, q\}$, then the theory with axiom $\vdash p$ is incomplete in Σ.

DEFINITION 1.3.14 Let T be a theory in Σ. The **deductive closure** of T, written $\text{Cn}(T)$, is the set of Σ sentences that is implied by T. If $T = \text{Cn}(T)$, then we say that T is **deductively closed**.

Example 1.3.15 Let $\Sigma = \{p\}$, and let $T = \{p\}$. Let $\Sigma' = \{p, q\}$, and let $T' = \{p\}$. Here we must think of T and T' as different theories, even though they consist of the same sentences – i.e., $T = T'$. One reason to think of these as different theories: T is complete, but T' is incomplete. Another reason to think of T and T' as distinct is that they have different deductive closures. For example, $q \vee \neg q$ is in the deductive closure of T', but not of T.

The point here turns out to be philosophically more important than one might think. Quine argued (correctly, we think) that choosing a theory is not just choosing axioms, but axioms in a particular language. Thus, one can't tell what theory a person accepts merely by seeing a list of the sentences that she believes to be true. ⌟

EXERCISE 1.3.16 Show that the theory T' from the previous example is not complete.

EXERCISE 1.3.17 Show that $\text{Cn}(\text{Cn}(T)) = \text{Cn}(T)$.

EXERCISE 1.3.18 Consider the signature $\Sigma = \{p\}$. How many complete theories are there in this signature? (We haven't been completely clear on the identity conditions of theories and, hence, on how to count theories. For this exercise, assume that theories are deductively closed, and two theories are equal just in case they contain exactly the same sentences.)

1.4 Translating between Theories

Philosophers constantly make claims about relations between theories – that they are equivalent, or inequivalent or one is reducible to the other, or one is stronger than another. What do all these claims mean? Now that we have a formal notion of a theory, we can consider how we might want to represent relations between theories. In fact, many of the relations that interest philosophers can be cashed out in terms of the notion of a **translation**.

There are many different kinds of translations between theories. Let's begin with the most trivial kind of translation – a change of notation. Imagine that at Princeton, a scientist is studying a theory T. Now, a scientist at Harvard manages to steal a copy of the Princeton scientist's file, in which she has been recording all the consequences of T. However, in order to avoid a charge of plagiarism, the Harvard scientist runs a "find and replace" on the file, replacing each occurence of the propositional constant p with the propositional constant h. Otherwise, the Harvard scientist's file is identical to the Princeton scientist's file.

What do you think: is the Harvard scientist's theory the same or different from the Princeton scientist's theory?

Most of us would say that the Princeton and Harvard scientists have the same theory. But it depends on what we mean by "same." These two theories aren't the same in the strictest sense, since one of the theories contains the letter "p," and the other doesn't. Nonetheless, in this case, we're likely to say that the theories are the same in the sense that they differ only in ways that are incidental to how they will be used. To borrow a phrase from Quine, we say that these two theories are **notational variants** of each other, and we assume that notational variants are equivalent.

Let's now try to make precise this notion of "notational variants" or, more generally, of **equivalent theories**. To do so, we will begin with the more general notion of a translation from one theory into another.

DEFINITION 1.4.1 Let Σ and Σ' be propositional signatures. A **reconstrual** from Σ to Σ' is a function from the set Σ to the set $\mathsf{Sent}(\Sigma')$.

A reconstrual f extends naturally to a function $\overline{f} : \mathsf{Sent}(\Sigma) \to \mathsf{Sent}(\Sigma')$, as follows:

1. For p in Σ, $\overline{f}(p) = f(p)$.
2. For any sentence ϕ, $\overline{f}(\neg\phi) = \neg\overline{f}(\phi)$.
3. For any sentences ϕ and ψ, $\overline{f}(\phi \circ \psi) = \overline{f}(\phi) \circ \overline{f}(\psi)$, where $\circ$ stands for an arbitrary binary connective.

When no confusion can result, we use f for $\overline{f}$.

THEOREM 1.4.2 (Substitution) *For any reconstrual* $f : \Sigma \to \Sigma'$, *if* $\phi \vdash \psi$ *then* $f(\phi) \vdash f(\psi)$.

Proof Since the family of sequents is constructed inductively, we will prove this result by induction.

(rule of assumptions) We have $\phi \vdash \phi$ by the rule of assumptions, and we also have $f(\phi) \vdash f(\phi)$.

($\wedge$ intro) Suppose that $\phi_1, \phi_2 \vdash \psi_1 \wedge \psi_2$ is derived from $\phi_1 \vdash \psi_1$ and $\phi_2 \vdash \psi_2$ by $\wedge$ intro, and assume that the result holds for the latter two sequents. That is, $f(\phi_1) \vdash f(\psi_1)$ and $f(\phi_2) \vdash f(\psi_2)$. But then $f(\phi_1), f(\phi_2) \vdash f(\psi_1) \wedge f(\psi_2)$ by $\wedge$ introduction. And since $f(\psi_1) \wedge f(\psi_2) = f(\psi_1 \wedge \psi_2)$, it follows that $f(\phi_1), f(\phi_2) \vdash f(\psi_1 \wedge \psi_2)$.

($\to$ intro) Suppose that $\theta \vdash \phi \to \psi$ is derived by conditional proof from $\theta, \phi \vdash \psi$. Now assume that the result holds for the latter sequent, i.e., $f(\theta), f(\phi) \vdash f(\psi)$. Then conditional proof yields $f(\theta) \vdash f(\phi) \to f(\psi)$. And since $f(\phi) \to f(\psi) = f(\phi \to \psi)$, it follows that $f(\theta) \vdash f(\phi \to \psi)$.

(reductio) Suppose that $\phi \vdash \neg\psi$ is derived by RAA from $\phi, \psi \vdash \bot$, and assume that the result holds for the latter sequent, i.e., $f(\phi), f(\psi) \vdash f(\bot)$. By the properties of f, $f(\bot) \vdash \bot$. Thus, $f(\phi), f(\psi) \vdash \bot$, and by RAA, $f(\phi) \vdash \neg f(\psi)$. But $\neg f(\psi) = f(\neg\psi)$, and, therefore, $f(\phi) \vdash f(\neg\psi)$, which is what we wanted to prove.

($\vee$ elim) We leave this step, and the others, as an exercise for the reader. □

DEFINITION 1.4.3 Let T be a theory in Σ, let T' be a theory in Σ', and let $f : \Sigma \to \Sigma'$ be a reconstrual. We say that f is a **translation** or **interpretation** of T into T', written $f : T \to T'$, just in case:

$$T \vdash \phi \quad \Longrightarrow \quad T' \vdash f(\phi).$$

Note that we have used the word "interpretation" here for a mapping from one theory to another, whereas we previously used that word for a mapping from a theory to a different sort of thing, viz. a set of truth values. However, there is no genuine difference between the two notions. We will soon see that an interpretation in the latter sense is just a special case of an interpretation in the former sense. We believe that it is a mistake to think that there is some other (mathematically precise) notion of interpretation where the targets are concrete (theory-independent) things.

DISCUSSION 1.4.4 Have we been too liberal by allowing translations to map elementary sentences, such as p, to complex sentences, such as $q \wedge r$? Could a "good" translation render a sentence that has no internal complexity as a sentence that does have internal complexity? Think about it.

We will momentarily propose a definition for an equivalence of theories. However, as motivation for our definition, consider the sorts of things that can happen in translating between natural languages. If I look up the word "car" in my English–German dictionary, then I find the word "Auto." But if I look up the word "Auto" in my German–English dictionary, then I find the word "automobile." This is as it should be – the English words "car" and "automobile" are synonymous and are equally good translations of "Auto." A good round-trip translation need not end where it started, but it needs to end at something that has the *same meaning* as where it started.

But how are we to represent this notion of "having the same meaning"? The convicted Quinean might want to cover his eyes now, as we propose that a theory defines its own internal notion of sameness of meaning. (Recall what we said in the preface: that first-order metatheory is chalk full of intensional concepts.) In particular, ϕ and ψ have the same meaning relative to T just in case $T \vdash \phi \leftrightarrow \psi$. With this notion in mind, we can also say that two translations $f : T \to T'$ and $g : T \to T'$ are synonymous just in case they agree up to synonymy in the target theory T'.

DEFINITION 1.4.5 (equality of translations) Let T and T' be theories, and let both f and g be translations from T to T'. We write $f \simeq g$ just in case $T' \vdash f(p) \leftrightarrow g(p)$ for each atomic sentence p in Σ.

With this looser notion of equality of translations, we are ready to propose a notion of an equivalence between theories.

DEFINITION 1.4.6 For each theory T, the identity translation $1_T : T \to T$ is given by the identity reconstrual on Σ. If $f : T \to T'$ and $g : T' \to T$ are translations, we let gf denote the translation from T to T given by $(gf)(p) = g(f(p))$, for each atomic sentence p of Σ. Theories T and T' are said to be **homotopy equivalent**, or simply **equivalent**, just in case there are translations $f : T \to T'$ and $g : T' \to T$ such that $gf \simeq 1_T$ and $fg \simeq 1_{T'}$.

EXERCISE 1.4.7 Prove that if v is a model of T', and $f : T \to T'$ is a translation, then $v \circ f$ is a model of T. Here $v \circ f$ is the interpretation of Σ obtained by applying f first, and then applying v.

EXERCISE 1.4.8 Prove that if $f : T \to T'$ is a translation, and T' is consistent, then T is consistent.

2 The Category of Sets

2.1 Introduction

In the previous chapter, we started to reason about theories (in propositional logic) without explicitly saying anything about the rules of reasoning that we would be permitted to use. Now we need to talk more explicitly about the theory we will use to talk about theories, i.e., our *metatheory*. We want our metatheory M to be able to describe theories, which we can take in the first instance to be "collections of sentences," or better, "structured collections of sentences." What's more, sentences themselves are structured collections of symbols. Fortunately, we won't need to press the inquiry further into the question of the nature of symbols. It will suffice to assume that there are enough symbols and that there is some primitive notion of identity of symbols. For example, I assume that you understand that "p" is the same symbol as "p" and is different from "q."

Fortunately, there is a theory of collections of things lying close to hand, namely "the theory of sets." At the beginning of the twentieth century, much effort was given to clarifying the theory of sets, since it was intended to serve as a foundation for all of mathematics. Amazingly, the theory of sets can be formalized in first-order logic with only one nonlogical symbol, viz. a binary relation symbol "$\in$." In the resulting first-order theory – usually called Zermelo–Frankel set theory – the quantifiers can be thought of as ranging over sets, and the relation symbol $\in$ can be used to define further notions such as subset, Cartesian products of sets, functions from one set to another, etc.

Set theory can be presented informally (sometimes called "naive set theory") or formally ("axiomatic set theory"). In both cases, the relation $\in$ is primitive. However, we're going to approach things from a different angle. We're not concerned as much with what sets *are*, but with what we can *do* with them. Thus, I'll present a version of ETCS, the elementary theory of the category of sets. Here "elementary theory" indicates that this theory can be formalized in elementary (i.e., first-order) logic. The phrase "category of sets" indicates that this theory treats the collection of sets as a structured object – a category consisting of sets and functions between them.

Axiom 1: Sets Is a Category

Sets is a **category**, i.e., it consists of two kinds of things: objects, which we call **sets**, and arrows, which we call **functions**. To say that **Sets** is a category means that

1. Every function f has a domain set $d_0 f$ and a codomain set $d_1 f$. We write $f : X \to Y$ to indicate that $X = d_0 f$ and $Y = d_1 f$.
2. Compatible functions can be composed. For example, if $f : X \to Y$ and $g : Y \to Z$ are functions, then $g \circ f : X \to Z$ is a function. (We frequently abbreviate $g \circ f$ as gf.)
3. Composition of functions is associative:

$$h \circ (g \circ f) = (h \circ g) \circ f$$

when all these compositions are defined.

4. For each set X, there is a function $1_X : X \to X$ that acts as a left and right identity relative to composition.

DISCUSSION 2.1.1 If our goal was to formalize ETCS rigorously in first-order logic, we might use two-sorted logic, with one sort for sets and one sort for functions. We will introduce the apparatus of many-sorted logic in Chapter 5. The primitive vocabulary of this theory would include symbols $\circ, d_0, d_1, 1$, but it would *not* include the symbol $\in$. In other words, containment is *not* a primitive notion of ETCS.

Set theory makes frequent use of bracket notation, such as

$$\{n \in N \mid n > 17\}.$$

These symbols should be read as "the set of n in N such that $n > 17$." Similarly, $\{x, y\}$ designates a set consisting of elements x and y. But so far, we have no rules for reasoning about such sets. In the following sections, we will gradually add axioms until it becomes clear which rules of inference are permitted vis-á-vis sets.

Suppose for a moment that we understand the bracket notation, and suppose that X and Y are sets. Then, given an element $x \in X$ and an element $y \in Y$, we can take the set $\{x, \{x, y\}\}$ as an "ordered pair" consisting of x and y. The pair is ordered because x and y play asymmetric roles: the element x occurs by itself, as well as with the element y. If we could then gather together these ordered pairs into a single set, we would designate it by $X \times Y$, which we call the **Cartesian product** of X and Y. The Cartesian product construction should be familiar from high school mathematics. For example, the plane (with x and y coordinates) is the Cartesian product of two copies of the real number line.

In typical presentations of set theory, the existence of product sets is derived from other axioms. Here we will proceed in the opposite direction: we will take the notion of a product set as primitive.

Axiom 2: Cartesian Products

For any two sets X and Y, there is a set $X \times Y$ and functions $\pi_0 : X \times Y \to X$ and $\pi_1 : X \times Y \to Y$, such that for any other set Z and functions $f : Z \to X$ and $g : Z \to Y$, there is a unique function $\langle f, g \rangle : Z \to X \times Y$, such that $\pi_0 \langle f, g \rangle = f$ and $\pi_1 \langle f, g \rangle = g$.

Here the angle brackets $\langle f, g \rangle$ are not intended to indicate anything about the internal structure of the denoted function. This notation is chosen merely to indicate that $\langle f, g \rangle$ is uniquely determined by f and g.

The defining conditions of a product set can be visualized by means of an arrow diagram.

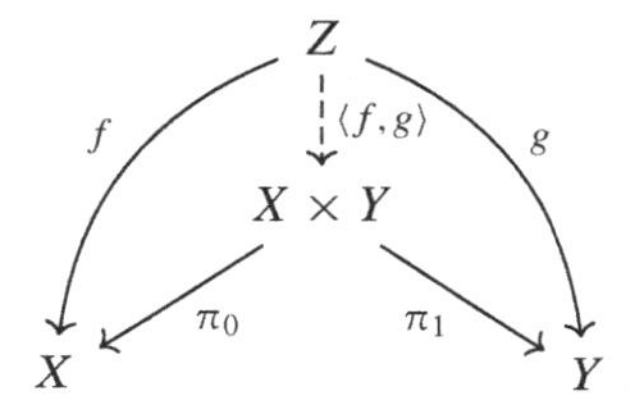

Here each node represents a set, and arrows between nodes represent functions. The dashed arrow is meant to indicate that the axiom asserts the existence of such an arrow (dependent on the existence of the other arrows in the diagram).

DISCUSSION 2.1.2 There is a close analogy between the defining conditions of a Cartesian product and the introduction and elimination rules for conjunction. If $\phi \wedge \psi$ is a conjunction, then there are arrows (i.e., derivations) $\phi \wedge \psi \to \phi$ and $\phi \wedge \psi \to \psi$. That's the $\wedge$ elimination rule. Moreover, for any sentence θ, if there are derivations $\theta \to \phi$ and $\theta \to \psi$, then there is a unique derivation $\theta \to \phi \wedge \psi$. That's the $\wedge$ introduction rule.

DEFINITION 2.1.3 Let γ and γ' be paths of arrows in a diagram that begin and end at the same node. We say that γ and γ' **commute** just in case the composition of the functions along γ is equal to the composition of the functions along γ'. We say that the diagram as a whole **commutes** just in case any two paths between nodes are equal. Thus, for example, the preceding product diagram commutes.

The functions $\pi_0 : X \times Y \to X$ and $\pi_1 : X \times Y \to Y$ are typically called **projections** of the product. What features do these projections have? Before we say more on that score, let's pause to talk about features of functions.

You may have heard before of some properties of functions such as being one-to-one, or onto, or continuous, etc. For bare sets, there is no notion of continuity of functions, per se. And with only the first two axioms in place, we do not yet have the means to define what it means for a function to be one-to-one or onto. Indeed, recall that a function $f : X \to Y$ is typically said to be one-to-one just in case $f(x) = f(y)$ implies $x = y$ for any two "points" x and y of X. But we don't yet have a notion of points!

Nonetheless, there are point-free surrogates for the notions of being one-to-one and onto.

DEFINITION 2.1.4 A function $f : X \to Y$ is said to be a **monomorphism** just in case for any two functions $g, h : Z \rightrightarrows X$, if $fg = fh$, then $g = h$.

DEFINITION 2.1.5 A function $f : X \to Y$ is said to be a **epimorphism** just in case for any two functions $g, h : Y \to Z$, if $gf = hf$, then $g = h$.

We will frequently say, "... is monic" as shorthand for "... is a monomorphism," and "... is epi" for "... is an epimorphism."

DEFINITION 2.1.6 A function $f : X \to Y$ is said to be an **isomorphism** just in case there is a function $g : Y \to X$ such that $gf = 1_X$ and $fg = 1_Y$. If there is an isomorphism $f : X \to Y$, we say that X and Y are **isomorphic**, and we write $X \cong Y$.

EXERCISE 2.1.7 Show the following:

1. If gf is monic, then f is monic.
2. If fg is epi, then f is epi.
3. If f and g are monic, then gf is monic.
4. If f and g are epi, then gf is epi.
5. If f is an isomorphism, then f is epi and monic.

PROPOSITION 2.1.8 *Suppose that both* (W, π_0, π_1) *and* (W', π_0', π_1') *are Cartesian products of* X *and* Y. *Then there is an isomorphism* $f : W \to W'$ *such that* $\pi_0' f = \pi_0$ *and* $\pi_1' f = \pi_1$.

Proof Since (W', π_0', π_1') is a Cartesian product of X and Y, there is a unique function $f : W \to W'$ such that $\pi_0' f = \pi_0$ and $\pi_1' f = \pi_1$. Since (W, π_0, π_1) is also a product of X and Y, there is a unique function $g : W' \to W$ such that $\pi_0 g = \pi_0'$ and $\pi_1 g = \pi_1'$. We claim that f and g are inverse to each other. Indeed,

$$\pi_i' \circ (f \circ g) = \pi_i \circ g = \pi_i'$$

for $i = 0, 1$. Thus, by the uniqueness clause in the definition of Cartesian products, $f \circ g = 1_{W'}$. A similar argument shows that $g \circ f = 1_W$. □

DEFINITION 2.1.9 If X is a set, we let $\delta : X \to X \times X$ denote the unique arrow $\langle 1_X, 1_X \rangle$ given by the definition of $X \times X$. We call δ the **diagonal** of X, or the **equality relation** on X. Note that δ is monic, since $\pi_0 \delta = 1_X$ is monic.

DEFINITION 2.1.10 Suppose that $f : W \to Y$ and $g : X \to Z$ are functions. Consider the following diagram:

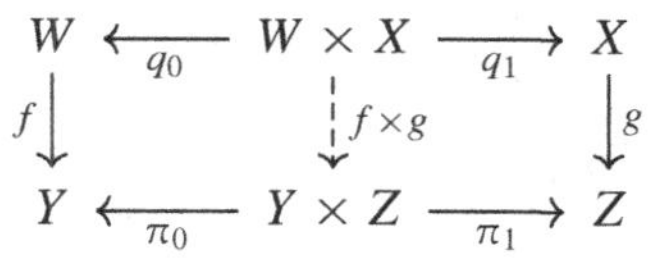

$$\begin{array}{ccccc} W & \xleftarrow[q_0]{} & W \times X & \xrightarrow[q_1]{} & X \\ {\scriptstyle f}\downarrow & & \downarrow{\scriptstyle f \times g} & & \downarrow{\scriptstyle g} \\ Y & \xleftarrow[\pi_0]{} & Y \times Z & \xrightarrow[\pi_1]{} & Z \end{array}$$

We let $f \times g = \langle f q_0, g q_1 \rangle$ be the unique function from $W \times X$ to $Y \times Z$ such that

$$\pi_0(f \times g) = f q_0, \qquad \pi_1(f \times g) = g q_1.$$

Recall here that, by the definition of products, a function into $Y \times Z$ is uniquely defined by its compositions with the projections π_0 and π_1.

PROPOSITION 2.1.11 *Suppose that $f : A \to B$ and $g : B \to C$ are functions. Then* $1_X \times (g \circ f) = (1_X \times g) \circ (1_X \times f)$.

Proof Consider the following diagram

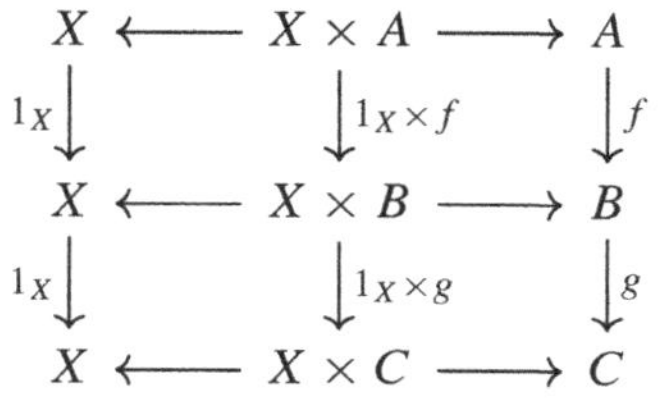

$$\begin{array}{ccccc} X & \longleftarrow & X \times A & \longrightarrow & A \\ {\scriptstyle 1_X}\downarrow & & \downarrow{\scriptstyle 1_X \times f} & & \downarrow{\scriptstyle f} \\ X & \longleftarrow & X \times B & \longrightarrow & B \\ {\scriptstyle 1_X}\downarrow & & \downarrow{\scriptstyle 1_X \times g} & & \downarrow{\scriptstyle g} \\ X & \longleftarrow & X \times C & \longrightarrow & C \end{array}$$

where $1_X \times f$ and $1_X \times g$ are constructed as in Definition 2.1.10. Since the top and bottom squares both commute, the entire diagram commutes. But then the composite arrow $(1_X \times g) \circ (1_X \times f)$ satisfies the defining properties of $1_X \times (g \circ f)$. □

EXERCISE 2.1.12 Show that $1_X \times 1_Y = 1_{X \times Y}$.

DEFINITION 2.1.13 Let X be a fixed set. Then X induces two mappings, as follows:

1. A mapping $Y \mapsto X \times Y$ of sets to sets.
2. A mapping $f \mapsto 1_X \times f$ of functions to functions. That is, if $f : Y \to Z$ is a function, then $1_X \times f : X \times Y \to X \times Z$ is a function.

By the previous results, the second mapping is compatible with the composition structure on arrows. In this case, we call the pair of mappings a **functor** from **Sets** to **Sets**.

EXERCISE 2.1.14 Suppose that $f : X \to Y$ is a function. Show that the following diagram commutes.

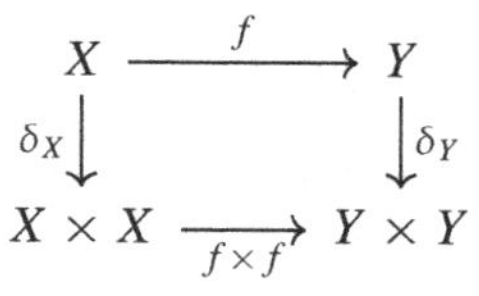

$$\begin{array}{ccc} X & \xrightarrow{f} & Y \\ {\scriptstyle \delta_X}\downarrow & & \downarrow{\scriptstyle \delta_Y} \\ X \times X & \xrightarrow[f \times f]{} & Y \times Y \end{array}$$

We will now recover the idea that sets consist of points by requiring the existence of a single-point set 1, which plays the privileged role of determining identity of functions.

Axiom 3: Terminal Object

There is a set 1 with the following two features:

1. For any set X, there is a unique function

   $$X \xrightarrow{\beta_X} 1.$$

   In this case, we say that 1 is a **terminal object** for **Sets**.
2. For any sets X and Y, and functions $f, g : X \rightrightarrows Y$, if $f \circ x = g \circ x$ for all functions $x : 1 \to X$, then $f = g$. In this case, we say that 1 is a **separator** for **Sets**.

The reader may wish to note that for a general category, a **terminal object** is required only to have the first of the two properties. So we are not merely requiring that **Sets** has a terminal object; we are requiring that it has a terminal object that also serves as a separator for functions.

EXERCISE 2.1.15 Show that if X and Y are terminal objects in a category, then $X \cong Y$.

DEFINITION 2.1.16 We write $x \in X$ to indicate that $x : 1 \to X$ is a function, and we say that x is an **element** of X. We say that X is **nonempty** just in case it has at least one element. If $f : X \to Y$ is a function, we sometimes write $f(x)$ for $f \circ x$. With this notation, the statement that 1 is a separator says: $f = g$ if and only if $f(x) = g(x)$, for all $x \in X$.

DISCUSSION 2.1.17 In ZF set theory, equality between functions is completely determined by equality between sets. Indeed, in ZF, functions $f, g : X \rightrightarrows Y$ are defined to be certain subsets of $X \times Y$; and subsets of $X \times Y$ are defined to be equal just in case they contain the same elements. In the ETCS approach to set theory, equality between functions is primitive, and Axiom 3 stipulates that this equality can be detected by checking elements.

Some might see this difference as arguing in favor of ZF; it is more parsimonious, because it derives $f = g$ from something more fundamental. However, the defender of ETCS might claim in reply that her theory defines $x \in y$ from something more fundamental. Which is *really* more fundamental, equality between arrows (functions) or containment of objects (sets)? We'll leave that for other philosophers to think about.

EXERCISE 2.1.18 Show that any function $x : 1 \to X$ is monic.

PROPOSITION 2.1.19 *A set X has exactly one element if and only if $X \cong 1$.*

Proof The terminal object 1 has exactly one element, since there is a unique function $1 \to 1$.

Suppose now that X has exactly one element $x : 1 \to X$. We will show that X is a terminal object. First, for any set Y, there is a function $x \circ \beta_Y$ from Y to X. Now suppose that f, g are functions from Y to X such that $f \neq g$. By Axiom 3, there is an element $y \in Y$ such that $fy \neq gy$. But then X has more than one element, a contradiction. Therefore, there is a unique function from Y to X, and X is a terminal object. □

PROPOSITION 2.1.20 *In any category with a terminal object* 1*, any object* X *is itself a Cartesian product of* X *and* 1.

Proof We have the obvious projections $\pi_0 = 1_X : X \to X$ and $\pi_1 = \beta_X : X \to 1$. Now let Y be an object, and let $f : Y \to X$ and $g : Y \to 1$ be arrows. We claim that $f : Y \to X$ is the unique arrow such that $1_X f = f$ and $\beta_X f = g$. To see that f satisfies this condition, note that $g : Y \to 1$ must be β_Y, the unique arrow from Y to the terminal object. If h is another arrow that satisfies this condition, then $h = 1_X h = f$. □

PROPOSITION 2.1.21 *Let* a *and* b *be elements of* $X \times Y$. *Then* $a = b$ *if and only if* $\pi_0(a) = \pi_0(b)$ *and* $\pi_1(a) = \pi_1(b)$.

Proof Suppose that $\pi_0(a) = \pi_0(b)$ and $\pi_1(a) = \pi_1(b)$. By the uniqueness property of the product, there is a unique function $c : 1 \to X \times Y$ such that $\pi_0(c) = \pi_0(a)$ and $\pi_1(c) = \pi_1(a)$. Since a and b both satisfy this property, $a = b$. □

NOTE 2.1.22 The previous proposition justifies the use of the notation

$$X \times Y = \{\langle x, y \rangle \mid x \in X, y \in Y\}.$$

Here the identity condition for ordered pairs is given by

$$\langle x, y \rangle = \langle x', y' \rangle \qquad \text{iff} \qquad x = x' \text{ and } y = y'.$$

PROPOSITION 2.1.23 *Let* $(X \times Y, \pi_0, \pi_1)$ *be the Cartesian product of* X *and* Y. *If* Y *is nonempty, then* π_0 *is an epimorphism.*

Proof Suppose that Y is nonempty, and that $y : 1 \to Y$ is an element. Let $\beta_X : X \to 1$ be the unique map, and let $f = y \circ \beta_X$. Then $\langle 1_X, f \rangle : X \to X \times Y$ such that $\pi_0 \langle 1_X, f \rangle = 1_X$. Since 1_X is epi, π_0 is epi. □

DEFINITION 2.1.24 We say that $f : X \to Y$ is **injective** just in case: for any $x, y \in X$ if $f(x) = f(y)$, then $x = y$. Written more formally:

$$\forall x \forall y [f(x) = f(y) \to x = y].$$

NOTE 2.1.25 "Injective" is synonymous with "one-to-one."

EXERCISE 2.1.26 Let $f : X \to Y$ be a function. Show that if f is monic, then f is injective.

PROPOSITION 2.1.27 *Let* $f : X \to Y$ *be a function. If* f *is injective, then* f *is monic.*

Proof Suppose that f is injective, and let $g, h : A \to X$ be functions such that $f \circ g = f \circ h$. Then for any $a \in A$, we have $f(g(a)) = f(h(a))$. Since f is injective, $g(a) = h(a)$. Since a was an arbitrary element of A, Axiom 3 entails that $g = h$. Therefore, f is monic. □

DEFINITION 2.1.28 Let $f : X \to Y$ be a function. We say that f is **surjective** just in case: for each $y \in Y$, there is an $x \in X$ such that $f(x) = y$. Written formally:

$$\forall y \exists x [f(x) = y].$$

And in diagrammatic form:

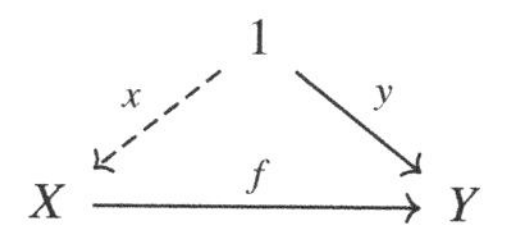

NOTE 2.1.29 "Surjective" is synonymous with "onto."

EXERCISE 2.1.30 Show that if $f : X \to Y$ is surjective, then f is an epimorphism.

We will eventually establish that all epimorphisms are surjective. However, first we need a couple more axioms. Given a set X, and some definable condition ϕ on X, we would like to be able to construct a subset consisting of those elements in X that satisfy ϕ. The usual notation here is $\{x \in X \mid \phi(x)\}$, which we read as "the x in X such that $\phi(x)$." But the important question is: which features ϕ do we allow? As an example of a definable condition ϕ, consider the condition of "having the same value under the functions f and g," – that is, $\phi(x)$ just in case $f(x) = g(x)$. We call the subset $\{x \in X \mid f(x) = g(x)\}$ the **equalizer** of f and g.

Axiom 4: Equalizers

Suppose that $f, g : X \rightrightarrows Y$ are functions. Then there is a set E and a function $m : E \to X$ with the following property: $fm = gm$, and for any other set F and function $h : F \to X$, if $fh = gh$, then there is a unique function $k : F \to E$ such that $mk = h$.

$$\begin{array}{ccccc} E & \xrightarrow{m} & X & \overset{f}{\underset{g}{\rightrightarrows}} & Y \\ {\scriptstyle k}\uparrow & \nearrow_{h} & & & \\ F & & & & \end{array}$$

We call (E, m) an **equalizer** of f and g. If we don't need to mention the object E, we will call the arrow m the equalizer of f and g.

EXERCISE 2.1.31 Suppose that (E, m) and (E', m') are both equalizers of f and g. Show that there is an isomorphism $k : E \to E'$.

DEFINITION 2.1.32 Let A, B, C be sets, and let $f : A \to C$ and $g : B \to C$ be functions. We say that g **factors through** f just in case there is a function $h : B \to A$ such that $fh = g$.

EXERCISE 2.1.33 Let $f, g : X \rightrightarrows Y$, and let $m : E \to X$ be the equalizer of f and g. Let $x \in X$. Show that x factors through m if and only if $f(x) = g(x)$.

PROPOSITION 2.1.34 *In any category, if (E, m) is the equalizer of f and g, then m is a monomorphism.*

Proof Let $x, y : Z \to E$ such that $mx = my$. Since $fmx = gmx$, there is a unique arrow $z : Z \to E$ such that $mz = mx$. Since both $mx = mx$ and $my = mx$, it follows that $x = y$. Therefore, m is monic. □

DEFINITION 2.1.35 Let $f : X \to Y$ be a function. We say that f is a **regular monomorphism** just in case f is the equalizer (up to isomorphism) of a pair of arrows $g, h : Y \rightrightarrows Z$.

EXERCISE 2.1.36 Show that if f is an epimorphism and a regular monomorphism, then f is an isomorphism.

In other approaches to set theory, one uses $\in$ to define a relation of inclusion between sets:

$$X \subseteq Y \iff \forall x(x \in X \to x \in Y).$$

We cannot define this exact notion in our approach since, for us, elements are attached to some particular set. However, for typical applications, every set under consideration will come equipped with a canonical monomorphism $m : X \to U$, where U is some fixed set. Thus, it will suffice to consider a relativized notion.

DEFINITION 2.1.37 A **subobject** or **subset** of a set X is a set B and a monomorphism $m : B \to X$, called the **inclusion** of B in X. Given two subsets $m : B \to X$ and $n : A \to X$, we say that B is a subset of A (relative to X), written $B \subseteq_X A$ just in case there is a function $k : B \to A$ such that $nk = m$. When no confusion can result, we omit X and write $B \subseteq A$.

Let $m : B \to Y$ be monic, and let $f : X \to Y$. Consider the diagram

$$f^{-1}(B) \xrightarrow{k} X \times B \underset{mp_1}{\overset{fp_0}{\rightrightarrows}} Y,$$

where $f^{-1}(B)$ is defined as the equalizer of $f\pi_0$ and mp_1. Intuitively, we have

$$\begin{aligned} f^{-1}(B) &= \{\langle x, y\rangle \in X \times B \mid f(x) = y\} \\ &= \{\langle x, y\rangle \in X \times Y \mid f(x) = y \text{ and } y \in B\} \\ &= \{x \in X \mid f(x) \in B\}. \end{aligned}$$

Now we verify that $f^{-1}(B)$ is a subset of X.

PROPOSITION 2.1.38 *The function $p_0 k : f^{-1}(B) \to X$ is monic.*

Proof To simplify notation, let $E = f^{-1}(B)$. Let $x, y : Z \to E$ such that $p_0kx = p_0ky$. Then $fp_0kx = fp_0ky$, and, hence, $mp_1kx = mp_1ky$. Since m is monic, $p_1kx = p_1ky$. Thus, $kx = ky$. (The identity of a function into $X \times B$ is determined by the identity of its projections onto X and B.) Since k is monic, $x = y$. Therefore, $p_0 k$ is monic. □

DEFINITION 2.1.39 Let $m : B \to X$ be a subobject, and let $x : 1 \to X$. We say that $x \in B$ just in case x factors through m as follows:

$$\begin{array}{ccc} & \nearrow & B \\ & & \downarrow m \\ 1 & \xrightarrow{x} & X \end{array}$$

PROPOSITION 2.1.40 *Let $A \subseteq B \subseteq X$. If $x \in A$ then $x \in B$.*

Proof

$$\begin{array}{ccccc} & \nearrow & A & \xrightarrow{k} & B \\ & & \downarrow & & \downarrow \\ 1 & \xrightarrow[x]{} & X & \xrightarrow[1_X]{} & X \end{array}$$

□

Recall that $x \in f^{-1}(B)$ means: $x : 1 \to X$ factors through the inclusion of $f^{-1}(B)$ in X. Consider the following diagram:

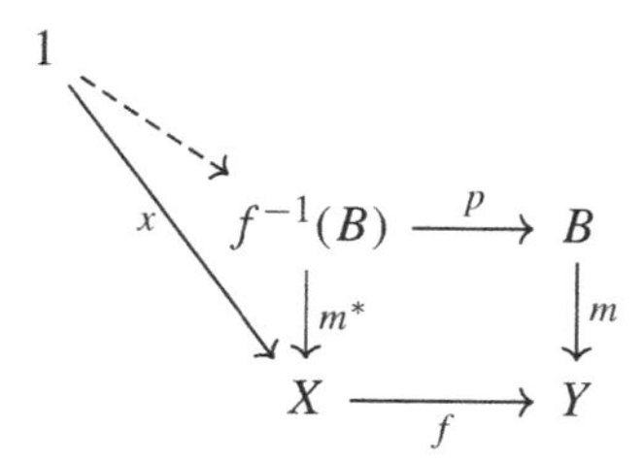

First look just at the lower-right square. This square commutes, in the sense that following the arrows from $f^{-1}(B)$ clockwise gives the same answer as following the arrows from $f^{-1}(B)$ counterclockwise. The square has another property: for any set Z, and functions $g : Z \to X$ and $h : Z \to B$, there is a unique function $k : Z \to f^{-1}(B)$ such that $m^*k = g$ and $pk = h$. When a commuting square has this property, then it's said to be a **pullback**.

PROPOSITION 2.1.41 *Let $f : X \to Y$, and let $B \subseteq Y$. Then $x \in f^{-1}(B)$ if and only if $f(x) \in B$.*

Proof If $x \in f^{-1}(B)$, then there is an arrow $\hat{x} : 1 \to f^{-1}(B)$ such that $m^*\hat{x} = x$. Thus, $fx = mp\hat{x}$, which entails that the element $f(x) \in Y$ factors through B, i.e., $f(x) \in B$. Conversely, if $f(x) \in B$, then, since the square is a pullback, $x : 1 \to X$ factors through $f^{-1}(B)$, i.e., $x \in f^{-1}(B)$. □

DEFINITION 2.1.42 Given functions $f : X \to Z$ and $g : Y \to Z$, we define

$$X \times_Z Y \;=\; \{\langle x,y\rangle \in X \times Y \mid f(x) = g(y)\}.$$

In other words, $X \times_Z Y$ is the equalizer of $f\pi_0$ and $g\pi_1$. The set $X \times_Z Y$, together with the functions $\pi_0 : X \times_Z Y \to X$ and $\pi_1 : X \times_Z Y \to Y$ is called the pullback of f and g, alternatively, the **fibered product** of f and g.

The pullback of f and g has the following distinguishing property: for any set A, and functions $h : A \to X$ and $k : A \to Y$ such that $fh = gk$, there is a unique function $j : A \to X \times_Z Y$ such that $\pi_0 j = h$ and $\pi_1 j = k$.

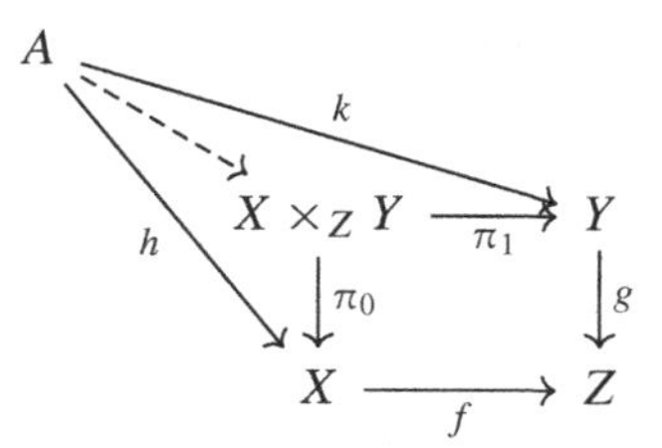

The following is an interesting special case of a pullback.

DEFINITION 2.1.43 Let $f : X \to Y$ be a function. Then the **kernel pair** of f is the pullback $X \times_Y X$, with projections $p_0 : X \times_Y X \to X$ and $p_1 : X \times_Y X \to X$. Intuitively, $X \times_Y X$ is the relation, "having the same image under f." Written in terms of braces,

$$X \times_Y X \;=\; \{\langle x,x'\rangle \in X \times X \mid f(x) = f(x')\}.$$

In particular, f is injective if and only if "having the same image under f" is coextensive with the equality relation on X. That is, $X \times_Y X = \{\langle x,x\rangle \mid x \in X\}$, which is the diagonal of X.

EXERCISE 2.1.44 Let $f : X \to Y$ be a function, and let $p_0, p_1 : X \times_Y X \rightrightarrows X$ be the kernel pair of f. Show that the following are equivalent:

1. f is a monomorphism.
2. p_0 and p_1 are isomorphisms.
3. $p_0 = p_1$.

2.2 Truth Values and Subsets

Axiom 5: Truth-Value Object

There is a set Ω with the following features:

1. Ω has exactly two elements, which we denote by $\mathsf{t} : 1 \to \Omega$ and $\mathsf{f} : 1 \to \Omega$.

2. For any set X, and subobject $m : B \to X$, there is a unique function $\chi_B : X \to \Omega$ such that the following diagram is a pullback:

$$\begin{array}{ccc} B & \longrightarrow & 1 \\ {\scriptstyle m}\downarrow & & \downarrow{\scriptstyle \mathsf{t}} \\ X & \xrightarrow[\chi_B]{} & \Omega \end{array}$$

In other words, $B = \{x \in X \mid \chi_B(x) = \mathsf{t}\}$.

Intuitively speaking, the first part of Axiom 5 says that Ω is a two-element set, say $\Omega = \{\mathsf{f}, \mathsf{t}\}$. The second part of Axiom 5 says that Ω classifies the subobjects of a set X. That is, each subobject $m : B \to X$ corresponds to a unique **characteristic function** $\chi_B : X \to \{\mathsf{f}, \mathsf{t}\}$ such that $\chi_B(x) = \mathsf{t}$ if and only if $x \in B$.

The terminal object 1 is a set with one element. Thus, it should be the case that 1 has two subsets, the empty set and 1 itself.

PROPOSITION 2.2.1 *The terminal object* 1 *has exactly two subobjects.*

Proof By Axiom 5, subobjects of 1 correspond to functions $1 \to \Omega$, that is, to elements of Ω. By Axiom 5, Ω has exactly two elements. Therefore, 1 has exactly two subobjects. □

Obviously the function $\mathsf{t} : 1 \to \Omega$ corresponds to the subobject $\mathrm{id}_1 : 1 \to 1$. Can we say more about the subobject $m : A \to 1$ corresponding to the function $\mathsf{f} : 1 \to \Omega$? Intuitively, we should have $A = \{x \in 1 \mid \mathsf{t} = \mathsf{f}\}$ – in other words, the empty set. To confirm this intuition, consider the pullback diagram:

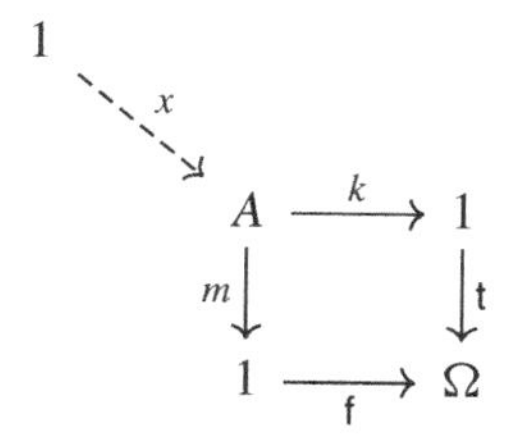

Note that m and k must both be the unique function from A to 1 – that is, $m = k = \beta_A$. Suppose that A is nonempty – i.e., there is a function $x : 1 \to A$. Then $\beta_A \circ x$ is the identity $1 \to 1$ and, since the square commutes, $\mathsf{t} = \mathsf{f}$, a contradiction. Therefore, A has no elements.

EXERCISE 2.2.2 Show that $\Omega \times \Omega$ has exactly four elements.

We now use the existence of a truth-value object in **Sets** to demonstrate further properties of functions.

EXERCISE 2.2.3 Show that, in any category, if $f : X \to Y$ is a regular monomorphism, then f is monic.

PROPOSITION 2.2.4 *Every monomorphism between sets is regular – i.e., an equalizer of a pair of parallel arrows.*

Proof Let $m : B \to X$ be monic. By Axiom 5, the following is a pullback diagram:

$$\begin{array}{ccc} B & \longrightarrow & 1 \\ {\scriptstyle m}\downarrow & & \downarrow{\scriptstyle t} \\ X & \underset{\chi_B}{\longrightarrow} & \Omega \end{array}$$

A straightforward verification shows that m is the equalizer of $X \overset{\beta_X}{\to} 1 \overset{t}{\to} \Omega$ and $\chi_B : X \to \Omega$. Therefore, m is regular monic. □

Students with some background in mathematics might assume that if a function $f : X \to Y$ is both a monomorphism and an epimorphism, then it is an isomorphism. However, that isn't true in all categories! (For example, in the category of monoids, the inclusion $i : \mathbb{N} \to \mathbb{Z}$ is epi and monic, but not an isomorphism.) Nonetheless, **Sets** is a special category, and in this case we have the result:

PROPOSITION 2.2.5 *In* **Sets**, *if a function is both a monomorphism and an epimorphism, then it is an isomorphism.*

Proof In any category, if m is regular monic and epi, then m is an isomorphism (Exercise 2.1.36). □

DEFINITION 2.2.6 Let $f : X \to Y$ be a function, and let $y \in Y$. The **fiber** over y is the subset $f^{-1}\{y\}$ of X given by the following pullback:

$$\begin{array}{ccc} f^{-1}\{y\} & \longrightarrow & 1 \\ \downarrow & & \downarrow{\scriptstyle y} \\ X & \overset{f}{\longrightarrow} & Y \end{array}$$

PROPOSITION 2.2.7 *Let $p : X \to Y$. If p is not a surjection, then there is a $y_0 \in Y$ such that the fiber $p^{-1}\{y_0\}$ is empty.*

Proof Since p is not a surjection, there is a $y_0 \in Y$ such that for all $x \in X$, $p(x) \neq y_0$. Now consider the pullback:

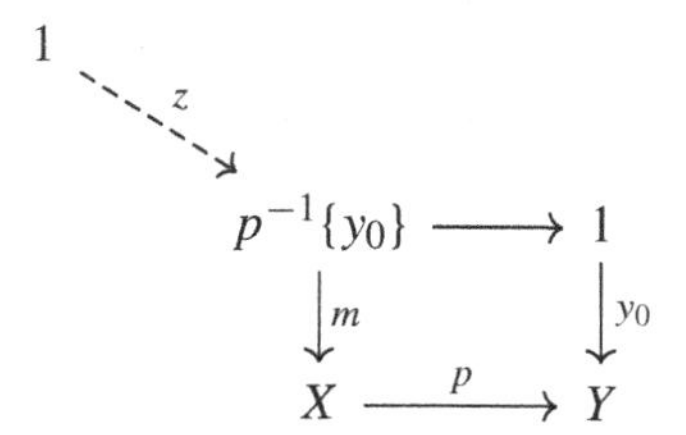

If there were a morphism $z : 1 \to p^{-1}\{y_0\}$, then we would have $p(m(z)) = y_0$, a contradiction. Therefore, $p^{-1}\{y_0\}$ is empty. □

PROPOSITION 2.2.8 *In* **Sets**, *epimorphisms are surjective.*

Proof Suppose that $p : X \to Y$ is not a surjection. Then there is a $y_0 \in Y$ such that for all $x \in X$, $p(x) \neq y_0$. Since 1 is terminal, the morphism $y_0 : 1 \to Y$ is monic. Consider the following diagram:

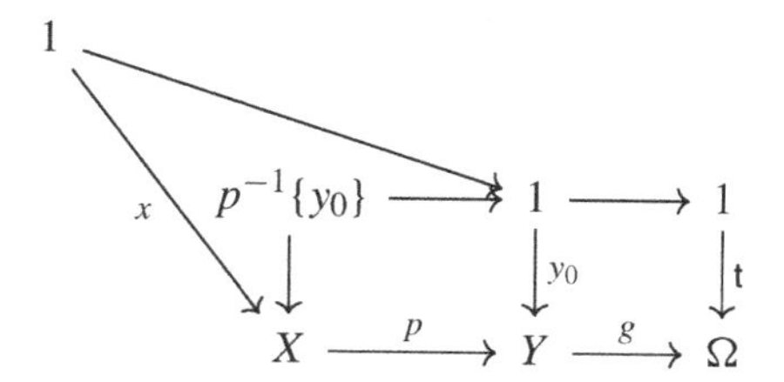

Here g is the characteristic function of $\{y_0\}$; by Axiom 5, g is the unique function that makes the right-hand square a pullback. Let $x \in X$ be arbitrary. If we had $g(p(x)) = \mathsf{t}$, then there would be an element $x' \in p^{-1}\{y_0\}$, in contradiction with the fact that the latter is empty (Proposition 2.2.7). By Axiom 5, either $g(p(x)) = \mathsf{t}$ or $g(p(x)) = \mathsf{f}$; therefore, $g(p(x)) = \mathsf{f}$. Now let h be the composite $Y \to 1 \xrightarrow{\mathsf{f}} \Omega$. Then, for any $x \in X$, we have $h(p(x)) = \mathsf{f}$. Since $g \circ p$ and $h \circ p$ agree on arbitrary $x \in X$, we have $g \circ p = h \circ p$. Since $g \neq h$, it follows that p is not an epimorphism. □

In a general category, there is no guarantee that an epimorphism pulls back to an epimorphism. However, in **Sets**, we have the following:

PROPOSITION 2.2.9 *In* **Sets**, *the pullback of an epimorphism is an epimorphism.*

Proof Suppose that $f : Y \to Z$ is epi, and let $x \in X$. Consider the pullback diagram:

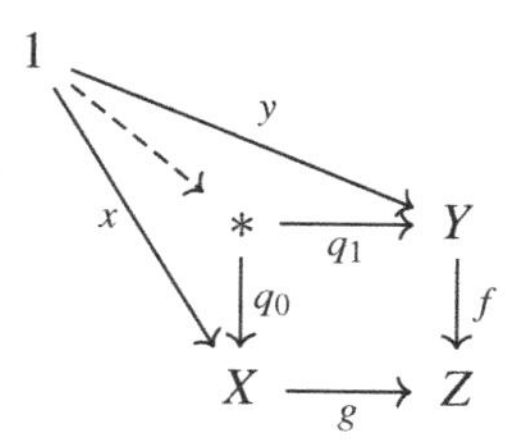

By Proposition 2.2.8, f is surjective. In particular, there is a $y \in Y$ such that $f(y) = g(x)$. Since the diagram is a pullback, there is a unique $\langle x, y\rangle : 1 \to *$ such that $q_0\langle x, y\rangle = x$ and $q_1\langle x, y\rangle = y$. Therefore, q_0 is surjective and, hence, epi. □

PROPOSITION 2.2.10 *If $f : X \to Y$ and $g : W \to Z$ are epimorphisms, then so is $f \times g : X \times W \to Y \times Z$.*

Proof Since $f \times g = (f \times 1) \circ (1 \times g)$, it will suffice to show that $f \times 1$ is epi when f is epi. Now, the following diagram is a pullback:

$$\begin{array}{ccc} X \times W & \xrightarrow{p_0} & X \\ {\scriptstyle f\times 1}\downarrow & & \downarrow{\scriptstyle f} \\ Y \times W & \xrightarrow{p_0} & Y \end{array}$$

By Proposition 2.2.9, if f is epi, then $f \times 1$ is epi. □

Suppose that $f : X \to Y$ is a function and that $p_0, p_1 : X \times_Y X \rightrightarrows X$ is the kernel pair of f. Suppose also that $h : E \to Y$ is a function, that $q_0, q_1 : E \times_Y E \rightrightarrows E$ is the kernel pair of h, and that $g : X \twoheadrightarrow E$ is an epimorphism. Then there is a unique function $b : X \times_Y X \to E \times_Y E$, such that $q_0 b = g p_0$ and $q_1 b = g p_1$.

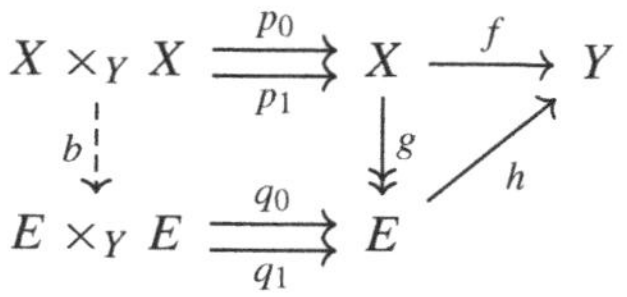

$$
\begin{array}{ccccc}
X \times_Y X & \overset{p_0}{\underset{p_1}{\rightrightarrows}} & X & \xrightarrow{f} & Y \\
{\scriptstyle b}\downarrow & & \downarrow{\scriptstyle g} & \nearrow{\scriptstyle h} & \\
E \times_Y E & \overset{q_0}{\underset{q_1}{\rightrightarrows}} & E & &
\end{array}
$$

An argument similar to the preceding argument shows that b is an epimorphism. We will use this fact to describe the properties of epimorphisms in **Sets**.

2.3 Relations

Equivalence Relations and Equivalence Classes

A relation R on a set X is a subset of $X \times X$ – i.e., a set of ordered-pairs. A relation is said to be an **equivalence relation** just in case it is reflexive, symmetric, and transitive. One particular way that equivalence relations on X arise is from functions with X as domain: given a function $f : X \to Y$, let's say that $\langle x, y \rangle \in R$ just in case $f(x) = f(y)$. (Sometimes we say that "x and y lie in the same fiber over Y.") Then R is an equivalence relation on X.

Given an equivalence relation R on X, and some element $x \in X$, let $[x] = \{y \in X \mid \langle x, y \rangle \in R\}$ denote the set of all elements of X that are equivalent to X. We say that $[x]$ is the **equivalence class** of x. It's straightforward to show that for any $x, y \in X$, either $[x] = [y]$ or $[x] \cap [y] = \emptyset$. Moreover, for any $x \in X$, we have $x \in [x]$. Thus, the equivalence classes form a **partition** of X into disjoint subsets.

We'd like now to be able to talk about the set of these equivalence classes – i.e., something that might intuitively be written as $\{[x] \mid x \in X\}$. The following axiom guarantees the existence of such a set, called X/R, and a canonical mapping $q : X \to X/R$ that takes each element $x \in X$ to its equivalence class $[x] \in X/R$.

Axiom 6: Equivalence Classes

Let R be an equivalence relation on X. Then there is a set X/R and a function $q : X \to X/R$ with the properties:

1. $\langle x, y \rangle \in R$ if and only if $q(x) = q(y)$.
2. For any set Y and function $f : X \to Y$ that is constant on equivalence classes, there is a unique function $\overline{f} : X/R \to Y$ such that $\overline{f} \circ q = f$.

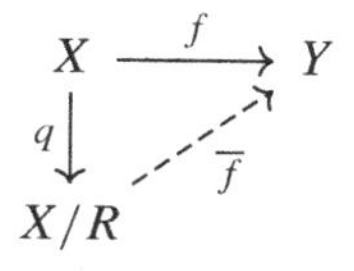

Here f is constant on equivalence classes just in case $f(x) = f(y)$ whenever $\langle x, y\rangle \in R$.

An equivalence relation R can be thought of as a subobject of $X \times X$, i.e., a subset of ordered pairs. Accordingly, there are two functions $p_0 : R \to X$ and $p_1 : R \to X$, given by $p_0\langle x, y\rangle = x$ and $p_1\langle x, y\rangle = y$. Then condition (1) in Axiom 6 says that $q \circ p_0 = q \circ p_1$. And condition (2) says that for any function $f : X \to Y$ such that $f \circ p_0 = f \circ p_1$, there is a unique function $\overline{f} : X/R \to Y$ such that $\overline{f} \circ q = f$. In this case, we say that q is a **coequalizer** of p_0 and p_1.

EXERCISE 2.3.1 Show that in any category, coequalizers are unique up to isomorphism.

EXERCISE 2.3.2 Show that in any category, a coequalizer is an epimorphism.

EXERCISE 2.3.3 For a function $f : X \to Y$, let $R = \{\langle x, y\rangle \in X \times X \mid f(x) = f(y)\}$. That is, R is the kernel pair of f. Show that R is an equivalence relation.

DEFINITION 2.3.4 A function $f : X \to Y$ is said to be a **regular epimorphism** just in case f is a coequalizer.

EXERCISE 2.3.5 Show that in any category, if $f : X \to Y$ is both a monomorphism and a regular epimorphism, then f is an isomorphism.

PROPOSITION 2.3.6 *Every epimorphism in* **Sets** *is regular. In particular, every epimorphism is the coequalizer of its kernel pair.*

Proof Let $f : X \to Y$ be an epimorphism. Let $p_0, p_1 : X \times_Y X \rightrightarrows X$ be the kernel pair of f. By Axiom 6, the coequalizer $g : X \to E$ of p_0 and p_1 exists; and since f also coequalizes p_0 and p_1, there is a unique function $m : E \to Y$ such that $f = mg$.

$$\begin{array}{ccccc} X \times_Y X & \overset{p_0}{\underset{p_1}{\rightrightarrows}} & X & \xrightarrow{f} & Y \\ \downarrow b & & \downarrow g & \nearrow m & \\ E \times_Y E & \overset{q_0}{\underset{q_1}{\rightrightarrows}} & E & & \end{array}$$

Here $E \times_Y E$ is the kernel pair of m. Since $mgp_0 = fp_0 = fp_1 = mgp_1$, there is a unique function $b : X \times_Y X \to E \times_Y E$ such that $gp_0 = q_0 b$ and $gp_1 = q_1 b$. By the considerations at the end of the previous section, b is an epimorphism. Furthermore,

$$q_0 b = gp_0 = gp_1 = q_1 b$$

and, therefore, $q_0 = q_1$. By Exercise 2.1.44, m is a monomorphism. Since $f = mg$, and f is epi, m is also epi. Therefore, by Proposition 2.2.5, m is an isomorphism. □

This last proposition actually shows that **Sets** is what is known as a **regular category**. In general, a category **C** is said to be **regular** just in case it has all finite limits and all coequalizers of kernel pairs and regular epimorphisms are stable under pullback. Now, it's known that if a category has products and equalizers, then it has all finite limits (Mac Lane, 1971, p. 113). Thus, **Sets** has all finite limits. Our most recent axiom says that **Sets** has coequalizers of kernel pairs. And, finally, all epimorphisms in **Sets** are regular, and epimorphisms in **Sets** are stable under pullback; therefore, regular epimorphisms are stable under pullback.

Regular categories have several nice features that will prove quite useful. In the remainder of this section, we will discuss one such feature: factorization of functions into a regular epimorphism followed by a monomorphism.

The Epi–Monic Factorization

Let $f : X \to Y$ be a function, and let $p_0, p_1 : X \times_Y X \rightrightarrows X$ be the kernel pair of f. By Axiom 6, the kernel pair has a coequalizer $g : X \twoheadrightarrow E$. Since f also coequalizes p_0 and p_1, there is a unique function $m : E \to Y$ such that $f = mg$.

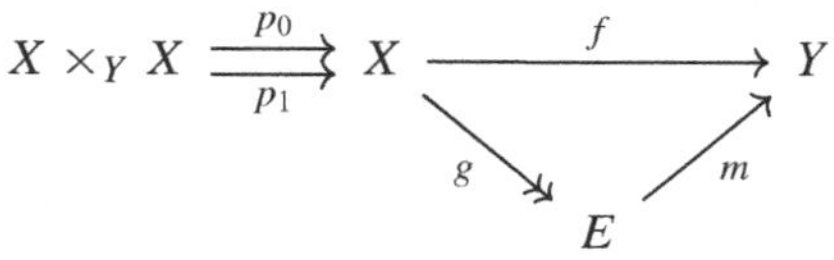

$$X \times_Y X \underset{p_1}{\overset{p_0}{\rightrightarrows}} X \xrightarrow{f} Y, \qquad X \xrightarrow{g} E \xrightarrow{m} Y$$

An argument similar to the one in Proposition 2.3.6 shows that m is a monomorphism. Thus, (E, m) is a subobject of Y, which we call the **image** of X under f, and we write $E = f(X)$. The pair (g, m) is called the **epi–monic factorization** of f. Since epis are surjections, and monics are injections, (g, m) can also be called the surjective–injective factorization.

DEFINITION 2.3.7 Suppose that A is a subset of X, in particular, $n : A \to X$ is monic. Then $f \circ n : A \to Y$, and we let $f(A)$ denote the image of A under $f \circ n$.

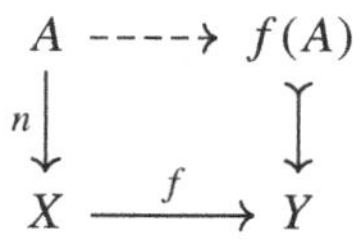

$$\begin{array}{ccc} A & \dashrightarrow & f(A) \\ {\scriptstyle n}\downarrow & & \downarrow \\ X & \xrightarrow{f} & Y \end{array}$$

We also use the suggestive notation

$$f(A) = \exists_f(A) = \{y \in Y \mid \exists x \in A. f(x) = y\}.$$

PROPOSITION 2.3.8 *Let $f : X \to Y$ be a function, and let A be a subobject of X. The image $f(A)$ is the smallest subobject of Y through which f factors.*

Proof Let $e : X \to Q$ and $m : Q \to Y$ be the epi–monic factorization of f. Suppose that $n : B \to Y$ is a subobject, and that f factors through n, say $f = ng$. Consider the following diagram.

$$
\begin{array}{ccccc}
E & \overset{p_0}{\underset{p_1}{\rightrightarrows}} & X & \xrightarrow{f} & Y \\
 & & {\scriptstyle e}\downarrow & \searrow^{g} & \uparrow{\scriptstyle n} \\
 & & Q & \overset{k}{\dashrightarrow} & B
\end{array}
$$

Then $ngp_0 = fp_0 = fp_1 = ngp_1$, since p_0, p_1 is the kernel pair of f. Since n is monic, $gp_0 = gp_1$ – i.e., g coequalizes p_0 and p_1. Since $e : X \to Q$ is the coequalizer of p_0 and p_1, there is a unique function $k : Q \to B$ such that $ke = g$. By uniqueness of the epi–monic factorization, $nk = m$. Therefore, $Q \subseteq B$. □

PROPOSITION 2.3.9 *For any $A \subseteq X$ and $B \subseteq Y$, we have*

$$A \subseteq f^{-1}(B) \quad \text{if and only if} \quad \exists_f(A) \subseteq B.$$

Proof Suppose first that $A \subseteq f^{-1}(B)$, in particular that $k : A \to f^{-1}(B)$. Consider the following diagram:

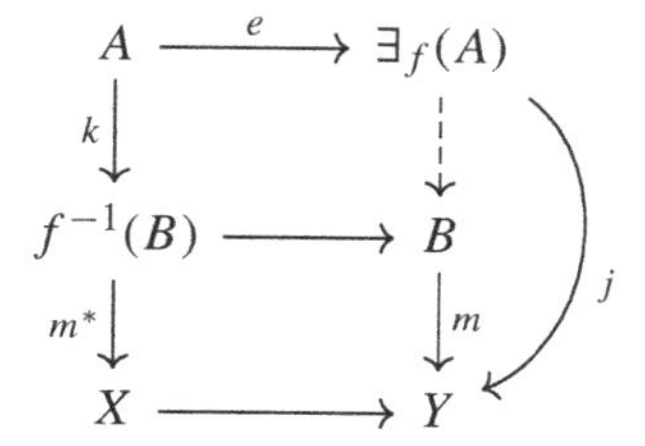

By definition, je is the epi–monic factorization of fm^*k. Since fm^*k also factors through $m : B \to Y$, we have $\exists_f(A) \subseteq B$, by Proposition 2.3.8.

Suppose now that $\exists_f(A) \subseteq B$. Using the fact that the lower square in the diagram is a pullback, we see that there is an arrow $k : A \to f^{-1}(B)$ such that m^*k is the inclusion of A in X. That is, $A \subseteq f^{-1}(B)$. □

EXERCISE 2.3.10 Use the previous result to show that $A \subseteq f^{-1}(\exists_f(A))$, for any subset A of X.

Functional Relations

DEFINITION 2.3.11 A relation $R \subseteq X \times Y$ is said to be **functional** just in case for each $x \in X$ there is a unique $y \in Y$ such that $\langle x, y \rangle \in R$.

DEFINITION 2.3.12 Suppose that $f : X \to Y$ is a function. We let $\text{graph}(f) = \{\langle x, y \rangle \mid f(x) = y\}$.

EXERCISE 2.3.13 Show that graph(f) is a functional relation.

The following result is helpful for establishing the existence of arrows $f : X \to Y$.

PROPOSITION 2.3.14 *Let $R \subseteq X \times Y$ be a functional relation. Then there is a unique function $f : X \to Y$ such that $R = $* graph($f$).

The proof of this result is somewhat complicated, and we omit it (for the time being).

2.4 Colimits

Axiom 7: Coproducts

For any two sets X, Y, there is a set $X \amalg Y$ and functions $i_0 : X \to X \amalg Y$ and $i_1 : Y \to X \amalg Y$ with the feature that for any set Z and functions $f : X \to Z$ and $g : Y \to Z$, there is a unique function $f \amalg g : X \amalg Y \to Z$ such that $(f \amalg g) \circ i_0 = f$ and $(f \amalg g) \circ i_1 = g$.

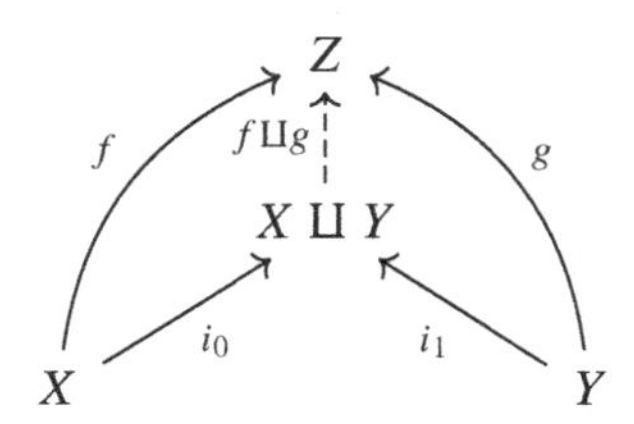

We call $X \amalg Y$ the **coproduct** of X and Y. We call i_0 and i_1 the coprojections of the coproduct.

Intuitively speaking, the coproduct $X \amalg Y$ is the disjoint union of the sets X and Y. What we mean by "disjoint" here is that if X and Y share elements in common (which doesn't make sense in our framework but does in some frameworks), then these elements are disidentified before the union is taken. For example, in terms of elements, we could think of $X \amalg Y$ as consisting of elements of the form $\langle x, 0\rangle$, with $x \in X$, and elements of the form $\langle y, 1\rangle$, with $y \in Y$. Thus, if x is contained in both X and Y, then $X \amalg Y$ contains two separate copies of x, namely $\langle x, 0\rangle$ and $\langle x, 1\rangle$.

We now show that that the inclusions $i_0 : X \to X \amalg Y$ and $i_1 : Y \to X \amalg Y$ do, in fact, have disjoint images.

PROPOSITION 2.4.1 *Coproducts in* **Sets** *are disjoint. In other words, if $i_0 : X \to X \amalg Y$ and $i_1 : Y \to X \amalg Y$ are the coprojections, then $i_0(x) \neq i_1(y)$ for all $x \in X$ and $y \in Y$.*

Proof Suppose for reductio ad absurdum that $i_0(x) = i_1(y)$. Let $g : X \to \Omega$ be the unique map that factors through $\mathsf{t} : 1 \to \Omega$. Let $h : Y \to \Omega$ be the unique map that factors through $\mathsf{f} : 1 \to \Omega$. By the universal property of the coproduct, there is a unique function $g \amalg h : X \amalg Y \to \Omega$ such that $(g \amalg h)i_0 = g$ and $(g \amalg h)i_1 = h$. Thus, we have

$$\mathsf{t} = g(x) = (g \amalg h)i_0 x = (g \amalg h)i_1 y = h(y) = \mathsf{f},$$

a contradiction. Therefore, $i_0(x) \neq i_1(y)$, and the ranges of i_0 and i_1 are disjoint. □

PROPOSITION 2.4.2 *The coprojections $i_0 : X \to X \amalg Y$ and $i_1 : Y \to X \amalg Y$ are monomorphisms.*

Proof We will show that i_0 is monic; the result then follows by symmetry. Suppose first that X has no elements. Then i_0 is trivially injective, hence monic by Proposition 2.1.27.

Suppose now that X has an element $x : 1 \to X$. Let $g = x \circ \beta_Y$, where $\beta_Y : Y \to 1$. Then $(1_X \amalg g)i_0 = 1_X$, and Exercise 2.1.7 entails that i_0 is monic. □

PROPOSITION 2.4.3 *The coprojections are jointly surjective. That is, for each $z \in X \amalg Y$, either there is an $x \in X$ such that $z = i_0(x)$, or there is a $y \in Y$ such that $z = i_1(y)$.*

Proof Suppose for reductio ad absurdum that z is neither in the image of i_0 nor in the image of i_1. Let $g : (X \amalg Y) \to \Omega$ be the characteristic function of $\{z_0\}$. Then for all $x \in X$, $g(i_0(x)) = f$. And for all $y \in Y$, $g(i_1(y)) = f$. Now let $h : (X \amalg Y) \to \Omega$ be the constant f function, i.e., $h(z) = f$ for all $z \in X \amalg Y$. Then $gi_0 = hi_0$ and $gi_1 = hi_1$. Since functions from $X \amalg Y$ are determined by their coprojections, $g = h$, a contradiction. Therefore, all $z \in X \amalg Y$ are either in the range of i_0 or in the range of i_1. □

PROPOSITION 2.4.4 *The function* $\mathsf{t} \amalg \mathsf{f} : 1 \amalg 1 \to \Omega$ *is an isomorphism.*

Proof Consider the diagram:

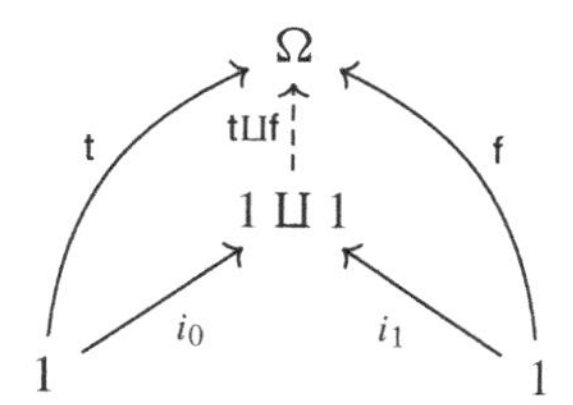

Then $\mathsf{t} \amalg \mathsf{f}$ is monic, since every element of $1 \amalg 1$ factors through either i_0 or i_1 (Proposition 2.4.3), and since $\mathsf{t} \neq \mathsf{f}$. Furthermore, $\mathsf{t} \amalg \mathsf{f}$ is epi since t and f are the only elements of Ω. By Proposition 2.2.5, $\mathsf{t} \amalg \mathsf{f}$ is an isomorphism. □

PROPOSITION 2.4.5 *Let X be a set, and let B be a subset of X. Then the inclusion $B \amalg X\backslash B \to X$ is an isomorphism.*

Proof Using the fact that Ω is Boolean, for every $x \in X$, either $x \in B$ or $x \in X\backslash B$. Thus the inclusion $B \amalg X\backslash B \to X$ is a bijection, hence an isomorphism. □

Axiom 8: Empty Set

There is a set $\emptyset$ with the following properties:

1. For any set X, there is a unique function

   $$\emptyset \xrightarrow{\alpha_X} X.$$

   In this case, we say that $\emptyset$ is an **initial object** in **Sets**.
2. $\emptyset$ is empty – i.e., there is no function $x : 1 \to \emptyset$.

EXERCISE 2.4.6 Show that in any category with coproducts, if A is an initial object, then $X \amalg A \cong X$, for any object X.

PROPOSITION 2.4.7 *Any function $f : X \to \emptyset$ is an isomorphism.*

Proof Since $\emptyset$ has no elements, f is trivially surjective. We now claim that X has no elements. Indeed, if $x : 1 \to X$ is an element of X, then $f(x)$ is an element of $\emptyset$. Since X has no elements, f is trivially injective. By Proposition 2.2.5, f is an isomorphism. □

PROPOSITION 2.4.8 *A set X has no elements if and only if $X \cong \emptyset$.*

Proof By Axiom 8, the set $\emptyset$ has no elements. Thus, if $X \cong \emptyset$, then X has no elements.

Suppose now that X has no elements. Since $\emptyset$ is an initial object, there is a unique arrow $\alpha_X : \emptyset \to X$. Since X has no elements, α_X is trivially surjective. Since $\emptyset$ has no elements, α_X is trivially injective. By Proposition 2.2.5, f is an isomorphism. □

2.5 Sets of Functions and Sets of Subsets

(Note: The following section is highly technical and can be skipped on a first reading.)

One distinctive feature of the category of sets is its ability to model almost any mathematical construction. One such construction is gathering together old things into a new set. For example, given two sets A and X, can we form a set X^A of all functions from A to X? Similarly, given a set X, can we form a set $\mathscr{P}X$ of all subsets of X?

As usual, we won't be interested in hard questions about what it takes to be a set. Rather, we're interested in hypothetical questions: if such a set existed, what would it be like? The crucial features of X^A seem to be captured by the following axiom:

Axiom 9: Exponential Objects

Suppose that A and X are sets. Then there is a set X^A, and a function $e_X : A \times X^A \to X$ such that for any set Z and function $f : A \times Z \to X$, there is a unique function $f^\sharp : Z \to X^A$ such that $e_X \circ (1_A \times f^\sharp) = f$.

$$\begin{array}{ccc} A \times X^A & \xrightarrow{e_X} & X \\ {\scriptstyle 1_A \times f^\sharp}\uparrow & \nearrow_{f} & \\ A \times Z & & \end{array}$$

The set X^A is called an **exponential object**, and the function $f^\sharp : Z \to X^A$ is called the **transpose** of $f : A \times Z \to X$.

The way to remember this axiom is to think of Y^X as the set of functions from X to Y, and to think of $e : X \times Y^X \to Y$ as a metafunction that takes an element $f \in Y^X$ and an element $x \in X$ and returns the value $e(f,x) = f(x)$. For this reason, $e : X \times Y^X \to Y$ is sometimes called the **evaluation function**.

Note further that if $f : X \times Z \to Y$ is a function, then for each $z \in Z$, $f(-,z)$ is a function from $X \to Y$. In other words, f corresponds uniquely to a function from Z to functions from Y to X. This latter function is the transpose $f^\sharp : Z \to Y^X$ of f.

We have written Axiom 9 in first-order fashion, but it might help to think of it as stating that there is a one-to-one correspondence between two sets:

$$\hom(X \times Z, Y) \cong \hom(Z, Y^X),$$

where $\hom(A, B)$ is thought of as the set of functions from A to B. As a particular case, when $Z = 1$, the terminal object, we have

$$\hom(X, Y) \cong \hom(1, Y^X).$$

In other words, elements of Y^X in the "internal sense" correspond to elements of $\hom(X, Y)$ in the "external sense."

Consider now the following special case of the construction:

$$\begin{array}{ccc} A \times X^A & \xrightarrow{e_X} & X^A \\ {\scriptstyle 1\times e^\sharp}\uparrow & \nearrow_{e_X} & \\ A \times X^A & & \end{array}$$

Thus, $e_X^\sharp = 1_{X^A}$.

DEFINITION 2.5.1 Suppose that $g : Y \to Z$ is a function. We let $g^A : X^A \to Y^A$ denote the transpose of the function:

$$A \times Y^A \xrightarrow{e_Y} Y \xrightarrow{g} Z.$$

That is, $g^A = (g \circ e_Y)^\sharp$, and the following diagram commutes:

$$\begin{array}{ccc} A \times Z^A & \xrightarrow{e_Z} & Z \\ {\scriptstyle 1\times g^A}\uparrow & & \uparrow{\scriptstyle g} \\ A \times Y^A & \xrightarrow{e_Y} & Y \end{array}$$

PROPOSITION 2.5.2 *Let $f : A \times X \to Y$ and $g : Y \to Z$ be functions. Then $(g \circ f)^\sharp = g^A \circ f^\sharp$.*

Proof Consider the following diagram:

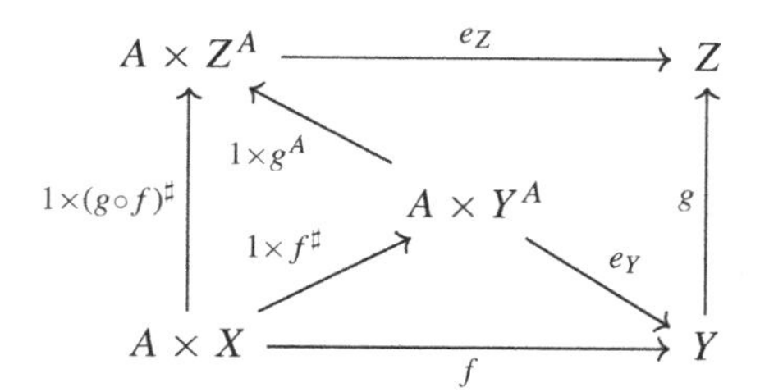

The bottom triangle commutes by the definition of $f^\sharp$. The upper-right triangle commutes by the definition of g^A. And the outer square commutes by the definition of $(g \circ f)^\sharp$. It follows that

$$e_Z \circ (1 \times (g^A \circ f^\sharp)) = g \circ f,$$

and hence $g^A \circ f^\sharp = (g \circ f)^\sharp$. □

Consider now the following particular case:

$$\begin{array}{ccc} A \times (A \times X)^A & \xrightarrow{e} & A \times X \\ {\scriptstyle 1\times p}\uparrow & \nearrow_{1} & \\ A \times X & & \end{array}$$

Here $p = 1^\sharp$ is the unique function such that $e(1_A \times p) = 1_{A\times X}$. Intuitively, we can think of p as the function that takes an element $x \in X$ and returns the function $p_x : A \to A \times X$ such that $p_x(a) = \langle a, x\rangle$. Thus, $(1 \times p)\langle a, x\rangle = \langle a, p_x\rangle$, and $e(1 \times p)\langle a, x\rangle = p_x(a) = \langle a, x\rangle$.

DEFINITION 2.5.3 Suppose that $f : Z \to X^A$ is a function. We define $f^\flat : Z \times A \to X$ to be the following composite function:

$$A \times Z \xrightarrow{1\times f} A \times X^A \xrightarrow{e_X} X.$$

PROPOSITION 2.5.4 *Let $f : X \to Y$ and $g : Y \to Z^A$ be functions. Then $(g \circ f)^\flat = g^\flat \circ (1_A \times f)$.*

Proof By definition,

$$(g \circ f)^\flat = e_X \circ (1 \times (g \circ f)) = e_X \circ (1 \times g) \circ (1 \times f) = g^\flat \circ (1 \times f).$$

□

PROPOSITION 2.5.5 *For any function $f : A \times Z \to X$, we have $(f^\sharp)^\flat = f$.*

Proof By the definitions, we have

$$(f^\sharp)^\flat = e_X \circ (1 \times f^\sharp) = f.$$

□

PROPOSITION 2.5.6 *For any function $f : Z \to X^A$, we have $(f^\flat)^\sharp = f$.*

Proof By definition, $(f^\flat)^\sharp$ is the unique function such that $e_X \circ (1 \times (f^\flat)^\sharp) = f^\flat$. But also $e_X \circ (1 \times f) = f^\flat$. Therefore, $(f^\flat)^\sharp = f$. □

PROPOSITION 2.5.7 *For any set X, we have $X^1 \cong X$.*

Proof Let $e : 1 \times X^1 \to X$ be the evaluation function from Axiom 9. We claim that e is a bijection. Recall that there is a natural isomorphism $i : 1 \times 1 \to 1$. Consider the following diagram:

$$\begin{array}{ccc} 1 \times X^1 & \xrightarrow{e} & X \\ {\scriptstyle 1\times x^\sharp}\uparrow & & \uparrow{\scriptstyle x} \\ 1 \times 1 & \xrightarrow{i} & 1 \end{array}$$

That is, for any element $x : 1 \to X$, there is a unique element $x^\sharp$ of X^1 such that $e(1 \times x^\sharp) = x$. Thus, e is a bijection, and $X \cong 1 \times X^1$ is isomorphic to X. □

PROPOSITION 2.5.8 *For any set X, we have $X^\emptyset \cong 1$.*

Proof Elements of $X^\emptyset$ correspond to functions $\emptyset \to X$. There is exactly one such function, hence $X^\emptyset$ has exactly one element $x : 1 \to X^\emptyset$. Thus, x is a bijection, and $X^\emptyset \cong 1$. □

PROPOSITION 2.5.9 *For any sets A, X, Y, we have $(X \times Y)^A \cong X^A \times Y^A$.*

Proof An elegant proof of this proposition would note that $(-)^A$ is a functor, and is right adjoint to the functor $A \times (-)$. Since right adjoints preserve products, $(X \times Y)^A \cong X^A \times Y^A$. Nonetheless, we will go into further detail.

By uniqueness of Cartesian products, it will suffice to show that $(X \times Y)^A$ is a Cartesian product of X^A and Y^A, with projections π_0^A and π_1^A. Let Z be an arbitrary set, and let $f : Z \to X^A$ and $g : Z \to Y^A$ be functions. Now take $\gamma = \langle f^\flat, g^\flat \rangle^\sharp$, where $f^\flat : A \times Z \to X$ and $g^\flat : A \times Z \to Y$.

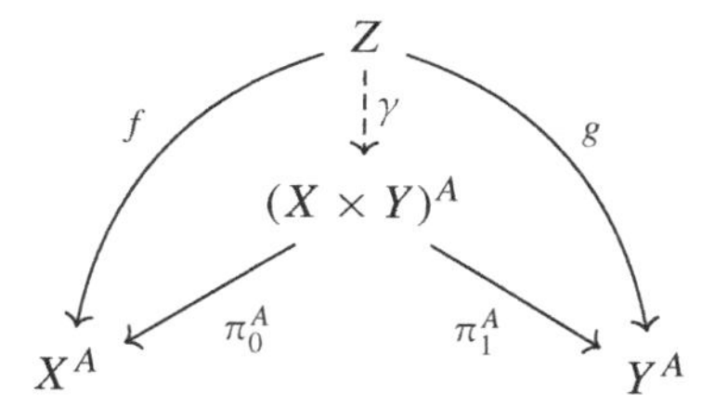

We claim that $\pi_0^A \gamma = f$ and $\pi_1^A \gamma = g$. Indeed,

$$\pi_0^A \circ \gamma = \pi_0^A \circ \langle f^\flat, g^\flat \rangle^\sharp = (\pi_0 \circ \langle f^\flat, g^\flat \rangle)^\sharp = (f^\flat)^\sharp = f.$$

Thus, $\pi_0^A \gamma = f$, and, similarly, $\pi_1^A \gamma = g$.

Suppose now that $h : Z \to (X \times Y)^A$ such that $\pi_0^A h = f$ and $\pi_1^A h = g$. Then

$$f = \pi_0^A \circ (h^\flat)^\sharp = (\pi_0 \circ h^\flat)^\sharp.$$

Hence, $\pi_0 \circ h^\flat = f^\flat$, and, similarly, $\pi_1 \circ h^\flat = g^\flat$. That is, $h^\flat = \langle f^\flat, g^\flat \rangle$, and $h = \langle f^\flat, g^\flat \rangle^\sharp = \gamma$. □

PROPOSITION 2.5.10 *For any sets A, X, Y, we have $A \times (X \amalg Y) \cong (A \times X) \amalg (A \times Y)$.*

Proof Even without Axiom 9, there is always a canonical function from $(A \times X) \amalg (A \times Y)$ to $A \times (X \amalg Y)$, namely $\phi := (1_A \times i_0) \amalg (1_A \times i_1)$, where i_0 and i_1 are the coproduct inclusions of $X \amalg Y$. That is,

$$\phi \circ j_0 = 1_A \times i_0, \qquad \text{and} \qquad \phi \circ j_1 = 1_A \times i_1,$$

where j_0 and j_1 are the coproduct inclusions of $(A \times X) \amalg (A \times Y)$.

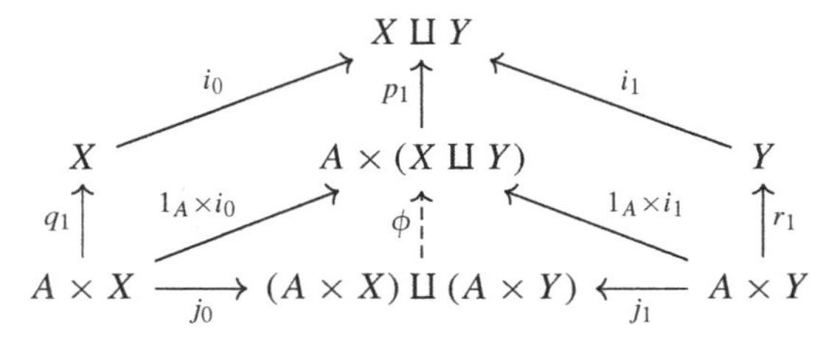

We will show that Axiom 9 entails that ϕ is invertible.

Let $g : A \times (X \amalg Y) \to A \times (X \amalg Y)$ be the identity, i.e., $g = 1_{A \times (X \amalg Y)}$. Then $g^\sharp : X \amalg Y \to (A \times (X \amalg Y))^A$ is the unique function such that $e(1_A \times g^\sharp) = g$. By Proposition 2.5.4,

$$(g^\sharp \circ i_0)^\flat = g \circ (1_A \times i_0) = 1_A \times i_0.$$

Similarly, $(g^\sharp \circ i_1)^\flat = 1_A \times i_1$. Thus,

$$g^\sharp \;=\; (1_A \times i_0)^\sharp \amalg (1_A \times i_1)^\sharp.$$

We also have $(1_A \times i_0)^\sharp = (\phi \circ j_0)^\sharp = \phi^A \circ j_0^\sharp$, and $(1^A \times i_1)^\sharp = \phi^A \circ j_1^\sharp$. Hence

$$g^\sharp \;=\; (\phi^A \circ j_0^\sharp) \amalg (\phi^A \circ j_1^\sharp) \;=\; \phi^A \circ (j_0^\sharp \amalg j_1^\sharp).$$

Now, for the inverse of ϕ, we take $\psi = (j_0^\sharp \amalg j_1^\sharp)^\flat$.

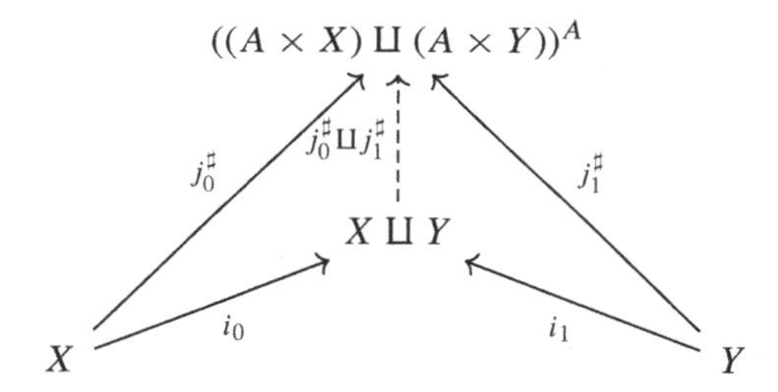

It then follows that

$$(\phi \circ \psi)^\sharp \;=\; \phi^A \circ (j_0^\sharp \amalg j_1^\sharp) \;=\; g^\sharp,$$

and, therefore, $\phi \circ \psi = 1_{A \times (X \amalg Y)}$. Similarly,

$$(\psi \circ \phi \circ j_0)^\sharp = \psi^A \circ (\phi \circ j_0)^\sharp = \psi^A \circ g^\sharp \circ i_0 = \psi^\sharp \circ i_0 = j_0^\sharp.$$

Thus, $\psi \circ \phi \circ j_0 = j_0$, and a similar calculation shows that $\psi \circ \phi \circ j_1 = j_1$. It follows that $\psi \circ \phi = 1_{(A \times X) \amalg (A \times Y)}$. Thus, ψ is a two-sided inverse for ϕ, and $A \times (X \amalg Y)$ is isomorphic to $(A \times X) \amalg (A \times Y)$. □

DEFINITION 2.5.11 (Powerset) If X is a set, we let $\mathscr{P}X = \Omega^X$.

Intuitively speaking, $\mathscr{P}X$ is the set of all subsets of X. For example, if $X = \{a, b\}$, then $\mathscr{P}X = \{\emptyset, \{a\}, \{b\}, \{a, b\}\}$. More rigorously, each element of Ω^X corresponds to a function $1 \to \Omega^X$, which in turn corresponds to a function $X \cong 1 \times X \to \Omega$, which corresponds to a subobject of X. Thus, we can think of $\mathscr{P}X$ as another name for Sub(X), although Sub(X) is not really an object in **Sets**.

2.6 Cardinality

When mathematics was rigorized in the nineteenth century, one of the important advances was a rigorous definition of "infinite set." It came as something of a surprise that there are different sizes of infinity and that some infinite sets (e.g., the real numbers) are strictly larger than the natural numbers. In this section, we define "finite" and "infinite." We then add an axiom that says there is a specific set N that behaves like the natural numbers; in particular, N is infinite. Finally, we show that the powerset $\mathscr{P}X$ of a set X is always larger than X.

DEFINITION 2.6.1 A set X is said to be **finite** if and only if for any function $m : X \to X$, if m is monic, then m is an isomorphism. A set X is said to be **infinite** if and only if there is a function $m : X \to X$ that is monic and not surjective.

We are already guaranteed the existence of finite sets: for example, the terminal object 1 is finite, as is the subobject classifier Ω. But the axioms we have stated thus far do not guarantee the existence of any infinite sets. We won't know that there are infinite sets until we add the natural number object (NNO) axiom (Axiom 10).

DEFINITION 2.6.2 We say that Y is at least as large as X, written $|X| \leq |Y|$, just in case there is a monomorphism $m : X \to Y$.

PROPOSITION 2.6.3 $|X| \leq |X \amalg Y|$.

Proof Proposition 2.4.2 shows that $i_0 : X \to X \amalg Y$ is monic. □

PROPOSITION 2.6.4 *If Y is nonempty, then* $|X| \leq |X \times Y|$.

Proof Consider the function $\langle 1_X, f \rangle : X \to X \times Y$, where $f : X \to 1 \to Y$. □

Axiom 10: Natural Number Object

There is an object N, and functions $z : 1 \to N$ and $s : N \to N$ such that for any other set X with functions $q : 1 \to X$ and $f : X \to X$, there is a unique function $u : N \to X$ such that the following diagram commutes:

$$\begin{array}{ccccc} 1 & \xrightarrow{z} & N & \xrightarrow{s} & N \\ & \searrow^{q} & \downarrow{u} & & \downarrow{u} \\ & & X & \xrightarrow{f} & X \end{array}$$

The set N is called a **natural number object**.

EXERCISE 2.6.5 Let N' be a set, and let $z' : 1 \to N'$ and $s' : N' \to N'$ be functions that satisfy the conditions in Axiom 10. Show that N' is isomorphic to N.

PROPOSITION 2.6.6 *$z \amalg s : 1 \amalg N \to N$ is an isomorphism.*

Proof Let $i_0 : 1 \to 1 \amalg N$ and $i_1 : N \to 1 \amalg N$ be the coproduct inclusions. Using the NNO axiom, there is a unique function $g : N \to 1 \amalg N$ such that the following diagram commutes:

$$\begin{array}{ccccc} 1 & \xrightarrow{z} & N & \xrightarrow{s} & N \\ & \searrow^{i_0} & \downarrow g & & \downarrow g \\ & & 1 \amalg N & \xrightarrow[i_1 z \amalg i_1 s]{} & 1 \amalg N \end{array}$$

We will show that g is a two-sided inverse of $z \amalg s$. To this end, we first establish that $g \circ s = i_1$. Consider the following diagram:

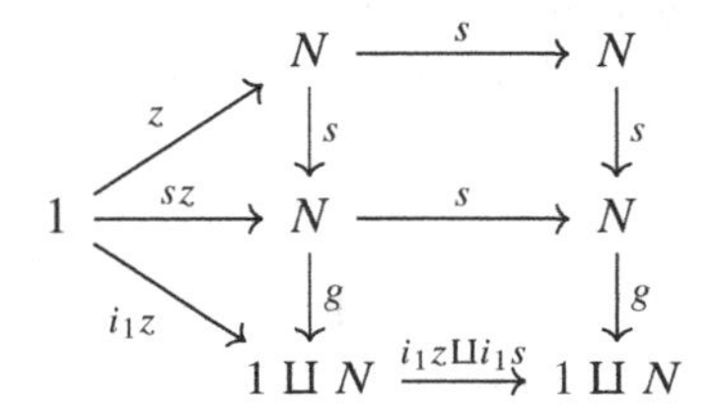

The lower triangle commutes because of the commutativity of the previous diagram. Thus, the entire diagram commutes. The outer triangle and square would also commute with i_1 in place of $g \circ s$. By the NNO axiom, $g \circ s = i_1$. Now, to see that $(z \amalg s) \circ g = 1_N$, note first that

$$(z \amalg s) \circ g \circ z = (z \amalg s) \circ i_0 = z.$$

Furthermore,

$$(z \amalg s) \circ g \circ s = (z \amalg s) \circ i_1 = s.$$

Thus, the NNO axiom entails that $(z \amalg s) \circ g = \mathrm{id}_N$. Finally, to see that $g \circ (z \amalg s) = \mathrm{id}_{1 \amalg N}$, we calculate

$$g \circ (z \amalg s) \circ i_0 = g \circ z = i_0.$$

Furthermore,

$$g \circ (z \amalg s) \circ i_1 = g \circ s = i_1.$$

Therefore, $g \circ (z \amalg s) = \mathrm{id}_{1 \amalg N}$. This establishes that g is a two-sided inverse of $z \amalg s$, and $1 \amalg N$ is isomorphic to N. □

PROPOSITION 2.6.7 *The function $s : N \to N$ is injective but not surjective. Thus, N is infinite.*

Proof By Proposition 2.4.2, the function $i_1 : N \to 1 \amalg N$ is monic. Since the images of i_0 and i_1 are disjoint, i_0 is not surjective. Since $z \amalg s$ is an isomorphism, $(z \amalg s) \circ i_1 = s$ is monic but not surjective. Therefore, N is infinite. □

PROPOSITION 2.6.8 *If $m : B \to X$ is a nonempty subobject, then there is an epimorphism $f : X \to B$.*

Proof Since B is nonempty, there is a function $g : X \backslash B \to B$. By Proposition 2.4.5, $B \cong B \amalg X \backslash B$. Finally, $1_B \amalg g : B \amalg X \backslash B \to B$ is an epimorphism, since 1_B is an epimorphism. □

DEFINITION 2.6.9 We say that a set X is **countable** just in case there is an epimorphism $f : N \to X$, where N is the natural numbers.

PROPOSITION 2.6.10 *$N \times N$ is countably infinite.*

Sketch of proof We will give two arguments: one quick and one slow (but hopefully more illuminating). For the quick argument, define a function $g : N \times N \to N$ by $g(x, y) = 2^x 3^y$. If $\langle x, y \rangle \neq \langle x', y' \rangle$, then either $x \neq x'$ or $y \neq y'$. In either case, unique factorizability of integers gives $2^x 3^y \neq 2^{x'} 3^{y'}$. Therefore, $g : N \times N \to N$ is monic. Since $N \times N$ is nonempty, Proposition 2.6.8 entails that there is an epimorphism $f : N \twoheadrightarrow N \times N$. Therefore, $N \times N$ is countable.

Now for the slow argument. Imagine writing down all elements in $N \times N$ in an infinite table, whose first few elements look like this:

$$\left(\begin{array}{cccc} \langle 0,0 \rangle & \langle 1,0 \rangle & \langle 2,0 \rangle & \cdots \\ \langle 0,1 \rangle & \langle 1,1 \rangle & \langle 2,1 \rangle & \cdots \\ \langle 0,2 \rangle & \langle 1,2 \rangle & \langle 2,2 \rangle & \cdots \\ \vdots & \vdots & \vdots & \vdots \end{array} \right.$$

Now imagine running a thread diagonally through the numbers: begin with $\langle 0, 0 \rangle$, then move down to $\langle 0, 1 \rangle$ and up to $\langle 1, 0 \rangle$, then over to $\langle 2, 0 \rangle$ and down its diagonal, etc. This process defines a function $f : N \to N \times N$ whose first few values are

$$\begin{aligned} f(0) &= \langle 0,0 \rangle \\ f(1) &= \langle 0,1 \rangle \\ f(2) &= \langle 1,0 \rangle \\ &\vdots \end{aligned}$$

It is not difficult to show that f is surjective, and so $N \times N$ is countable. □

EXERCISE 2.6.11 Show that if A and B are countable, then $A \cup B$ is countable.

We're now going to show that exponentiation creates sets of larger and larger size. In the case of finite sets A and X, it's easy to see that the following equation holds:

$$|A^X| = |A|^{|X|},$$

where $|X|$ denotes the number of elements in X. In particular, Ω^X can be thought of as the set of binary sequences indexed by X. We're now going to show that for any set X, the set Ω^X is larger than X.

DEFINITION 2.6.12 Let $g : A \to A$ be a function. We say that $a \in A$ is a **fixed point** of g just in case $g(a) = a$. We say that A has the **fixed-point property** just in case any function $g : A \to A$ has a fixed point.

THEOREM 2.6.13 (Lawvere's fixed-point theorem) *Let A and X be sets. If there is a surjective function $p : X \to A^X$, then A has the fixed point property.*

Proof Suppose that $p : X \to A^X$ is surjective. That is, for any function $f : X \to A$, there is an $x_f \in X$ such that $f = p(x_f)$. Let $\phi = p^\flat$, so that $f = \phi(x_f, -)$. Now let $g : A \to A$ be any function. We need to show that g has a fixed point. Consider the function $f : X \to A$ defined by $f = g \circ \phi \circ \delta_X$, where $\delta_X : X \to X \times X$ is the diagonal map. Then we have

$$g\phi(x,x) = f(x) = \phi(x_f, x),$$

for all $x \in X$. In particular, $g\phi(x_f, x_f) = \phi(x_f, x_f)$, which means that $a = \phi(x_f, x_f)$ is a fixed point of g. Since $g : A \to A$ was arbitrary, it follows that A has the fixed point property. □

THEOREM 2.6.14 (Cantor's theorem) *There is no surjective function $X \to \Omega^X$.*

Proof The function $\Omega \to \Omega$ that permutes t and f has no fixed points. The result then follows from Lawvere's fixed-point theorem. □

EXERCISE 2.6.15 Show that there is an injective function $X \to \Omega^X$. (The proof is easy if you simply think of Ω^X as functions from X to $\{\mathsf{t}, \mathsf{f}\}$. For a bigger challenge, try to prove that it's true using the definition of the exponential set Ω^X.)

COROLLARY 2.6.16 *For any set X, the set $\mathscr{P}X$ of its subsets is strictly larger than X.*

There are several other facts about cardinality that are important for certain parts of mathematics – in our case, they will be important for the study of topology. For example, if X is an infinite set, then the set $\mathscr{F}X$ of all finite subsets of a set X has the same cardinality as X. Similarly, a countable coproduct of countable sets is countable. However, these facts – well known from ZF set theory – are not obviously provable in ETCS.

DISCUSSION 2.6.17 Intuitively speaking, X^N is the set of all sequences with values in X. Thus, we should have something like

$$X^N \cong X \times X \times \ldots$$

However, we don't have any axiom telling us that **Sets** has infinite products such as the one on the right-hand side. Can it be proven that X^N satisfies the definition of an infinite product? In other words, are there projections $\pi_i : X^N \to X$ that satisfy an appropriate universal property?

2.7 The Axiom of Choice

In recent years, it has become routine to supplement set theory with a further axiom, the so-called **axiom of choice**. (The axiom of choice is regularly used in fields such as functional analysis, e.g., to prove the existence of an orthonormal basis for Hilbert spaces of arbitrarily large dimension.) While the name of this axiom suggests that it has something to do with our choices, in fact it really just asserts the existence of further sets. Following our typical procedure in this chapter, we will provide a structural version of the axiom.

DEFINITION 2.7.1 Let $f : X \to Y$ be a function. We say that f is a **split epimorphism** just in case there is a function $s : Y \to X$ such that $fs = 1_Y$. In this case, we say that s is a **section** of f.

EXERCISE 2.7.2 Prove that if f is a split epimorphism, then f is a regular epimorphism. Prove that if s is a section, then s is a regular monomorphism.

Axiom 11: Axiom of Choice

Every epimorphism in **Sets** has a section.

A more typical formulation of the axiom of choice might say that for any set-indexed collection of nonempty sets, say $\{X_i \mid i \in I\}$, the product set $\prod_{i \in I} X_i$ is nonempty. To translate that version of the axiom of choice into our version, suppose that the sets X_i are stacked side by side, and that f is the map that projects each $x \in X_i$ to the value i. Then a section s of f is a function with domain I that returns an element $s(i) \in X_i$ for each $i \in I$. If such a function exists, then $\prod_{i \in I} X_i$ is nonempty.

In this book, we will never use the full axiom of choice. However, we will use a couple of weaker versions of it, specifically in the proofs of the completeness theorems for propositional and predicate logic. For propositional logic, we will assume the Boolean prime ideal theorem; and for predicate logic, we will use a version of the axiom of dependent choices to prove the Baire category theorem.

2.8 Notes

- There are many good books on category theory. The classic reference is Mac Lane (1971), but it can be difficult going for those without extensive mathematical training. We also find the following useful: Borceux (1994); Awodey (2010); Van Oosten (2002). The latter two are good entry points for people with some background in formal logic.
- The elementary theory of the category of sets (ETCS) was first presented by Lawvere (1964). For pedagogical presentations, see Lawvere and Rosebrugh (2003); Leinster (2014).

3 The Category of Propositional Theories

One of the primary goals of this book is to provide a formal model of "the universe of all scientific theories." In the twentieth century, mathematics stepped up another level of abstraction, and it began to talk of structured collections of mathematical objects – e.g., the category of groups, topological spaces, manifolds, Hilbert spaces, or sets. This maneuver can be a little bit challenging for foundationally oriented thinkers, viz. philosophers, because we are now asked to consider collections that are bigger than any set. However, mathematicians know very well how to proceed in this manner without falling into contradictions (e.g., by availing themselves of Grothendieck universes).

We want to follow the lead of the mathematicians, but instead of talking about the category of groups, or manifolds, or Hilbert spaces, etc., we want to talk about the **category of all theories**. In the present chapter, we work out one special case: the category of all propositional theories. Of course, this category is too simple to serve as a good model for the category of all scientific theories. However, already for predicate logic, the category of theories becomes extremely complex, almost to the point of mathematical intractability. In subsequent chapters, we will make some headway with that case; for the remainder of this chapter, we restrict ourselves to the propositional case.

After defining the relevant category **Th** of propositional theories, we will show that **Th** is equivalent to the category **Bool** of Boolean algebras. We then prove a version of the famous Stone duality theorem, which shows that **Bool** is dual to a certain category **Stone** of topological spaces. This duality shows that each propositional theory corresponds to a unique topological space, viz. the space of its models, and each translation between theories corresponds to a continuous mapping between their spaces of models.

3.1 Basics

DEFINITION 3.1.1 We let **Th** denote the category whose objects are propositional theories and whose arrows are translations between theories. We say that two translations $f,g : T \rightrightarrows T'$ are equal, written $f \simeq g$, just in case $T' \vdash f(\phi) \leftrightarrow g(\phi)$ for every $\phi \in \mathsf{Sent}(\Sigma)$. (Note well: equality between translations is weaker than set-theoretic equality.)

DEFINITION 3.1.2 We say that a translation $f : T \to T'$ is **conservative** just in case, for any $\phi \in \mathsf{Sent}(\Sigma)$, if $T' \vdash f(\phi)$ then $T \vdash \phi$.

PROPOSITION 3.1.3 *A translation $f : T \to T'$ is conservative if and only if f is a monomorphism in the category* **Th**.

Proof Suppose first that f is conservative, and let $g, h : T'' \to T$ be translations such that $f \circ g = f \circ h$. That is, $T' \vdash fg(\phi) \leftrightarrow fh(\phi)$ for every sentence ϕ of Σ''. Since f is conservative, $T \vdash g(\phi) \leftrightarrow h(\phi)$ for every sentence ϕ of Σ''. Thus, $g = h$, and f is a monomorphism in **Th**.

Conversely, suppose that f is a monomorphism in the category **Th**. Let ϕ be a Σ sentence such that $T' \vdash f(\phi)$. Thus, $T' \vdash f(\phi) \leftrightarrow f(\psi)$, where ψ is any Σ sentence such that $T \vdash \psi$. Now let T'' be the empty theory in signature $\Sigma'' = \{p\}$. Define $g : \Sigma'' \to \mathsf{Sent}(\Sigma)$ by $g(p) = \phi$, and define $h : \Sigma'' \to \mathsf{Sent}(\Sigma)$ by $h(p) = \psi$. It's easy to see then that $f \circ g = f \circ h$. Since f is monic, $g = h$, which means that $T \vdash g(p) \leftrightarrow h(p)$. Therefore, $T \vdash \phi$, and f is conservative. □

DEFINITION 3.1.4 We say that a translation $f : T \to T'$ is **essentially surjective** just in case for any sentence ϕ of Σ', there is a sentence ψ of Σ such that $T' \vdash \phi \leftrightarrow f(\psi)$. (Sometimes we use the abbreviation "eso" for essentially surjective.)

PROPOSITION 3.1.5 *If $f : T \to T'$ is essentially surjective, then f is an epimorphism in* **Th**.

Proof Suppose that $f : T \to T'$ is eso. Let $g, h : T' \rightrightarrows T''$ such that $g \circ f = h \circ f$. Let ϕ be an arbitrary Σ' sentence. Since f is eso, there is a sentence ψ of Σ such that $T' \vdash \phi \leftrightarrow f(\psi)$. But then $T'' \vdash g(\phi) \leftrightarrow h(\phi)$. Since ϕ was arbitrary, $g = h$. Therefore, f is an epimorphism. □

What about the converse of this proposition? Are all epimorphisms in **Th** essentially surjective? The answer is yes, but the result is not easy to prove. We'll prove it later on, by means of the correspondence that we establish between theories, Boolean algebras, and Stone spaces.

PROPOSITION 3.1.6 *Let $f : T \to T'$ be a translation. If f is conservative and essentially surjective, then f is a homotopy equivalence.*

Proof Let $p \in \Sigma'$. Since f is eso, there is some $\phi_p \in \mathsf{Sent}(\Sigma)$ such that $T' \vdash p \leftrightarrow f(\phi_p)$. Define a reconstrual $g : \Sigma' \to \mathsf{Sent}(\Sigma)$ by setting $g(p) = \phi_p$. As usual, g extends naturally to a function from $\mathsf{Sent}(\Sigma')$ to $\mathsf{Sent}(\Sigma)$, and it immediately follows that $T' \vdash \psi \leftrightarrow fg(\psi)$, for every sentence ψ of Σ'.

We claim now that g is a translation from T' to T. Suppose that $T' \vdash \psi$. Since $T' \vdash \psi \leftrightarrow fg(\psi)$, it follows that $T' \vdash fg(\psi)$. Since f is conservative, $T \vdash g(\psi)$. Thus, for all sentences ψ of Σ', if $T' \vdash \psi$ then $T \vdash g(\psi)$, which means that $g : T' \to T$ is a translation. By the previous paragraph, $1_{T'} \simeq fg$.

It remains to show that $1_T \simeq gf$. Let ϕ be an arbitrary sentence of Σ. Since f is conservative, it will suffice to show that $T' \vdash f(\phi) \leftrightarrow fgf(\phi)$. But by the previous paragraph, $T' \vdash \psi \leftrightarrow fg(\psi)$ for all sentences ψ of Σ'. Therefore, $1_T \simeq gf$, and f is a homotopy equivalence. □

Before proceeding, let's remind ourselves of some of the motivations for these technical investigations.

The category **Sets** is, without a doubt, extremely useful. However, a person who is familiar with **Sets** might have developed some intuitions that could be misleading when applied to other categories. For example, in **Sets**, if there are injections $f : X \to Y$ and $g : Y \to X$, then there is a bijection between X and Y. Thus, it's tempting to think, for example, that if there are embeddings $f : T \to T'$ and $g : T' \to T$ of theories, then T and T' are equivalent. (Here an embedding between theories is a monomorphism in **Th**, i.e., a conservative translation.) Similarly, in **Sets**, if there is an injection $f : X \to Y$ and a surjection $g : X \to Y$, then there is a bijection between X and Y. However, in **Th**, the analogous result fails to hold.

TECHNICAL ASIDE 3.1.7 For those familiar with the category **Vect** of vector spaces: **Vect** is similar to **Sets** in that mutually embeddable vector spaces are isomorphic. That is, if $f : V \to W$ and $g : W \to V$ are monomorphisms (i.e., injective linear maps), then V and W have the same dimension and, hence, are isomorphic. The categories **Sets** and **Vect** share in common the feature that the objects can be classified by cardinal numbers. In the case of sets, if $|X| = |Y|$, then $X \cong Y$. In the case of vector spaces, if $\dim(V) = \dim(W)$, then $V \cong W$.

In Exercise 1.4.7, you showed that if $f : T \to T'$ is a translation, and if v is a model of T', then $v \circ f$ is a model of T. Let $M(T)$ be the set of all models of T, and define a function $f^* : M(T') \to M(T)$ by setting $f^*(v) = f \circ v$.

PROPOSITION 3.1.8 *Let $f : T \to T'$ be a translation. If $f^* : M(T') \to M(T)$ is surjective, then f is conservative.*

Proof Suppose that f^* is surjective, and suppose that ϕ is a sentence of Σ such that $T \nvdash \phi$. Then there is a $v \in M(T)$ such that $v(\phi) = 0$. (Here we have invoked the completeness theorem, but we haven't proven it yet. Note that our proof of the completeness theorem, page 79, does not cite this result or any that depend on it.) Since f^* is surjective, there is a $w \in M(T')$ such that $f^*(w) = v$. But then

$$w(f(\phi)) = f^*w(\phi) = v(\phi) = 0,$$

from which it follows that $T' \nvdash f(\phi)$. Therefore, f is conservative. □

Example 3.1.9 Let $\Sigma = \{p_0, p_1, \dots\}$, and let T be the empty theory in Σ. Let $\Sigma' = \{q_0, q_1, \dots\}$, and let T' be the theory with axioms $q_0 \to q_i$, for $i = 0, 1, \dots$. We will show that there are conservative translations $f : T \to T'$ and $g : T' \to T$.

Define $f : \Sigma \to \mathsf{Sent}(\Sigma')$ by $f(p_i) = q_{i+1}$. Since T is the empty theory, f is a translation. Then for any valuation v of Σ', we have

$$f^*v(p_i) = v(f(p_i)) = v(q_{i+1}).$$

Furthermore, for any sequence of zeros and ones, there is a valuation v of Σ' that assigns that sequence to $q_1, q_2, \dots$. Thus, f^* is surjective, and f is conservative.

Now define $g : \Sigma' \to \mathsf{Sent}(\Sigma)$ by setting $g(q_i) = p_0 \vee p_i$. Since $T \vdash p_0 \vee p_0 \to p_0 \vee p_i$, it follows that g is a translation. Furthermore, for any valuation v of Σ, we have

$$g^*v(q_i) = v(g(q_i)) = v(p_0 \vee p_i).$$

Recall that $M(T')$ splits into two parts: (1) a singleton set containing the valuation z where $z(q_i) = 1$ for all i, and (2) the infinitely many other valuations that assign 0 to q_0. Clearly, $z = g^*v$, where v is any valuation such that $v(p_0) = 1$. Furthermore, for any valuation w of Σ' such that $w(p_0) = 0$, we have $w = g^*v$, where $v(p_i) = w(q_i)$. Therefore, g^* is surjective, and g is conservative. ⌟

EXERCISE 3.1.10 In Example 3.1.9, show that f and g are not essentially surjective.

Example 3.1.11 Let T and T' be as in the previous example. Now we'll show that there are essentially surjective (eso) translations $k : T \to T'$ and $h : T' \to T$. The first is easy: the translation $k(p_i) = q_i$ is obviously eso. For the second, define $h(q_0) = \bot$, where $\bot$ is some contradiction, and define $h(q_i) = p_{i-1}$ for $i > 0$. ⌟

Let's pause to think about some of the questions we might want to ask about theories. We arrange these in roughly decreasing order of technical tractability.

1. Does **Th** have the **Cantor–Bernstein property**? That is, if there are monomorphisms $f : T \to T'$ and $g : T' \to T$, then is there an isomorphism $h : T \to T'$?
2. Is **Th** balanced, in the sense that if $f : T \to T'$ is both a monomorphism and an epimorphism, then f is an isomorphism?
3. If there is both a monomorphism $f : T \to T'$ and an epimorphism $g : T' \to T$, then are T and T' homotopy equivalent?
4. Can an arbitrary theory T be embedded into a theory T_0 that has no axioms? Quine and Goodman (1940) present a proof of this claim – and they argue that it undercuts the analytic-synthetic distinction. They are right about the technical claim (see 3.7.10), but have perhaps misconstrued its philosophical implications.
5. If theories have the same number of models, then are they equivalent? If not, then can we determine whether T and T' are equivalent by inspecting $M(T)$ and $M(T')$?
6. How many theories (up to isomorphism) are there with n models?
7. (Does supervenience imply reduction?) Suppose that the truth value of a sentence ψ **supervenes** on the truth value of some other sentences $\phi_1, \dots, \phi_n$, i.e., for any valuations v, w of the propositional constants occurring in $\phi_1, \dots, \phi_n, \psi$, if $v(\phi_i) = w(\phi_i)$, for $i = 1, \dots, n$, then $v(\psi) = w(\psi)$. Does it follow then that $\vdash \psi \leftrightarrow \theta$, where θ contains only the propositional constants that occur in $\phi_1, \dots, \phi_n$? We will return to this issue in Section 6.7.
8. Suppose that $f : T \to T'$ is conservative. Suppose also that every model of T extends uniquely to a model of T'. Does it follow that $T \cong T'$?
9. Suppose that T and T' are consistent in the sense that there is no sentence θ in $\Sigma \cap \Sigma'$ such that $T \vdash \theta$ and $T' \vdash \neg\theta$. Is there a unified theory T'' that extends both T and T'? (The answer is yes, as shown by **Robinson's theorem**.)
10. What does it mean for one theory to be **reducible** to another? Can we explicate this notion in terms of a certain sort of translation between the relevant theories? Some philosophers have claimed that the reduction relation ought to be treated

semantically, rather than syntactically. In other words, they would have us consider functions from $M(T')$ to $M(T)$, rather than translations from T to T'. In light of the Stone duality theorem proved later in the chapter, it appears that syntactic and semantic approaches are equivalent to each other.

11. Consider various formally definable notions of theoretical equivalence. What are the advantages and disadvantages of the various notions? Is homotopy equivalence too liberal? Is it too conservative?

3.2 Boolean Algebras

DEFINITION 3.2.1 A **Boolean algebra** is a set B together with a unary operation $\neg$, two binary operations $\wedge$ and $\vee$, and designated elements $0 \in B$ and $1 \in B$, which satisfy the following equations:

1. Top and Bottom

$$a \wedge 1 = a \vee 0 = a$$

2. Idempotence

$$a \wedge a = a \vee a = a$$

3. De Morgan's Rules

$$\neg(a \wedge b) = \neg a \vee \neg b, \quad \neg(a \vee b) = \neg a \wedge \neg b$$

4. Commutativity

$$a \wedge b = b \wedge a, \quad a \vee b = b \vee a$$

5. Associativity

$$(a \wedge b) \wedge c = a \wedge (b \wedge c), \quad (a \vee b) \vee c = a \vee (b \vee c)$$

6. Distribution

$$a \wedge (b \vee c) = (a \wedge b) \vee (a \wedge c), \quad a \vee (b \wedge c) = (a \vee b) \wedge (a \vee c)$$

7. Excluded Middle

$$a \wedge \neg a = 0, \quad a \vee \neg a = 1$$

Here we are implicitly universally quantifying over a, b, c.

Example 3.2.2 Let 2 denote the Boolean algebra of subsets of a one-point set. Note that 2 looks just like the truth-value set Ω. Indeed, Ω is equipped with operations $\wedge, \vee$, and $\neg$ that make it into a Boolean algebra. ⌟

Example 3.2.3 Let $\Sigma = \{p\}$. Define an equivalence relation $\simeq$ on sentences of Σ by $\phi \simeq \psi$ just in case $\vdash \phi \leftrightarrow \psi$. If we let F denote the set of equivalence classes, then it's not hard to see that F has four elements: $0, 1, [p], [\neg p]$. Define $[\phi] \wedge [\psi] = [\phi \wedge \psi]$, where the $\wedge$ on the right is the propositional connective, and the $\wedge$ on the left is a newly defined binary function on F. Perform a similar construction for the other logical connectives. Then F is a Boolean algebra. ⌟

We now derive some basic consequences from the axioms for Boolean algebras. The first two results are called the **absorption laws**.

1. $a \wedge (a \vee b) = a$

$$a \wedge (a \vee b) = (a \vee 0) \wedge (a \vee b) = a \vee (0 \wedge b) = a \vee 0 = a.$$

2. $a \vee (a \wedge b) = a$

$$a \vee (a \wedge b) = (a \wedge 1) \vee (a \wedge b) = a \wedge (1 \vee b) = a \wedge 1 = a.$$

3. $a \vee 1 = 1$

$$a \vee 1 = a \vee (a \vee \neg a) = a \vee \neg a = 1.$$

4. $a \wedge 0 = 0$

$$a \wedge 0 = a \wedge (a \wedge \neg a) = a \wedge \neg a = 0.$$

DEFINITION 3.2.4 If B is a Boolean algebra and $a, b \in B$, we write $a \leq b$ when $a \wedge b = a$.

Since $a \wedge 1 = a$, it follows that $a \leq 1$, for all $a \in B$. Since $a \wedge 0 = 0$, it follows that $0 \leq a$, for all $a \in B$. Now we will show that $\leq$ is a partial order, i.e., reflexive, transitive, and asymmetric.

PROPOSITION 3.2.5 *The relation $\leq$ on a Boolean algebra B is a partial order.*

Proof (Reflexive) Since $a \wedge a = a$, it follows that $a \leq a$.

(Transitive) Suppose that $a \wedge b = a$ and $b \wedge c = b$. Then

$$a \wedge c = (a \wedge b) \wedge c = a \wedge (b \wedge c) = a \wedge b = a,$$

which means that $a \leq c$.

(Asymmetric) Suppose that $a \wedge b = a$ and $b \wedge a = b$. By commutativity of $\wedge$, it follows that $a = b$. □

We now show how $\leq$ interacts with $\wedge, \vee$, and $\neg$. In particular, we show that if $\leq$ is thought of as implication, then $\wedge$ behaves like conjunction, $\vee$ behaves like disjunction, $\neg$ behaves like negation, 1 behaves like a tautology, and 0 behaves like a contradiction.

PROPOSITION 3.2.6 *$c \leq a \wedge b$ iff $c \leq a$ and $c \leq b$.*

Proof Since $a \wedge (a \wedge b) = a \wedge b$, it follows that $a \wedge b \leq a$. By similar reasoning, $a \wedge b \leq b$. Thus, if $c \leq a \wedge b$, then transitivity of $\leq$ entails that both $c \leq a$ and $c \leq b$.

Now suppose that $c \leq a$ and $c \leq b$. That is, $c \wedge a = c$ and $c \wedge b = c$. Then $c \wedge (a \wedge b) = (c \wedge a) \wedge (c \wedge b) = c \wedge c = c$. Therefore, $c \leq a \wedge b$. □

Notice that $\leq$ and $\wedge$ interact precisely as implication and conjunction interact in propositional logic. The elimination rule says that $a \wedge b$ implies a and b. Hence, if c implies $a \wedge b$, then c implies a and b. The introduction rule says that a and b imply $a \wedge b$. Hence, if c implies a and b, then c implies $a \wedge b$.

PROPOSITION 3.2.7 *$a \leq c$ and $b \leq c$ iff $a \vee b \leq c$*

Proof Suppose first that $a \leq c$ and $b \leq c$. Then

$$(a \vee b) \wedge c = (a \wedge c) \vee (b \wedge c) = a \vee b.$$

Therefore, $a \vee b \leq c$.

Suppose now that $a \vee b \leq c$. By the absorption law, $a \wedge (a \vee b) = a$, which implies that $a \leq a \vee b$. By transitivity, $a \leq c$. Similarly, $b \leq a \vee b$, and by transitivity, $b \leq c$. □

Now we show that the connectives $\wedge$ and $\vee$ are monotonic.

PROPOSITION 3.2.8 *If $a \leq b$, then $a \wedge c \leq b \wedge c$, for any $c \in B$.*

Proof

$$(a \wedge c) \wedge (b \wedge c) = (a \wedge b) \wedge c = a \wedge c.$$

□

PROPOSITION 3.2.9 *If $a \leq b$, then $a \vee c \leq b \vee c$, for any $c \in B$.*

Proof

$$(a \vee c) \wedge (b \vee c) = (a \wedge b) \vee c = a \vee c.$$

□

PROPOSITION 3.2.10 *If $a \wedge b = a$ and $a \vee b = a$, then $a = b$.*

Proof $a \wedge b = a$ means that $a \leq b$. We now claim that $a \vee b = a$ iff $b \wedge a = b$ iff $b \leq a$. Indeed, if $a \vee b = a$, then

$$b \wedge a = b \wedge (a \vee b) = (0 \vee b) \wedge (a \vee b) = (0 \wedge a) \vee b = b.$$

Conversely, if $b \wedge a = b$, then

$$a \vee b = a \vee (a \wedge b) = (a \wedge 1) \vee (a \wedge b) = a \wedge (1 \vee b) = a.$$

Thus, if $a \wedge b = a$ and $a \vee b = a$, then $a \leq b$ and $b \leq a$. By asymmetry of $\leq$, it follows that $a = b$. □

We now show that $\neg a$ is the unique complement of a in B.

PROPOSITION 3.2.11 *If $a \wedge b = 0$ and $a \vee b = 1$, then $b = \neg a$.*

Proof Since $b \vee a = 1$, we have

$$b = b \vee 0 = b \vee (a \wedge \neg a) = (b \vee a) \wedge (b \vee \neg a) = b \vee \neg a.$$

Since $b \wedge a = 0$, we also have

$$b = b \wedge 1 = b \wedge (a \vee \neg a) = (b \wedge a) \vee (b \wedge \neg a) = b \wedge \neg a.$$

By the preceding proposition, $b = \neg a$. □

PROPOSITION 3.2.12 $\neg 1 = 0$.

Proof We have $1 \wedge 0 = 0$ and $1 \vee 0 = 1$. By the preceding proposition, $0 = \neg 1$. □

PROPOSITION 3.2.13 *If $a \leq b$, then $\neg b \leq \neg a$.*

Proof Suppose that $a \leq b$, which means that $a \wedge b = a$, and, equivalently, $a \vee b = b$. Thus, $\neg a \wedge \neg b = \neg(a \vee b) = \neg b$, which means that $\neg b \leq \neg a$. □

PROPOSITION 3.2.14 $\neg\neg a = a$.

Proof We have $\neg a \vee \neg\neg a = 1$ and $\neg a \wedge \neg\neg a = 1$. By Proposition 3.2.11, it follows that $\neg\neg a = a$. □

DEFINITION 3.2.15 Let A and B be Boolean algebras. A **homomorphism** is a map $\phi : A \to B$ such that $\phi(0) = 0$, $\phi(1) = 1$, and for all $a, b \in A$, $\phi(\neg a) = \neg\phi(a)$, $\phi(a \wedge b) = \phi(a) \wedge \phi(b)$ and $\phi(a \vee b) = \phi(a) \vee \phi(b)$.

It is easy to see that if $\phi : A \to B$ and $\psi : B \to C$ are homomorphisms, then $\psi \circ \phi : A \to C$ is also a homomorphism. Moreover, $1_A : A \to A$ is a homomorphism, and composition of homomorphisms is associative.

DEFINITION 3.2.16 We let **Bool** denote the category whose objects are Boolean algebras and whose arrows are homomorphisms of Boolean algebras.

Since **Bool** is a category, we have notions of **monomorphisms**, **epimorphisms**, **isomorphisms**, etc. Once again, it is easy to see that an injective homomorphism is a monomorphism and a surjective homomorphism is an epimorphism.

PROPOSITION 3.2.17 *Monomorphisms in* **Bool** *are injective.*

Proof Let $f : A \to B$ be a monomorphism, and let $a, b \in A$. Let F denote the Boolean algebra with four elements, and let p denote one of the two elements in F that is neither 0 nor 1. Define $\hat{a} : F \to A$ by $\hat{a}(p) = a$, and define $\hat{b} : F \to A$ by $\hat{b}(p) = b$. It is easy to see that $\hat{a}$ and $\hat{b}$ are uniquely defined by these conditions, and that they are Boolean homomorphisms. Suppose now that $f(a) = f(b)$. Then $f\hat{a} = f\hat{b}$, and, since f is a monomorphism, $\hat{a} = \hat{b}$, and, therefore, $a = b$. Therefore, f is injective. □

It is also true that epimorphisms in **Bool** are surjective. However, proving that fact is no easy task. We will return to it later in the chapter.

PROPOSITION 3.2.18 *If $f : A \to B$ is a homomorphism of Boolean algebras, then $a \leq b$ only if $f(a) \leq f(b)$.*

Proof $a \leq b$ means that $a \wedge b = a$. Thus,

$$f(a) \wedge f(b) = f(a \wedge b) = f(a),$$

which means that $f(a) \leq f(b)$. □

DEFINITION 3.2.19 A homomorphism $\phi : B \to 2$ is called a **state** of B.

3.3 Equivalent Categories

We now have two categories on the table: the category **Th** of theories and the category **Bool** of Boolean algebras. Our next goal is to show that these categories are **structurally identical**. But what do we mean by this? What we mean is that they are **equivalent categories**. In order to explain what that means, we need a few more definitions.

DEFINITION 3.3.1 Suppose that $\mathbf{C}$ and $\mathbf{D}$ are categories. We let $\mathbf{C}_0$ denote the objects of $\mathbf{C}$, and we let $\mathbf{C}_1$ denote the arrows of $\mathbf{C}$. A (covariant) **functor** $F : \mathbf{C} \to \mathbf{D}$ consists of a pair of maps: $F_0 : \mathbf{C}_0 \to \mathbf{D}_0$, and $F_1 : \mathbf{C}_1 \to \mathbf{D}_1$ with the following properties:

1. F_0 and F_1 are compatible in the sense that if $f : X \to Y$ in $\mathbf{C}$, then $F_1(f) : F_0(X) \to F_0(Y)$ in $\mathbf{D}$.
2. F_1 preserves identities and composition in the following sense: $F_1(1_X) = 1_{F_0(X)}$, and $F_1(g \circ f) = F_1(g) \circ F_1(f)$.

When no confusion can result, we simply use F in place of F_0 and F_1.

NOTE 3.3.2 There is also a notion of a **contravariant functor**, where F_1 reverses the direction of arrows: if $f : X \to Y$ in $\mathbf{C}$, then $F_1(f) : F_0(Y) \to F_0(X)$ in $\mathbf{D}$. Contravariant functors will be especially useful for examining the relation between a theory and its set of models. We've already seen that a translation $f : T \to T'$ induces a function $f^* : M(T') \to M(T)$. In Section 3.7, we will see that $f \mapsto f^*$ is part of a contravariant functor.

Example 3.3.3 For any category $\mathbf{C}$, there is a functor $1_{\mathbf{C}}$ that acts as the identity on both objects and arrows. That is, for any object X of $\mathbf{C}$, $1_{\mathbf{C}}(X) = X$. And for any arrow f of $\mathbf{C}$, $1_{\mathbf{C}}(f) = f$. ⌟

DEFINITION 3.3.4 Let $F : \mathbf{C} \to \mathbf{D}$ and $G : \mathbf{C} \to \mathbf{D}$ be functors. A **natural transformation** $\eta : F \Rightarrow G$ consists of a family $\{\eta_X : F(X) \to G(X) \mid X \in \mathbf{C}_0\}$ of arrows in $\mathbf{D}$, such that for any arrow $f : X \to Y$ in $\mathbf{C}$, the following diagram commutes:

$$\begin{array}{ccc} F(X) & \xrightarrow{F(f)} & F(Y) \\ \downarrow{\scriptstyle \eta_X} & & \downarrow{\scriptstyle \eta_Y} \\ G(X) & \xrightarrow{G(f)} & G(Y) \end{array}$$

DEFINITION 3.3.5 A natural transformation $\eta : F \Rightarrow G$ is said to be a **natural isomorphism** just in case each arrow $\eta_X : F(X) \to G(X)$ is an isomorphism. In this case, we write $F \cong G$.

DEFINITION 3.3.6 Let $F : \mathbf{C} \to \mathbf{D}$ and $G : \mathbf{D} \to \mathbf{C}$ be functors. We say that F and G are a **categorical equivalence** just in case $GF \cong 1_{\mathbf{C}}$ and $FG \cong 1_{\mathbf{D}}$.

3.4 Propositional Theories Are Boolean Algebras

In this section, we show that there is a one-to-one correspondence between theories (in propositional logic) and Boolean algebras. We first need some preliminaries.

DEFINITION 3.4.1 Let Σ be a propositional signature (i.e., a set), let B be a Boolean algebra, and let $f : \Sigma \to B$ be an arbitrary function. (Here we use $\cap, \cup$ and $-$ for the Boolean operations in order to avoid confusion with the logical connectives $\wedge, \vee$ and $\neg$.) Then f naturally extends to a map $f : \mathsf{Sent}(\Sigma) \to B$ as follows:

1. $f(\phi \wedge \psi) = f(\phi) \cap f(\psi)$
2. $f(\phi \vee \psi) = f(\phi) \cup f(\psi)$
3. $f(\neg\phi) = -f(\phi)$.

Now let T be a theory in Σ. We say that f is an **interpretation** of T in B just in case: for all sentences ϕ, if $T \vdash \phi$ then $f(\phi) = 1$.

DEFINITION 3.4.2 Let $f : T \to B$ be an interpretation. We say that

1. f is **conservative** just in case: for all sentences ϕ, if $f(\phi) = 1$ then $T \vdash \phi$.
2. f **surjective** just in case: for each $a \in B$, there is a $\phi \in \mathsf{Sent}(\Sigma)$ such that $f(\phi) = a$.

LEMMA 3.4.3 *Let $f : T \to B$ be an interpretation. Then the following are equivalent:*

1. *f is conservative.*
2. *For any $\phi, \psi \in \mathsf{Sent}(\Sigma)$, if $f(\phi) = f(\psi)$ then $T \vdash \phi \leftrightarrow \psi$.*

Proof Note first that $f(\phi) = f(\psi)$ if and only if $f(\phi \leftrightarrow \psi) = 1$. Suppose then that f is conservative. If $f(\phi) = f(\psi)$, then $f(\phi \leftrightarrow \psi) = 1$, and hence $T \vdash \phi \leftrightarrow \psi$. Suppose now that (2) holds. If $f(\phi) = 1$, then $f(\phi) = f(\phi \vee \neg\phi)$, and hence $T \vdash (\phi \vee \neg\phi) \leftrightarrow \phi$. Therefore, $T \vdash \phi$, and f is conservative. □

LEMMA 3.4.4 *If $f : T \to B$ is an interpretation, and $g : B \to A$ is a homomorphism, then $g \circ f$ is an interpretation.*

Proof This is almost obvious. □

LEMMA 3.4.5 *If $f : T \to B$ is an interpretation, and $g : T' \to T$ is a translation, then $f \circ g : T' \to B$ is an interpretation.*

Proof This is almost obvious. □

LEMMA 3.4.6 *Suppose that T is a theory, and $e : T \to B$ is a surjective interpretation. If $f, g : B \rightrightarrows A$ are homomorphisms such that $fe = ge$, then $f = g$.*

Proof Suppose that $fe = ge$, and let $a \in B$. Since e is surjective, there is a $\phi \in \mathsf{Sent}(\Sigma)$ such that $e(\phi) = a$. Thus, $f(a) = fe(\phi) = ge(\phi) = g(a)$. Since a was arbitrary, $f = g$. □

Let T' and T be theories, and let $f, g : T' \rightrightarrows T$ be translations. Recall that we defined identity between translations as follows: $f = g$ if and only if $T \vdash f(\phi) \leftrightarrow g(\phi)$ for all $\phi \in \mathsf{Sent}(\Sigma')$.

LEMMA 3.4.7 *Suppose that $m : T \to B$ is a conservative interpretation. If $f, g : T' \rightrightarrows T$ are translations such that $mf = mg$, then $f = g$.*

Proof Let $\phi \in \mathsf{Sent}(\Sigma')$, where Σ' is the signature of T'. Then $mf(\phi) = mg(\phi)$. Since m is conservative, $T \vdash f(\phi) \leftrightarrow g(\phi)$. Since this holds for all sentences, it follows that $f = g$. □

PROPOSITION 3.4.8 *For each theory T, there is a Boolean algebra $L(T)$ and a conservative, surjective interpretation $i_T : T \to L(T)$ such that for any Boolean algebra B and interpretation $f : T \to B$, there is a unique homomorphism $\overline{f} : L(T) \to B$ such that $\overline{f} i_T = f$.*

$$\begin{array}{ccc} T & \xrightarrow{i_T} & L(T) \\ & \searrow^{f} & \downarrow{\scriptstyle \overline{f}} \\ & & B \end{array}$$

We define an equivalence relation $\equiv$ on the sentences of Σ:

$$\phi \equiv \psi \quad \text{iff} \quad T \vDash \phi \leftrightarrow \psi,$$

and we let

$$E_\phi := \{\psi \mid \phi \equiv \psi\}.$$

Finally, let

$$L(T) := \{E_\phi \mid \phi \in \mathsf{Sent}(\Sigma)\}.$$

We now equip $L(T)$ with the structure of a Boolean algebra. To this end, we need the following facts, which correspond to easy proofs in propositional logic.

FACT 3.4.9 *If $E_\phi = E_{\phi'}$ and $E_\psi = E_{\psi'}$, then:*

1. $E_{\phi \wedge \psi} = E_{\phi' \wedge \psi'}$
2. $E_{\phi \vee \psi} = E_{\phi' \vee \psi'}$
3. $E_{\neg\phi} = E_{\neg\phi'}$.

We then define a unary operation $-$ on $L(T)$ by

$$-E_\phi := E_{\neg\phi},$$

and we define two binary operations on $L(T)$ by

$$E_\phi \cap E_\psi := E_{\phi \wedge \psi}, \qquad E_\phi \cup E_\psi := E_{\phi \vee \psi}.$$

Finally, let ϕ be an arbitrary Σ sentence, and let $0 = E_{\phi \wedge \neg\phi}$ and $1 = E_{\phi \vee \neg\phi}$. The proof that $\langle L(T), \cap, \cup, -, 0, 1\rangle$ is a Boolean algebra requires a series of straightforward verifications. For example, let's show that $1 \cap E_\psi = E_\psi$, for all sentences ψ. Recall that $1 = E_{\phi\vee\neg\phi}$ for some arbitrarily chosen sentence ϕ. Thus,

$$1 \cap E_\psi = E_{\phi\vee\neg\phi} \cap E_\psi = E_{(\phi\vee\neg\phi)\wedge\psi}.$$

Moreover, $T \vdash \psi \leftrightarrow ((\phi \vee \neg\phi) \wedge \psi)$, from which it follows that $E_{(\phi\vee\neg\phi)\wedge\psi} = E_\psi$. Therefore, $1 \cap E_\psi = E_\psi$.

Consider now the function $i_T : \Sigma \to L(T)$ given by $i_T(\phi) = E_\phi$, and its natural extension to $\mathsf{Sent}(\Sigma)$. A quick inductive argument, using the definition of the Boolean operations on $L(T)$, shows that $i_T(\phi) = E_\phi$ for all $\phi \in \mathsf{Sent}(\Sigma)$. The following shows that i_T is a conservative interpretation of T in $L(T)$.

PROPOSITION 3.4.10 *$T \vdash \phi$ if and only if $i_T(\phi) = 1$.*

Proof $T \vdash \phi$ iff $T \vdash (\psi \vee \neg\psi) \leftrightarrow \phi$ iff $i_T(\phi) = E_\phi = E_{\psi\vee\neg\psi} = 1$. □

Since $i_T(\phi) = E_\phi$, the interpretation i_T is also surjective.

PROPOSITION 3.4.11 *Let B be a Boolean algebra, and let $f : T \to B$ be an interpretation. Then there is a unique homomorphism $\overline{f} : L(T) \to B$ such that $\overline{f} i_T = f$.*

Proof If $E_\phi = E_\psi$, then $T \vdash \phi \leftrightarrow \psi$, and so $f(\phi) = f(\psi)$. Thus, we may define $\overline{f}(E_\phi) = f(\phi)$. It is straightforward to verify that $\overline{f}$ is a Boolean homomorphism, and it is clearly unique. □

DEFINITION 3.4.12 The Boolean algebra $L(T)$ is called the **Lindenbaum algebra** of T.

PROPOSITION 3.4.13 *Let B be a Boolean algebra. There is a theory T_B and a conservative, surjective interpretation $e_B : T_B \to B$ such that for any theory T and interpretation $f : T \to B$, there is a unique interpretation $\overline{f} : T \to T_B$ such that $e_B \overline{f} = f$.*

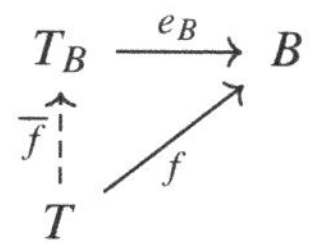

Proof Let $\Sigma_B = B$ be a signature. (Recall that a propositional signature is just a set where each element represents an elementary proposition.) We define $e_B : \Sigma_B \to B$ as the identity and use the symbol e_B also for its extension to $\mathsf{Sent}(\Sigma_B)$. We define a theory T_B on Σ_B by $T_B \vdash \phi$ if and only if $e_B(\phi) = 1$. Thus, $e_B : T_B \to B$ is automatically a conservative interpretation of T_B in B.

Now let T be some theory in signature Σ, and let $f : T \to B$ be an interpretation. Since $\Sigma_B = B$, f automatically gives rise to a reconstrual $f : \Sigma \to \Sigma_B$, which we will rename $\overline{f}$ for clarity. And since e_B is just the identity on $B = \Sigma_B$, we have $f = e_B \overline{f}$.

Finally, to see that $\overline{f} : T \to T_B$ is a translation, suppose that $T \vdash \phi$. Since f is an interpretation of T_B, $f(\phi) = 1$, which means that $e_B(\overline{f}(\phi)) = 1$. Since e_B is conservative, $T_B \vdash \overline{f}(\phi)$. Therefore, $\overline{f}$ is a translation. □

We have shown that each propositional theory T corresponds to a Boolean algebra $L(T)$ and each Boolean algebra B corresponds to a propositional theory T_B. We will now show that these correspondences are functorial. First we show that a morphism $f : B \to A$ in **Bool** naturally gives rise to a morphism $T(f) : T_B \to T_A$ in **Th**. Indeed, consider the following diagram:

$$\begin{array}{ccc} T_B & \xrightarrow{T(f)} & T_A \\ \downarrow{\scriptstyle e_B} & & \downarrow{\scriptstyle e_A} \\ B & \xrightarrow{f} & A \end{array}$$

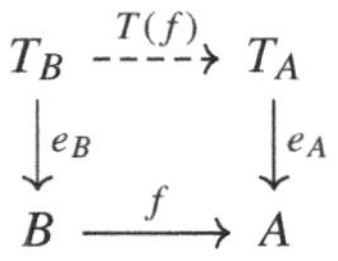

Since fe_B is an interpretation of T_B in A, Prop. 3.4.13 entails that there is a unique translation $T(f) : T_B \to T_A$ such that $e_A T(f) = fe_B$. The uniqueness clause also entails that T commutes with composition of morphisms, and maps identity morphisms to identity morphisms. Thus, T : **Bool** $\to$ **Th** is a functor.

Let's consider this translation $T(f) : T_B \to T_A$ more concretely. First of all, recall that translations from T_B to T_A are actually equivalence classes of maps from Σ_B to $\mathsf{Sent}(\Sigma_A)$. Thus, there's no sense to the question, "which function is $T(f)$?" However, there's a natural choice of a representative function. Indeed, consider f itself as a function from $\Sigma_B = B$ to $\Sigma_A = A$. Then, for $x \in \Sigma_B = B$, we have

$$(e_A \circ T(f))(x) = e_A(f(x)) = f(x) = f(e_B(x)),$$

since e_A is the identity on Σ_A, and e_B is the identity on Σ_B. In other words, $T(f)$ is the equivalence class of f itself. [But recall that translations, while initially defined on the signature Σ_B, extend naturally to all elements of $\mathsf{Sent}(\Sigma_B)$. From this point of view, $T(f)$ has a larger domain than f.]

A similar construction can be used to define the functor L : **Th** $\to$ **Bool**. In particular, let $f : T \to T'$ be a morphism in **Th**, and consider the following diagram:

$$\begin{array}{ccc} T & \xrightarrow{f} & T' \\ \downarrow{\scriptstyle i_T} & & \downarrow{\scriptstyle i_{T'}} \\ L(T) & \xrightarrow{L(f)} & L(T') \end{array}$$

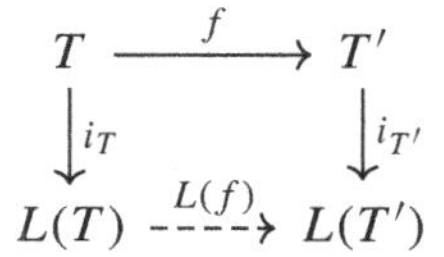

Since $i_{T'}f$ is an interpretation of T in $L(T')$, Prop. 3.4.8 entails that there is a unique homomorphism $L(f) : L(T) \to L(T')$ such that $L(f)i_T = i_{T'}f$.

More explicitly,

$$L(f)(E_\phi) = L(f)(i_T(\phi)) = i_{T'}f(\phi) = E_{f(\phi)}.$$

Recall, however, that identity of arrows in **Th** is *not* identity of the corresponding functions, in the set-theoretic sense. Rather, $f \simeq g$ just in case $T' \vdash f(\phi) \leftrightarrow g(\phi)$, for all $\phi \in \mathsf{Sent}(\Sigma)$. Thus, we must verify that if $f \simeq g$ in **Th**, then $L(f) = L(g)$. Indeed, since $i_{T'}$ is an interpretation of T', we have $i_{T'}(f(\phi)) = i_{T'}(g(\phi))$; and since the diagram above commutes, $L(f) \circ i_T = L(g) \circ i_T$. Since i_T is surjective, $L(f) = L(g)$. Thus, $f \simeq g$ only if $L(f) = L(g)$. Finally, the uniqueness clause in Prop. 3.4.8

entails that L commutes with composition and maps identities to identities. Therefore, $L : \mathbf{Th} \to \mathbf{Bool}$ is a functor.

We will soon show that the functor $L : \mathbf{Th} \to \mathbf{Bool}$ is an equivalence of categories, from which it follows that L preserves all categorically definable properties. For example, a translation $f : T \to T'$ is monic if and only if $L(f) : L(T) \to L(T')$ is monic, etc. However, it may be illuminating to prove some such facts directly.

PROPOSITION 3.4.14 *Let $f : T \to T'$ be a translation. Then f is conservative if and only if $L(f)$ is injective.*

Proof Suppose first that f is conservative. Let $E_\phi, E_\psi \in L(T)$ such that $L(f)(E_\phi) = L(f)(E_\psi)$. Using the definition of $L(f)$, we have $E_{f(\phi)} = E_{f(\psi)}$, which means that $T' \vdash f(\phi) \leftrightarrow f(\psi)$. Since f is conservative, $T \vdash \phi \leftrightarrow \psi$, from which $E_\phi = E_\psi$. Therefore, $L(f)$ is injective.

Suppose now that $L(f)$ is injective. Let ϕ be a Σ sentence such that $T' \vdash f(\phi)$. Since $f(\top) = \top$, we have $T' \vdash f(\top) \leftrightarrow f(\phi)$, which means that $L(f)(E_\top) = L(f)(E_\phi)$. Since $L(f)$ is injective, $E_\top = E_\phi$, from which $T \vdash \phi$. Therefore, f is conservative. □

PROPOSITION 3.4.15 *For any Boolean algebra B, there is a natural isomorphism $\eta_B : B \to L(T_B)$.*

Proof Let $e_B : T_B \to B$ be the interpretation from Prop. 3.4.13, and let $i_{T_B} : T_B \to L(T_B)$ be the interpretation from Prop. 3.4.8. Consider the following diagram:

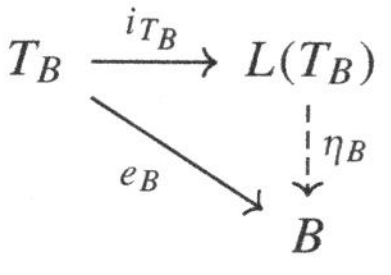

$$\begin{array}{ccc} T_B & \xrightarrow{i_{T_B}} & L(T_B) \\ & \searrow^{e_B} & \downarrow \eta_B \\ & & B \end{array}$$

By Prop. 3.4.8, there is a unique homomorphism $\eta_B : L(T_B) \to B$ such that $e_B = \eta_B i_{T_B}$. Since e_B is the identity on Σ_B,

$$\eta_B(E_x) = \eta_B i_{T_B}(x) = e_B(x) = x,$$

for any $x \in B$. Thus, if η_B has an inverse, it must be given by the map $x \mapsto E_x$. We claim that this map is a Boolean homomorphism. To see this, recall that $\Sigma_B = B$. Moreover, for $x, y \in B$, the Boolean meet $x \cap y$ is again an element of B, hence an element of the signature Σ_B. By the definition of T_B, we have $T_B \vdash (x \cap y) \leftrightarrow (x \wedge y)$, where the $\wedge$ symbol on the right is conjunction in $\mathsf{Sent}(\Sigma_B)$. Thus,

$$E_{x \cap y} = E_{x \wedge y} = E_x \cap E_y.$$

A similar argument shows that $E_{-x} = -E_x$. Therefore, $x \mapsto E_x$ is a Boolean homomorphism, and η_B is an isomorphism.

It remains to show that η_B is natural in B. Consider the following diagram:

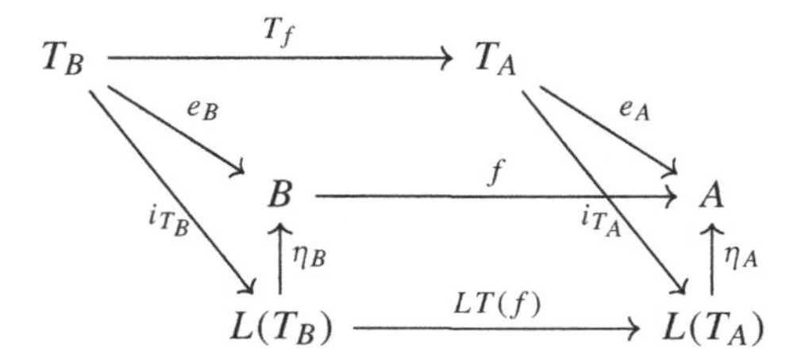

The top square commutes by the definition of the functor T. The triangles on the left and right commute by the definition of η. And the outmost square commutes by the definition of the functor L. Thus we have

$$\begin{aligned} f \circ \eta_B \circ i_{T_B} &= f \circ e_B \\ &= e_A \circ T_f \\ &= \eta_A \circ i_{T_A} \circ T_f \\ &= \eta_A \circ LT(f) \circ i_{T_B}. \end{aligned}$$

Since i_{T_B} is surjective, it follows that $f \circ \eta_B = \eta_A \circ LT(f)$, and, therefore, η is a natural transformation. □

DISCUSSION 3.4.16 Consider the algebra $L(T_B)$, which we have just proved is isomorphic to B. This result is hardly surprising. For any $x, y \in \Sigma_B$, we have $T_B \vdash x \leftrightarrow y$ if and only if $x = e_B(x) = e_B(y) = y$. Thus, the equivalence class E_x contains x and no other element from Σ_B. (That's why $\eta_B(E_x) = x$ makes sense.) We also know that for every $\phi \in \mathsf{Sent}(\Sigma_B)$, there is an $x \in \Sigma_B = B$ such that $T_B \vdash x \leftrightarrow \phi$. In particular, $T_B \vdash e_B(\phi) \leftrightarrow \phi$. Thus, $E_\phi = E_x$, and there is a natural bijection between elements of $L(T_B)$ and elements of B.

PROPOSITION 3.4.17 *For any theory T, there is a natural isomorphism $\varepsilon_T : T \to T_{L(T)}$.*

Proof Consider the following diagram:

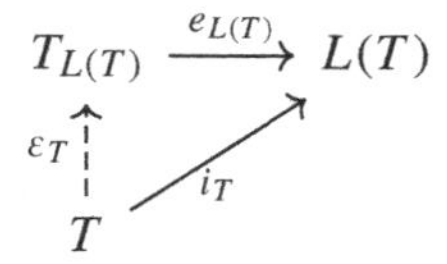

By Prop. 3.4.13, there is a unique interpretation $\varepsilon_T : T \to T_{L(T)}$ such that $e_{L(T)}\varepsilon_T = i_T$. We claim that ε_T is an isomorphism. To see that ϵ_T is conservative, suppose that $T_{L(T)} \vdash \epsilon_T(\phi)$. Since $e_{L(T)}$ is an interpretation, $e_{L(T)}\epsilon_T(\phi) = 1$ and hence $i_T(\phi) = 1$. Since i_T is conservative, $T \vdash \phi$. Therefore ϵ_T is conservative.

To see that ϵ_T is essentially surjective, suppose that $\psi \in \mathsf{Sent}(\Sigma_{L(T)})$. Since i_T is surjective, there is a $\phi \in \mathsf{Sent}(\Sigma)$ such that $i_T(\phi) = e_{L(T)}(\psi)$. Thus, $e_{L(T)}(\epsilon_T(\phi)) = e_{L(T)}(\psi)$. Since $e_{L(T)}$ is conservative, $T_{L(T)} \vdash \epsilon_T(\phi) \leftrightarrow \psi$. Therefore, ϵ_T is essentially surjective.

It remains to show that ϵ_T is natural in T. Consider the following diagram:

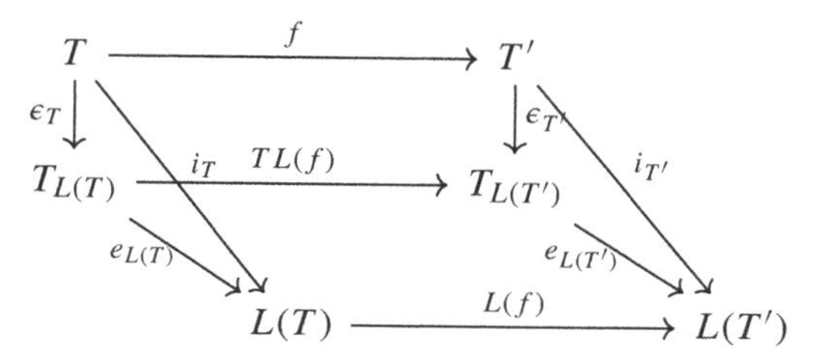

The triangles on the left and the right commute by the definition of ϵ. The top square commutes by the definition of L, and the bottom square commutes by the definition of T. Thus, we have

$$\begin{aligned} e_{L(T')} \circ \epsilon_{T'} \circ f &= i_{T'} \circ f \\ &= L(f) \circ i_T \\ &= L(f) \circ e_{L(T)} \circ \epsilon_T \\ &= e_{L(T')} \circ TL(f) \circ \epsilon_T. \end{aligned}$$

Since $e_{L(T')}$ is conservative, $\epsilon_{T'} \circ f = TL(f) \circ \epsilon_T$. Therefore, ϵ_T is natural in T. □

DISCUSSION 3.4.18 Recall that ϵ_T doesn't denote a unique function; it denotes an equivalence class of functions. One representative of this equivalence class is the function $\epsilon_T : \Sigma \to \Sigma_{L(T)}$ given by $\epsilon_T(p) = E_p$. In this case, a straightforward inductive argument shows that $T_{L(T)} \vdash E_\phi \leftrightarrow \epsilon_T(\phi)$, for all $\phi \in \mathsf{Sent}(\Sigma)$.

We know that ϵ_T has an inverse, which itself is an equivalence class of functions from $\Sigma_{L(T)}$ to $\mathsf{Sent}(\Sigma)$. We can define a representative f of this equivalence class by choosing, for each $E \in \Sigma_{L(T)} = L(T)$, some $\phi \in E$, and setting $f(E) = \phi$. Another straightforward argument shows that if we made a different set of choices, the resulting function f' would be equivalent to f – i.e., it would correspond to the same translation from $T_{L(T)}$ to T.

Since there are natural isomorphisms $\epsilon : 1_{\mathbf{Th}} \Rightarrow TL$ and $\eta : 1_{\mathbf{Bool}} \Rightarrow LT$, we have the following result:

Lindenbaum Theorem

The categories **Th** and **Bool** are equivalent.

3.5 Boolean Algebras Again

The Lindenbaum theorem would deliver everything we wanted – if we had a perfectly clear understanding of the category **Bool**. However, there remain questions about **Bool**.

For example, are all epimorphisms in **Bool** surjections? In order to shed even more light on **Bool**, and hence on **Th**, we will show that **Bool** is dual to a certain category of topological spaces. This famous result is called the **Stone duality theorem**. But in order to prove it, we need to collect a few more facts about Boolean algebras.

DEFINITION 3.5.1 Let B be a Boolean algebra. A subset $F \subseteq B$ is said to be a **filter** just in case

1. If $a, b \in F$, then $a \wedge b \in F$.
2. If $a \in F$ and $a \leq b$, then $b \in F$.

If, in addition, $F \neq B$, then we say that F is a **proper filter**. We say that F is an **ultrafilter** just in case F is maximal among proper filters – i.e., if $F \subseteq F'$ where F' is a proper filter, then $F = F'$.

DISCUSSION 3.5.2 Consider the Boolean algebra B as a theory. Then a filter $F \subseteq B$ can be thought of as supplying an update of information. The first condition says that if we learn a and b, then we've learned $a \wedge b$. The second condition says that if we learn a, and $a \leq b$, then we've learned b. In particular, an ultrafilter supplies maximal information.

EXERCISE 3.5.3 Let F be a filter. Show that F is proper if and only if $0 \notin F$.

DEFINITION 3.5.4 Let $F \subseteq B$ be a filter, and let $a \in B$. We say that a is **compatible** with F just in case $a \wedge x \neq 0$ for all $x \in F$.

LEMMA 3.5.5 *Let $F \subseteq B$ be a proper filter, and let $a \in B$. Then either a or $\neg a$ is compatible with F.*

Proof Suppose for reductio ad absurdum that neither a nor $\neg a$ is compatible with F. That is, there is an $x \in F$ such that $x \wedge a = 0$, and there is a $y \in F$ such that $y \wedge \neg a = 0$. Then

$$x \wedge y = (x \wedge y) \wedge (a \vee \neg a) = (x \wedge y \wedge a) \vee (x \wedge y \wedge \neg a) = 0.$$

Since $x, y \in F$, it follows that $0 = x \wedge y \in F$, contradicting the assumption that F is proper. Therefore, either a or $\neg a$ is compatible with F. □

PROPOSITION 3.5.6 *Let F be a proper filter on B. Then the following are equivalent:*

1. *F is an ultrafilter.*
2. *For all $a \in B$, either $a \in F$ or $\neg a \in F$.*
3. *For all $a, b \in B$, if $a \vee b \in F$, then either $a \in F$ or $b \in F$.*

Proof ($1 \Rightarrow 2$) Suppose that F is an ultrafilter. By Lemma 3.5.5, either a or $\neg a$ is compatible with F. Suppose first that a is compatible with F. Then the set

$$F' = \{y : x \wedge a \leq y, \text{ some } x \in F\},$$

is a proper filter that contains F and a. Since F is an ultrafilter, $F' = F$, and hence $a \in F$. By symmetry, if $\neg a$ is compatible with F, then $\neg a \in F$.

(2 $\Rightarrow$ 3) Suppose that $a \vee b \in F$. By 2, either $a \in F$ or $\neg a \in F$. If $\neg a \in F$, then $\neg a \wedge (a \vee b) \in F$. But $\neg a \wedge (a \vee b) \leq b$, and so $b \in F$.

(3 $\Rightarrow$ 1) Suppose that F' is a filter that contains F, and let $a \in F' - F$. Since $a \vee \neg a = 1 \in F$, it follows from (3) that $\neg a \in F$. But then $0 = a \wedge \neg a \in F'$; that is, $F' = B$. Therefore F is an ultrafilter. □

PROPOSITION 3.5.7 *There is a bijective correspondence between ultrafilters in B and homomorphisms from B into 2. In particular, for any homomorphism $f : B \to 2$, the subset $f^{-1}(1)$ is an ultrafilter in B.*

Proof Let U be an ultrafilter on B. Define $f : B \to 2$ by setting $f(a) = 1$ iff $a \in U$. Then

$$\begin{aligned} f(a \wedge b) = 1 \quad &\text{iff} \quad a \wedge b \in U \\ &\text{iff} \quad a \in U \text{ and } b \in U \\ &\text{iff} \quad f(a) = 1 \text{ and } f(b) = 1. \end{aligned}$$

Furthermore,

$$\begin{aligned} f(\neg a) = 1 \quad &\text{iff} \quad \neg a \in U \\ &\text{iff} \quad a \notin U \\ &\text{iff} \quad f(a) = 0. \end{aligned}$$

Therefore, f is a homomorphism.

Now suppose that $f : B \to 2$ is a homomorphism, and let $U = f^{-1}(1)$. Since $f(a) = 1$ and $f(b) = 1$ only if $f(a \wedge b) = 1$, it follows that U is closed under conjunction. Since $a \leq b$ only if $f(a) \leq f(b)$, it follows that U is closed under implication. Finally, since $f(a) = 0$ iff $f(\neg a) = 1$, it follows that $a \notin U$ iff $\neg a \in U$. □

DEFINITION 3.5.8 For $a, b \in B$, define

$$a \to b \; := \; \neg a \vee b,$$

and define

$$a \leftrightarrow b \; := \; (a \to b) \wedge (b \to a).$$

It's straightforward to check that $\to$ behaves like the conditional from propositional logic. The next lemma gives a Boolean algebra version of modus ponens.

LEMMA 3.5.9 *Let F be a filter. If $a \to b \in F$ and $a \in F$, then $b \in F$.*

Proof Suppose that $\neg a \vee b = a \to b \in F$ and $a \in F$. We then compute

$$b = b \vee 0 = b \vee (a \wedge \neg a) = (a \vee b) \wedge (\neg a \vee b).$$

Since $a \in F$ and $a \leq a \vee b$, we have $a \vee b \in F$. Since F is a filter, $b \in F$. □

EXERCISE 3.5.10

1. Let B be a Boolean algebra, and let $a, b, c \in B$. Show that the following hold:
   (a) $(a \to b) = 1$ iff $a \leq b$
   (b) $(a \wedge b) \leq c$ iff $a \leq (b \to c)$
   (c) $a \wedge (a \to b) \leq b$
   (d) $(a \leftrightarrow b) = (b \leftrightarrow a)$
   (e) $(a \leftrightarrow a) = 1$
   (f) $(a \leftrightarrow 1) = a$
2. Let $\mathscr{P}N$ be the powerset of the natural numbers, and let $\mathscr{U}$ be an ultrafilter on $\mathscr{P}N$. Show that if $\mathscr{U}$ contains a finite set F, then $\mathscr{U}$ contains a singleton set.

DEFINITION 3.5.11 Let B be a Boolean algebra, and let R be an equivalence relation on the underlying set of B. We say that R is a **congruence** just in case R is compatible with the operations on B in the following sense: if aRa' and bRb', then $(a \wedge b)R(a' \wedge b')$, and $(a \vee b)R(a' \vee b')$, and $(\neg a)R(\neg a')$.

In a category $\mathbf{C}$ with limits (products, equalizers, pullbacks, etc.), it's possible to formulate the notion of an equivalence relation in $\mathbf{C}$. Thus, in **Bool**, an equivalence relation R on B is a subalgebra R of $B \times B$ that satisfies the appropriate analogues of reflexivity, symmetry, and transitivity. Since R is a subalgebra of $B \times B$, it follows in particular that if $\langle a, b \rangle \in R$, and $\langle a', b' \rangle \in R$, then $\langle a \wedge a', b \wedge b' \rangle \in R$. Continuing this reasoning, it's not difficult to see that congruences, as defined earlier, are precisely the equivalence relations in the category **Bool** of Boolean algebras. Thus, in the remainder of this chapter, when we speak of an equivalence relation on a Boolean algebra B, we mean an equivalence relation in **Bool** – in other words, a congruence. (To be clear, not every equivalence relation on the set B is an equivalence relation on the Boolean algebra B.)

Now suppose that $\mathbf{C}$ is a category in which equivalence relations are definable, and let $p_0, p_1 : R \rightrightarrows B$ be an equivalence relation. (Here p_0 and p_1 are the projections of R, considered as a subobject of $B \times B$.) Then we can ask, do these two maps p_0 and p_1 have a coequalizer? That is, is there an object B/R, and a map $q : B \to B/R$, with the relevant universal property? In the case of **Bool**, a coequalizer can be constructed directly. We merely note that the Boolean operations on B can be used to induce Boolean operations on the set B/R of equivalence classes.

DEFINITION 3.5.12 (Quotient algebra) Suppose that R is an equivalence relation on B. For each $a \in B$, let E_a denote its equivalence class, and let $B/R = \{E_a \mid a \in B\}$. We then define $E_a \wedge E_b = E_{a \wedge b}$, and similarly for $E_a \vee E_b$ and $\neg E_a$. Since R is a congruence (i.e., an equivalence relation on **Bool**), these operations are well defined. It then follows immediately that B/R is a Boolean algebra, and the quotient map $q : B \to B/R$ is a surjective Boolean homomorphism.

LEMMA 3.5.13 *Let $R \subseteq B \times B$ be an equivalence relation. Then $q : B \to B/R$ is the coequalizer of the projection maps $p_0 : R \to B$ and $p_1 : R \to B$. In particular, q is a regular epimorphism.*

Proof It is obvious that $qp_0 = qp_1$. Now suppose that A is another Boolean algebra and $f : B \to A$ such that $fp_0 = fp_1$. Define $g : B/R \to A$ by setting $g(E_x) = f(x)$. Since $fp_0 = fp_1$, g is well defined. Furthermore,

$$g(E_x \wedge E_y) = g(E_{x \wedge y}) = f(x \wedge y) = f(x) \wedge f(y) = g(E_x) \wedge g(E_y).$$

Similarly, $g(\neg E_x) = \neg g(E_x)$. Therefore, g is a Boolean homomorphism. Since q is an epimorphism, g is the unique homomorphism such that $gq = f$. Therefore, $q : B \to B/R$ is the coequalizer of p_0 and p_1. □

The category **Bool** has further useful structure: there is a one-to-one correspondence between equivalence relations and filters.

LEMMA 3.5.14 *Suppose that $R \subseteq B \times B$ is an equivalence relation. Let $F = \{a \in B \mid aR1\}$. Then F is a filter, and $R = \{\langle a,b\rangle \in B \times B \mid a \leftrightarrow b \in F\}$.*

Proof Suppose that $a, b \in F$. That is, $aR1$ and $bR1$. Since R is a congruence, $(a \wedge b)R(1 \wedge 1)$ and, therefore, $(a \wedge b)R1$. That is, $a \wedge b \in F$. Now suppose that x is an arbitrary element of B such that $a \leq x$. That is, $x \vee a = x$. Since R is a congruence, $(x \vee a)R(x \vee 1)$ and so $(x \vee a)R1$, from which it follows that $xR1$. Therefore, $x \in F$, and F is a filter.

Now suppose that aRb. Since R is reflexive, $(a \vee \neg a)R1$, and, thus, $(b \vee \neg a)R1$. Similarly, $(a \vee \neg b)R1$, and, therefore, $(a \leftrightarrow b)R1$. That is, $a \leftrightarrow b \in F$. □

LEMMA 3.5.15 *Suppose that F is a filter on B. Let $R = \{\langle a,b\rangle \in B \times B \mid a \leftrightarrow b \in F\}$. Then R is an equivalence relation, and $F = \{a \in B \mid aR1\}$.*

Proof Showing that R is an equivalence relation requires several straightforward verifications. For example, $a \leftrightarrow a = 1$, and $1 \in F$; therefore, aRa. We leave the remaining verifications to the reader.

Now suppose that $a \in F$. Since $a = (a \leftrightarrow 1)$, it follows that $a \leftrightarrow 1 \in F$, which means that $aR1$. □

DEFINITION 3.5.16 (Quotient algebra) Let F be a filter on B. Given the correspondence between filters and equivalence relations, we write B/F for the corresponding algebra of equivalence classes.

PROPOSITION 3.5.17 *Let F be a proper filter on B. Then B/F is a two-element Boolean algebra if and only if F is an ultrafilter.*

Proof Suppose first that $B/F \cong 2$. That is, for any $a \in B$, either $a \leftrightarrow 1 \in F$ or $a \leftrightarrow 0 \in F$. But $a \leftrightarrow 1 = a$ and $a \leftrightarrow 0 = \neg a$. Therefore, either $a \in F$ or $\neg a \in F$, and F is an ultrafilter.

Suppose now that F is an ultrafilter. Then for any $a \in B$, either $a \in F$ or $\neg a \in F$. In the former case, $a \leftrightarrow 1 \in F$. In the latter case, $a \leftrightarrow 0 \in F$. Therefore, $B/F \cong 2$. □

EXERCISE 3.5.18 (This exercise presupposes knowledge of measure theory.) Let Σ be the Boolean algebra of Borel subsets of $[0, 1]$, and let μ be Lebesgue measure on $[0, 1]$. Let $\mathscr{F} = \{S \in \Sigma \mid \mu(S) = 1\}$. Show that $\mathscr{F}$ is a filter, and describe the equivalence relation on Σ corresponding to $\mathscr{F}$.

According to our motivating analogy, a Boolean algebra B is like a theory, and a homorphism $\phi : B \to 2$ is like a model of this theory. We say that the algebra B is **syntactically consistent** just in case $0 \neq 1$. (In fact, we defined Boolean algebras so as to require syntactic consistency.) We say that the algebra B is **semantically consistent** just in case there is a homomorphism $\phi : B \to 2$. Then semantic consistency clearly implies syntactic consistency. But does syntactic consistency imply semantic consistency?

It's at this point that we have to invoke a powerful theorem – or, more accurately, a powerful set-theoretic axiom. In short, if we use the axiom of choice, or some equivalent such as Zorn's lemma, then we can prove that every syntactically consistent Boolean algebra is semantically consistent. However, we do not actually need the full power of the Axiom of Choice. As set-theorists know, the Boolean ultrafilter axiom (UF for short) is strictly weaker than the Axiom of Choice.

PROPOSITION 3.5.19 *The following are equivalent:*

1. ***Boolean ultrafilter axiom (UF):*** *For any Boolean algebra B, there is a homomorphism $f : B \to 2$.*
2. *For any Boolean algebra B, and proper filter $F \subseteq B$, there is a homomorphism $f : B \to 2$ such that $f(a) = 1$ when $a \in F$.*
3. *For any Boolean algebra B, if $a, b \in B$ such that $a \neq b$, then there is a homomorphism $f : B \to 2$ such that $f(a) \neq f(b)$.*
4. *For any Boolean algebra B, if $\phi(a) = 1$ for all $\phi : B \to 2$, then $a = 1$.*
5. *For any two Boolean algebras A, B, and homomorphisms $f, g : A \rightrightarrows B$, if $\phi f = \phi g$ for all $\phi : B \to 2$, then $f = g$.*

Proof ($1 \Rightarrow 2$) Suppose that F is a proper filter in B. Then there is a homomorphism $q : B \to B/F$ such that $q(a) = 1$ for all $a \in F$. By UF, there is a homomorphism $\phi : B/F \to 2$. Therefore, $\phi \circ q : B \to 2$ is a homomorphism such that $(\phi \circ q)(a) = 1$ for all $a \in F$.

($1 \Rightarrow 3$) Suppose that $a, b \in B$ with $a \neq b$. Then either $\neg a \wedge b \neq 0$ or $a \wedge \neg b \neq 0$. Without loss of generality, we assume that $\neg a \wedge b \neq 0$. In this case, the filter F generated by $\neg a \wedge b$ is proper. By UF, there is a homomorphism $\phi : B \to 2$ such that $\phi(x) = 1$ when $x \in F$. In particular, $\phi(\neg a \wedge b) = 1$. But then $\phi(a) = 0$ and $\phi(b) = 1$.

($2 \Rightarrow 4$) Suppose that $\phi(a) = 1$ for all $\phi : B \to 2$. Now let F be the filter generated by $\neg a$. If F is proper, then by (2), there is a $\phi : B \to 2$ such that $\phi(\neg a) = 1$, a contradiction. Thus, $F = B$, which implies that $\neg a = 0$ and $a = 1$.

($4 \Rightarrow 5$) Let $f, g : A \to B$ be homomorphisms, and suppose that for all $\phi : B \to 2$, $\phi f = \phi g$. That is, for each $a \in A$, $\phi(f(a)) = \phi(g(a))$. But then $\phi(f(a) \leftrightarrow g(a)) = 1$ for all $\phi : B \to 2$. By (4), $f(a) \leftrightarrow g(a) = 1$ and, therefore, $f(a) = g(a)$.

($5 \Rightarrow 3$) Let B be a Boolean algebra, and $a, b \in B$. Suppose that $\phi(a) = \phi(b)$ for all $\phi : B \to 2$. Let F be the four element Boolean algebra, with generator p. Then there is a homomorphism $\hat{a} : F \to B$ such that $\hat{a}(p) = a$, and a homomorphism $\hat{b} : F \to B$ such that $\hat{b}(p) = b$. Thus, $\phi\hat{a} = \phi\hat{b}$ for all $\phi : B \to 2$. By (5), $\hat{a} = \hat{b}$, and therefore $a = b$.

(3 ⇒ 1) Let B be an arbitrary Boolean algebra. Since $0 \neq 1$, (3) implies that there is a homomorphism $\phi : B \to 2$. □

We are finally in a position to prove the completeness of the propositional calculus. The following result assumes the Boolean ultrafilter axiom (UF).

Completeness Theorem

If $T \models \phi$, then $T \vdash \phi$.

Proof Suppose that $T \nvdash \phi$. Then in the Lindenbaum algebra $L(T)$, we have $E_\phi \neq 1$. In this case, there is a homomorphism $h : L(T) \to 2$ such that $h(E_\phi) = 0$. Hence, $h \circ i_T$ is a model of T such that $(h \circ i_T)(\phi) = h(E_\phi) = 0$. Therefore, $T \not\models \phi$. □

EXERCISE 3.5.20 Let $\mathscr{P}N$ be the powerset of the natural numbers. We say that a subset E of N is **cofinite** just in case $N \backslash E$ is finite. Let $\mathscr{F} \subseteq \mathscr{P}N$ be the set of cofinite subsets of N. Show that $\mathscr{F}$ is a filter, and show that there are infinitely many ultrafilters containing $\mathscr{F}$.

3.6 Stone Spaces

If we're going to undertake an exact study of "possible worlds," then we need to make a proposal about what structure this space carries. But what do I mean here by "structure"? Isn't the collection of possible worlds just a bare set? Let me give you a couple of reasons why it's better to think of possible worlds as forming a **topological space**.

Suppose that there are infinitely many possible worlds, which we represent by elements of a set X. As philosophers are wont to do, we then represent **propositions** by subsets of X. But should we think that all $2^{|X|}$ subsets of X correspond to genuine propositions? What would warrant such a claim?

There is another reason to worry about this approach. For a person with training in set theory, it is not difficult to build a collection $C_1, C_2, \dots$ of subsets of X with the following features: (1) each C_i is nonempty, (2) $C_{i+1} \subseteq C_i$ for all i, and (3) $\bigcap_i C_i$ is empty. Intuitively speaking, $\{C_i \mid i \in \mathbb{N}\}$ is a family of propositions that are individually consistent (since nonempty) and that are becoming more and more specific, and yet there is no world in X that makes all C_i true. Why not? It seems that X is missing some worlds! Indeed, here's a description of a new world w that does not belong to X: for each proposition ϕ, let ϕ be true in w if and only if $\phi \cap C_i$ is nonempty for all i. It's not difficult to see that w is, in fact, a truth valuation on the set of all propositions – i.e., it is a possible world. But w is not represented by a point in X. What we have here is a mismatch between the set X of worlds and the set of propositions describing these worlds.

The idea behind logical topology is that not all subsets of X correspond to propositions. A designation of a topology on X is tantamount to saying which subsets of X correspond to propositions. However, the original motivation for the study of topology comes from geometry (and analysis), not from logic. Recall high school mathematics, where you learned that a continuous function is one where you don't have to lift your pencil from the paper in order to draw the graph. If your high school class was really good, or if you studied calculus in college, then you will have learned that there is a more rigorous definition of a continuous function – a definition involving epsilons and deltas. In the early twentieth century, it was realized that the essence of continuity is even more abstract than epsilons and deltas would suggest: all we need is a notion of nearness of points, which we can capture in terms of a notion of a neighborhood of a point. The idea then is that a function $f : X \to Y$ is continuous at a point x just in case for any neighborhood V of $f(x)$, there is some neighborhood U of x such that $f(U) \subseteq V$. Intuitively speaking, f preserves closeness of points.

Notice, however, that if X is an arbitrary set, then it's not obvious what "closeness" means. To be able to talk about closeness of points in X, we need specify which subsets of X count as the neighborhoods of points. Thus, a **topology** on X is a set of subsets of X that satisfies certain conditions.

DEFINITION 3.6.1 A **topological space** is a set X and a family $\mathscr{F}$ of subsets of X satisfying the following conditions:

1. $\emptyset \in \mathscr{F}$ and $X \in \mathscr{F}$.
2. If $U, V \in \mathscr{F}$ then $U \cap V \in \mathscr{F}$.
3. If $\mathscr{F}_0$ is a subfamily of $\mathscr{F}$, then $\bigcup_{U \in \mathscr{F}_0} U \in \mathscr{F}$.

The sets in $\mathscr{F}$ are called **open subsets** of the space $(X, \mathscr{F})$. If $p \in U$ with U an open subset, we say that U is a **neighborhood** of p.

There are many familiar examples of topological spaces. In many cases, however, we only know the open sets indirectly, by means of certain nice open sets. For example, in the case of the real numbers, not every open subset is an interval. However, every open subset is a union of intervals. In that case, we call the open intervals in $\mathbb{R}$ a **basis** for the topology.

PROPOSITION 3.6.2 *Let $\mathscr{B}$ be a family of subsets of X with the property that if $U, V \in \mathscr{B}$, then $U \cap V \in \mathscr{B}$. Then there is a unique smallest topology $\mathscr{F}$ on X containing $\mathscr{B}$.*

Proof Let $\mathscr{F}$ be the collection obtained by taking all unions of sets in $\mathscr{B}$, and then taking finite intersections of the resulting collection. Clearly $\mathscr{F}$ is a topology on X, and any topology on X containing $\mathscr{B}$ also contains $\mathscr{F}$. □

DEFINITION 3.6.3 If $\mathscr{B}$ is a family of subsets of X that is closed under intersection, and if $\mathscr{F}$ is the topology generated by $\mathscr{B}$, then we say that $\mathscr{B}$ is a **basis** for $\mathscr{F}$.

PROPOSITION 3.6.4 *Let $(X, \mathscr{F})$ be a topological space. Let $\mathscr{F}_0$ be a subfamily of $\mathscr{F}$ with the following properties: (1) $\mathscr{F}_0$ is closed under finite intersections, and (2) for* *each $x \in X$ and $U \in \mathscr{F}_0$ with $x \in U$, there is a $V \in \mathscr{F}_0$ such that $x \in V \subseteq U$. Then $\mathscr{F}_0$ is a basis for the topology $\mathscr{F}$.*

Proof We need only show that each $U \in \mathscr{F}$ is a union of elements in $\mathscr{F}_0$. And that follows immediately from the fact that if $x \in U$, then there is $V \in \mathscr{F}_0$ with $x \in V \subseteq U$. □

DEFINITION 3.6.5 Let X be a topological space. A subset C of X is called **closed** just in case $C = X \backslash U$ for some open subset U of X. The intersection of closed sets is closed. Hence, for each subset E of X, there is a unique smallest closed set $\overline{E}$ containing E, namely the intersection of all closed supersets of E. We call $\overline{E}$ the **closure** of E.

PROPOSITION 3.6.6 *Let $p \in X$ and let $S \subseteq X$. Then $p \in \overline{S}$ if and only if every open neighborhood U of p has nonempty intersection with S.*

Proof Exercise. □

DEFINITION 3.6.7 Let S be a subset of X. We say that S is **dense** in X just in case $\overline{S} = X$.

DEFINITION 3.6.8 Let $E \subseteq X$. We say that p is a **limit point** of E just in case for each open neighborhood U of p, $U \cap E$ contains some point besides p. We let E' denote the set of all limit points of E.

LEMMA 3.6.9 $E' \subseteq \overline{E}$.

Proof Let $p \in E'$, and let C be a closed set containing E. If $p \in X \backslash C$, then p is contained in an open set that has empty intersection with E. Thus, $p \in C$. Since C was an arbitrary closed superset of E, it follows that $p \in \overline{E}$. □

PROPOSITION 3.6.10 $\overline{E} = E \cup E'$.

Proof The previous lemma gives $E' \subseteq \overline{E}$. Thus, $E \cup E' \subseteq \overline{E}$.

Suppose now that $p \notin E$ and $p \notin E'$. Then there is an open neighborhood U of p such that $U \cap E$ is empty. Then $E \subseteq X \backslash U$, and since $X \backslash U$ is closed, $\overline{E} \subseteq X \backslash U$. Therefore, $p \notin \overline{E}$. □

DEFINITION 3.6.11 A topological space X is said to be

- T_1, or **Frechet**, just in case all singleton subsets are closed.
- T_2, or **Hausdorff**, just in case, for any $x, y \in X$, if $x \neq y$, then there are disjoint open neighborhoods of x and y.
- T_3, or **regular**, just in case for each $x \in X$, and for each closed $C \subseteq X$ such that $x \notin C$, there are open neighborhoods U of x, and V of C, such that $U \cap V = \emptyset$.
- T_4, or **normal**, just in case any two disjoint closed subsets of X can be separated by disjoint open sets.

Clearly we have the implications

$$(T_1 + T_4) \Rightarrow (T_1 + T_3) \Rightarrow T_2 \Rightarrow T_1.$$

A discrete space satisfies all of the separation axioms. A nontrivial indiscrete space satisfies none of the separation axioms. A useful heuristic here is that the stronger the separation axiom, the closer the space is to discrete. In this book, most of the spaces we consider are very close to discrete (which means that all subsets are open).

EXERCISE 3.6.12

1. Show that X is regular iff for each $x \in X$ and open neighborhood U of x, there is an open neighborhood V of x such that $\overline{V} \subseteq U$.
2. Show that if $E \subseteq F$, then $\overline{E} \subseteq \overline{F}$.
3. Show that $\overline{\overline{E}} = \overline{E}$.
4. Show that the intersection of two topologies is a topology.
5. Show that the infinite distributive law holds:

$$U \cap \left(\bigcup_{i \in I} V_i \right) = \bigcup_{i \in I} (U \cap V_i).$$

DEFINITION 3.6.13 Let $S \subseteq X$. A family $\mathscr{C}$ of open subsets of X is said to **cover** S just in case $S \subseteq \bigcup_{U \in \mathscr{C}} U$. We say that S is **compact** just in case for every open cover $\mathscr{C}$ of S, there is a finite subcollection $\mathscr{C}_0$ of $\mathscr{C}$ that also covers S. We say that the space X is compact just in case it's compact as a subset of itself.

DEFINITION 3.6.14 A collection $\mathscr{C}$ of subsets of X is said to satisfy the **finite intersection property** if for every finite subcollection $C_1, \dots, C_n$ of $\mathscr{C}$, the intersection $C_1 \cap \dots \cap C_n$ is nonempty.

DISCUSSION 3.6.15 Suppose that X is the space of possible worlds, so that we can think of subsets of X as propositions. If $A \cap B$ is nonempty, then the propositions A and B are consistent – i.e., there is a world in which they are both true. Thus, a collection $\mathscr{C}$ of propositions has the finite intersection property just in case it is finitely consistent.

Recall that compactness of propositional logic states that if a set $\mathscr{C}$ of propositions is finitely consistent, then $\mathscr{C}$ is consistent. The terminology here is no accident; a topological space is compact just in case finite consistency entails consistency.

PROPOSITION 3.6.16 *A space X is compact if and only if for every collection $\mathscr{C}$ of closed subsets of X, if $\mathscr{C}$ satisfies the finite intersection property, then $\bigcap \mathscr{C}$ is nonempty.*

Proof ($\Rightarrow$) Assume first that X is compact, and let $\mathscr{C}$ be a family of closed subsets of X. We will show that if $\mathscr{C}$ satisfies the finite intersection property, then the intersection of all sets in $\mathscr{C}$ is nonempty. Assume the negation of the consequent, i.e., that $\bigcap_{C \in \mathscr{C}} C$ is empty. Let $\mathscr{C}' = \{C' : C \in \mathscr{C}\}$, where $C' = X \backslash C$ is the complement of C in X. (Warning: this notation can be confusing. Previously we used E' to denote the set of limit points of E. This C' has nothing to do with limit points.) Each C' is open, and

$$\left(\bigcup_{C \in \mathscr{C}} C' \right)' = \bigcap_{C \in \mathscr{C}} C,$$

which is empty. It follows then that $\mathscr{C}'$ is an open cover of X. Since X is compact, there is a finite subcover $\mathscr{C}'_0$ of $\mathscr{C}'$. If we let $\mathscr{C}_0$ be the complements of sets in $\mathscr{C}'_0$, then $\mathscr{C}_0$ is a finite collection of sets in $\mathscr{C}$ whose intersection is empty. Therefore, $\mathscr{C}$ does not satisfy the finite intersection property.

($\Leftarrow$) Assume now that X is not compact. In particular, suppose that $\mathscr{U}$ is an open cover with no finite subcover. Let $\mathscr{C} = \{X\backslash U \mid U \in \mathscr{U}\}$. For any finite subcollection $X\backslash U_1, \dots, X\backslash U_n$ of $\mathscr{C}$, we have

$$U_1 \cup \cdots \cup U_n \neq X,$$

and hence

$$(X\backslash U_1) \cap \cdots \cap (X\backslash U_n) \neq \emptyset.$$

Thus, $\mathscr{C}$ has the fip. Nonetheless, since $\mathscr{U}$ covers X, the intersection of all sets in $\mathscr{C}$ is empty. □

PROPOSITION 3.6.17 *In a compact space, closed subsets are compact.*

Proof Let $\mathscr{C}$ be an open cover of S, and consider the cover $\mathscr{C}' = \mathscr{C} \cup \{X\backslash S\}$ of X. Since X is compact, there is a finite subcover $\mathscr{C}_0$ of $\mathscr{C}'$. Removing $X\backslash S$ from $\mathscr{C}_0$ gives a finite subcover of the original cover $\mathscr{C}$ of S. □

PROPOSITION 3.6.18 *Suppose that X is compact, and let U be an open set in X. Let $\{F_i\}_{i\in I}$ be a family of closed subsets of X such that $\bigcap_{i\in I} F_i \subseteq U$. Then there is a finite subset J of I such that $\bigcap_{i\in J} F_i \subseteq U$.*

Proof Let $C = X\backslash U$, which is closed. Thus, the hypotheses of the proposition say that the family $\mathscr{C} := \{C\} \cup \{F_i : i \in I\}$ has empty intersection. Since X is compact, $\mathscr{C}$ also fails to have the finite intersection property. That is, there are $i_1, \dots, i_k \in I$ such that $C \cap F_{i_1} \cap \cdots \cap F_{i_k} = \emptyset$. Therefore, $F_{i_1} \cap \cdots \cap F_{i_k} \subseteq U$. □

PROPOSITION 3.6.19 *If X is compact Hausdorff, then X is regular.*

Proof Let $x \in X$, and let $C \subseteq X$ be closed. For each $y \in C$, let U_y be an open neighborhood of x, and V_y an open neighborhood of y such that $U_y \cap V_y = \emptyset$. The V_y form an open cover of C. Since C is closed and X is compact, C is compact. Hence, there is a finite subcollection $V_{y_1}, \dots, V_{y_n}$ that cover C. But then $U = \cap_{i=1}^{n} U_{y_i}$ is an open neighborhood of x, and $V = \cup_{i=1}^{n} V_{y_i}$ is an open neighborhood of C, such that $U \cap V = \emptyset$. Therefore, X is regular. □

PROPOSITION 3.6.20 *In Hausdorff spaces, compact subsets are closed.*

Proof Let p be a point of X that is not in K. Since X is Hausdorff, for each $x \in K$, there are open neighborhoods U_x of x and V_x of p such that $U_x \cap V_x = \emptyset$. The family $\{U_x : x \in K\}$ covers K. Since K is compact, it is covered by a finite subcollection $U_{x_1}, \dots, U_{x_n}$. But then $\cap_{i=1}^{n} V_{x_i}$ is an open neighborhood of p that is disjoint from K. It follows that $X\backslash K$ is open, and K is closed. □

DEFINITION 3.6.21 Let X, Y be topological spaces. A function $f : X \to Y$ is said to be **continuous** just in case for each open subset U of Y, $f^{-1}(U)$ is an open subset of X.

Example 3.6.22 Let $f : \mathbb{R} \to \mathbb{R}$ be the function that is constantly 0 on $(-\infty, 0)$, and 1 on $[0, \infty)$. Then f is not continuous: $f^{-1}(\frac{1}{2}, \frac{3}{2}) = [0, \infty)$, which is not open. ⌟

In the exercises, you will show that a function f is continuous if and only if $f^{-1}(C)$ is closed whenever C is closed. Thus, in particular, if C is a clopen subset of Y, then $f^{-1}(C)$ is a clopen subset of X.

PROPOSITION 3.6.23 *Let* **Top** *consist of the class of topological spaces and continuous maps between them. For* $X \xrightarrow{f} Y \xrightarrow{g} Z$, *define* $g \circ f$ *to be the composition of* g *and* f. *Then* **Top** *is a category.*

Proof It needs to be confirmed that if f and g are continuous, then $g \circ f$ is continuous. We leave this to the exercises. Since composition is associative, **Top** is a category. □

PROPOSITION 3.6.24 *Suppose that* $f : X \to Y$ *is continuous. If* K *is compact in* X, *then* $f(K)$ *is compact in* Y.

Proof Let $\mathscr{G}$ be a collection of open subsets of Y that covers $f(K)$. Let

$$\mathscr{G}' = \{f^{-1}(U) : U \in \mathscr{G}\}.$$

When $\mathscr{G}'$ is an open cover of K. Since K is compact, $\mathscr{G}'$ has a finite subcover $f^{-1}(U_1), \ldots, f^{-1}(U_n)$. But then $U_1, \ldots, U_n$ is a finite subcover of $\mathscr{G}$. □

We remind the reader of the category theoretic definitions:

- f is a **monomorphism** just in case $fh = fk$ implies $h = k$.
- f is an **epimorphism** just in case $hf = kf$ implies $h = k$.
- f is an **isomorphism** just in case there is a $g : Y \to X$ such that $gf = 1_X$ and $fg = 1_Y$.

For historical reasons, isomorphisms in **Top** are usually called **homeomorphisms**. It is easy to show that a continuous map $f : X \to Y$ is monic if and only if f is injective. It is also true that $f : X \to Y$ is epi if and only if f is surjective (but the proof is somewhat subtle). In contrast, a continuous bijection is not necessarily an isomorphism in **Top**. For example, if we let X be a two-element set with the discrete topology, and if we let Y be a two-element set with the indiscrete topology, then any bijection $f : X \to Y$ is continuous but is not an isomorphism.

EXERCISE 3.6.25

1. Show that if f and g are continuous, then $g \circ f$ is continuous.
2. Suppose that $f : X \to Y$ is a surjection. Show that if E is dense in X, then $f(E)$ is dense in Y.
3. Show that $f : X \to Y$ is continuous if and only if $f^{-1}(C)$ is closed whenever C is closed.
4. Let Y be a Hausdorff space, and let $f, g : X \to Y$ be continuous. Show that if f and g agree on a dense subset of X, then $f = g$.

EXERCISE 3.6.26 Show that $f^{-1}(V) \subseteq U$ if and only if $V \subseteq Y \backslash f(X \backslash U)$.

DEFINITION 3.6.27 A continuous mapping $f : X \to Y$ is said to be **closed** just in case for every closed set $C \subseteq X$, the image $f(C)$ is closed in Y. Similarly, $f : X \to Y$ is said to be **open** just in case for every open set $U \subseteq X$, the image $f(U)$ is open in Y.

PROPOSITION 3.6.28 *Let $f : X \to Y$ be continuous. Then the following are equivalent.*

1. *f is closed.*
2. *For every open set $U \subseteq X$, the set $\{y \in Y \mid f^{-1}\{y\} \subseteq U\}$ is open.*
3. *For every $y \in Y$, and every neighborhood U of $f^{-1}\{y\}$, there is a neighborhood V of y such that $f^{-1}(V) \subseteq U$.*

Proof ($2 \Leftrightarrow 3$) The equivalence of (2) and (3) is straightforward, and we leave its proof as an exercise.

($3 \Rightarrow 1$) Suppose that f satisfies condition (3), and let C be a closed subset of X. To show that $f(C)$ is closed, assume that $y \in Y \backslash f(C)$. Then $f^{-1}\{y\} \subseteq X \backslash C$. Since $X \backslash C$ is open, there is a neighborhood V of y such that $f^{-1}(V) \subseteq U$. Then

$$V \subseteq Y \backslash f(X \backslash U) = Y \backslash f(C).$$

Since y was an arbitrary element of $Y \backslash f(C)$, it follows that $Y \backslash f(C)$ is open, and $f(C)$ is closed.

($1 \Rightarrow 3$) Suppose that f is closed. Let $y \in Y$, and let U be a neighborhood of $f^{-1}\{y\}$. Then $X \backslash U$ is closed, and $f(X \backslash U)$ is also closed. Let $V = Y \backslash f(X \backslash U)$. Then V is an open neighborhood of y and $f^{-1}(V) \subseteq U$. □

PROPOSITION 3.6.29 *Suppose that X and Y are compact Hausdorff. If $f : X \to Y$ is continuous, then f is a closed map.*

Proof Let B be a closed subset of X. By Proposition 3.6.17, B is compact. By Proposition 3.6.24, $f(B)$ is compact. And by Proposition 3.6.20, $f(B)$ is closed. Therefore, f is a closed map. □

PROPOSITION 3.6.30 *Suppose that X and Y are compact Hausdorff. If $f : X \to Y$ is a continuous bijection, then f is an isomorphism.*

Proof Let $f : X \to Y$ be a continuous bijection. Thus, there is function $g : Y \to X$ such that $gf = 1_X$ and $fg = 1_Y$. We will show that g is continuous. By Proposition 3.6.29, f is closed. Moreover, for any closed subset B of X, we have $g^{-1}(B) = f(B)$. Thus, g^{-1} preserves closed subsets, and hence g is continuous. □

DEFINITION 3.6.31 A topological space X is said to be **totally separated** if for any $x, y \in X$, if $x \neq y$ then there is a closed and open (clopen) subset of X containing x but not y.

DEFINITION 3.6.32 We say that X is a **Stone space** if X is compact and totally separated. We let **Stone** denote the full subcategory of **Top** consisting of Stone spaces.

To say that **Stone** is a full subcategory means that the arrows between two Stone spaces X and Y are just the arrows between X and Y considered as topological spaces – i.e., continuous functions.

NOTE 3.6.33 Let E be a clopen subset of X. Then there is a continuous function $f : X \to \{0, 1\}$ such that $f(x) = 1$ for $x \in E$, and $f(x) = 0$ for $x \in X \backslash E$. Here we are considering $\{0, 1\}$ with the discrete topology.

PROPOSITION 3.6.34 *Let X and Y be Stone spaces. If $f : X \to Y$ is an epimorphism, then f is surjective.*

Proof Suppose that f is not surjective. Since X is compact, the image $f(X)$ is compact in Y, hence closed. Since f is not surjective, there is a $y \in Y\backslash f(X)$. Since Y is a regular space, there is a clopen neighborhood U of y such that $U \cap f(X) = \emptyset$. Define $g : Y \to \{0, 1\}$ to be constantly 0. Define $h : Y \to \{0, 1\}$ to be 1 on U, and 0 on $Y\backslash U$. Then $g \circ f = h \circ f$, but $g \neq h$. Therefore, f is not an epimorphism. □

PROPOSITION 3.6.35 *Let X and Y be Stone spaces. If $f : X \to Y$ is both a monomorphism and an epimorphism, then f is an isomorphism.*

Proof By Proposition 3.6.34, f is surjective. Therefore, f is a continuous bijection. By Proposition 3.6.30, f is an isomorphism. □

3.7 Stone Duality

In this section, we show that the category **Bool** is dual to the category **Stone** of Stone spaces. To say that categories are "dual" means that the first is equivalent to the mirror image of the second.

DEFINITION 3.7.1 We say that categories $\mathbf{C}$ and $\mathbf{D}$ are **dual** just in case there are contravariant functors $F : \mathbf{C} \to \mathbf{D}$ and $G : \mathbf{D} \to \mathbf{C}$ such that $GF \cong 1_{\mathbf{C}}$ and $FG \cong 1_{\mathbf{D}}$. To see that this definition makes sense, note that if F and G are contravariant functors, then GF and FG are covariant functors. If $\mathbf{C}$ and $\mathbf{D}$ are dual, we write $\mathbf{C} \cong \mathbf{D}^{op}$, to indicate that $\mathbf{C}$ is equivalent to the opposite category of $\mathbf{D}$ – i.e., the category that has the same objects as $\mathbf{D}$, but arrows running in the opposite direction.

The Functor from Bool to Stone

We now define a contravariant functor S : **Bool** $\to$ **Stone**. For reasons that will become clear later, the functor S is sometimes called the **semantic functor**.

Consider the set $\hom(B, 2)$ of two-valued homomorphisms of the Boolean algebra B. For each $a \in B$, define

$$C_a = \{\phi \in \hom(B, 2) \mid \phi(a) = 1\}.$$

Clearly, the family $\{C_a \mid a \in B\}$ forms a basis for a topology on $\hom(B, 2)$. We let $S(B)$ denote the resulting topological space. Note that $S(B)$ has a basis of clopen sets. Thus, if $S(B)$ is compact, then $S(B)$ is a Stone space.

LEMMA 3.7.2 *If B is a Boolean algebra, then $S(B)$ is a Stone space.*

Proof Let $\mathscr{B} = \{C_a \mid a \in B\}$ denote the chosen basis for the topology on $S(B)$. To show that $S(B)$ is compact, it will suffice to show that for any subfamily $\mathscr{C}$ of $\mathscr{B}$, if $\mathscr{C}$ has the finite intersection property, then $\bigcap \mathscr{C}$ is nonempty. Now let F be the set of $b \in B$ such that

$$C_{a_1} \cap \cdots \cap C_{a_n} \subseteq C_b,$$

for some $C_{a_1}, \ldots, C_{a_n} \in \mathscr{C}$. Since $\mathscr{C}$ has the finite intersection property, F is a filter in B. Thus, UF entails that F is contained in an ultrafilter U. This ultrafilter U corresponds to a $\phi : B \to 2$, and we have $\phi(a) = 1$ whenever $C_a \in \mathscr{C}$. In other words, $\phi \in C_a$, whenever $C_a \in \mathscr{C}$. Therefore, $\bigcap \mathscr{C}$ is nonempty, and $S(B)$ is compact. □

Let $f : A \to B$ be a homomorphism, and let $S(f) : S(B) \to S(A)$ be given by $S(f) = \hom(f, 2)$; that is,

$$S(f)(\phi) = \phi \circ f, \qquad \forall \phi \in S(B).$$

We claim now that $S(f)$ is a continuous map. Indeed, for any basic open subset C_a of $S(A)$, we have

$$S(f)^{-1}(C_a) = \{\phi \in S(B) \mid \phi(f(a)) = 1\} = C_{f(a)}. \tag{3.1}$$

It is straightforward to verify that $S(1_A) = 1_{S(A)}$, and that $S(g \circ f) = S(f) \circ S(f)$. Therefore, $S : \mathbf{Bool} \to \mathbf{Stone}$ is a contravariant functor.

The Functor from Stone to Bool

Let X be a Stone space. Then the set $K(X)$ of clopen subsets of X is a Boolean algebra, and is a basis for the topology on X. We now show that K is the object part of a contravariant functor $K : \mathbf{Stone} \to \mathbf{Bool}$. For reasons that will become clear later, K is sometimes called the **syntactic functor**.

Indeed, if X, Y are Stone spaces, and $f : X \to Y$ is continuous, then for each clopen subset U of Y, $f^{-1}(U)$ is a clopen subset of X. Moreover, f^{-1} preserves union, intersection, and complement of subsets; thus $f^{-1} : K(Y) \to K(X)$ is a Boolean homomorphism. We define the mapping K on arrows by $K(f) = f^{-1}$. Obviously, $K(1_X) = 1_{K(X)}$, and $K(g \circ f) = K(f) \circ K(g)$. Therefore, K is a contravariant functor.

Now we will show that KS is naturally isomorphic to the identity on **Bool**, and SK is naturally isomorphic to the identity on **Stone**. For each Boolean algebra B, define $\eta_B : B \to KS(B)$ by

$$\eta_B(a) = C_a = \{\phi \in S(B) \mid \phi(a) = 1\}.$$

LEMMA 3.7.3 *The map $\eta_B : B \to KS(B)$ is an isomorphism of Boolean algebras.*

Proof We first verify that $a \mapsto C_a$ is a Boolean homomorphism. For $a, b \in B$, we have

$$\begin{aligned} C_{a \wedge b} &= \{\phi \mid \phi(a \wedge b) = 1\} \\ &= \{\phi \mid \phi(a) = 1 \text{ and } \phi(b) = 1\} \\ &= C_a \wedge C_b. \end{aligned}$$

A similar calculation shows that $C_{\neg a} = X \backslash C_a$. Therefore, $a \mapsto C_a$ is a Boolean homomorphism.

To show that $a \mapsto C_a$ is injective, it will suffice to show that $C_a = \emptyset$ only if $a = 0$. In other words, it will suffice to show that for each $a \in B$, if $a \neq 0$ then there is some $\phi : B \to 2$ such that $\phi(a) = 1$. Thus, the result follows from UF.

Finally, to see that η_B is surjective, let U be a clopen subset of $S(B)$. Since U is open, $U = \bigcup_{a \in I} C_a$, for some subset I of B. Since U is closed in the compact space $G(B)$, it follows that U is compact. Thus, there is a finite subset F of B such that $U = \bigcup_{a \in F} C_a$. And since $a \mapsto C_a$ is a Boolean homomorphism, $\bigcup_{a \in F} C_a = C_b$, where $b = \bigvee_{a \in F} a$. Therefore, η_B is surjective. □

LEMMA 3.7.4 *The family of maps* $\{\eta_A : A \to KS(A)\}$ *is natural in* A.

Proof Suppose that A and B are Boolean algebras and that $f : A \to B$ is a Boolean homomorphism. Consider the following diagram:

$$\begin{array}{ccc} A & \xrightarrow{f} & B \\ \downarrow{\scriptstyle \eta_A} & & \downarrow{\scriptstyle \eta_B} \\ KS(A) & \xrightarrow{KS(f)} & KS(B) \end{array}$$

For $a \in A$, we have $\eta_B(f(a)) = C_{f(a)}$, and $\eta_A(a) = C_a$. Furthermore,

$$KS(f)(C_a) = S(f)^{-1}(C_a) = C_{f(a)},$$

by Eqn. 3.1. Therefore, the diagram commutes, and η is a natural transformation. □

Now we define a natural isomorphism $\theta : 1_{\mathbf{S}} \Rightarrow SK$. For a Stone space X, $K(X)$ is the Boolean algebra of clopen subsets of X, and $SK(X)$ is the Stone space of $K(X)$. For each point $\phi \in X$, let $\hat{\phi} : K(X) \to 2$ be defined by

$$\hat{\phi}(C) = \begin{cases} 1 & \phi \in C, \\ 0 & \phi \notin C. \end{cases}$$

It's straightforward to verify that $\hat{\phi}$ is a Boolean homomorphism. We define $\theta_X : X \to SK(X)$ by $\theta_X(\phi) = \hat{\phi}$.

LEMMA 3.7.5 *The map* $\theta_X : X \to SK(X)$ *is a homeomorphism of Stone spaces.*

Proof It will suffice to show that θ_X is bijective and continuous. (Do you remember why? Hint: Stone spaces are compact Hausdorff.) To see that θ_X is injective, suppose that ϕ and ψ are distinct elements of X. Since X is a Stone space, there is a clopen set U of X such that $\phi \in U$ and $\psi \notin U$. But then $\hat{\phi} \neq \hat{\psi}$. Thus, θ_X is injective.

To see that θ_X is surjective, let $h : K(X) \to 2$ be a Boolean homomorphism. Let

$$\mathscr{C} = \{C \in K(X) \mid h(C) = 1\}.$$

In particular $X \in \mathscr{C}$; and since h is a homomorphism, $\mathscr{C}$ has the finite intersection property. Since X is compact, $\bigcap \mathscr{C}$ is nonempty. Let ϕ be a point in $\bigcap \mathscr{C}$. Then for any $C \in K(X)$, if $h(C) = 1$, then $C \in \mathscr{C}$ and $\phi \in C$, from which it follows that $\hat{\phi}(C) = 1$. Similarly, if $h(C) = 0$ then $X \backslash C \in \mathscr{C}$, and $\hat{\phi}(C) = 0$. Thus, $\theta_X(\phi) = \hat{\phi} = h$, and θ_X is surjective.

To see that θ_X is continuous, note that each basic open subset of $SK(X)$ is of the form

$$\hat{C} = \{h : K(X) \to 2 \mid h(C) = 1\},$$

for some $C \in K(X)$. Moreover, for any $\phi \in X$, we have $\hat{\phi} \in \hat{C}$ iff $\hat{\phi}(C) = 1$ iff $\phi \in C$. Therefore,

$$\theta_X^{-1}(\hat{C}) = \{\phi \in X \mid \hat{\phi}(C) = 1\} = C.$$

Therefore, θ_X is continuous. □

LEMMA 3.7.6 *The family of maps* $\{\theta_X : X \to SK(X)\}$ *is natural in* X.

Proof Let X, Y be Stone spaces, and let $f : X \to Y$ be continuous. Consider the diagram:

$$\begin{array}{ccc} X & \xrightarrow{f} & Y \\ \downarrow{\scriptstyle \theta_X} & & \downarrow{\scriptstyle \theta_Y} \\ SK(X) & \xrightarrow{SK(f)} & SK(Y) \end{array}$$

For arbitrary $\phi \in X$, we have $(\theta_Y \circ f)(\phi) = \widehat{f(\phi)}$. Furthermore.

$$SK(f) = \hom(K(f), 2) = \hom(f^{-1}, 2),$$

In other words, for a homomorphism $h : K(X) \to 2$, we have

$$SK(f)(h) = h \circ f^{-1}.$$

In particular, $SK(f)(\hat{\phi}) = \hat{\phi} \circ f^{-1}$. For any $C \in K(Y)$, we have

$$(\hat{\phi} \circ f^{-1})(C) = \begin{cases} 1 & f(\phi) \in C, \\ 0 & f(\phi) \notin C. \end{cases}$$

That is, $\hat{\phi} \circ f^{-1} = \widehat{f(\phi)}$. Therefore, the diagram commutes, and θ is a natural isomorphism. □

This completes the proof that K and S are quasi-inverse, and yields the famous theorem:

Stone Duality Theorem

The categories **Stone** and **Bool** are dual to each other. In particular, any Boolean algebra B is isomorphic to the field of clopen subsets of its state space $S(B)$.

PROPOSITION 3.7.7 *Let $A \subseteq B$, and let $a \in B$. Then the following are equivalent:*

1. *For any states f and g of B, if $f|_A = g|_A$ then $f(a) = g(a)$.*
2. *If h is a state of A, then any two extensions of h to B agree on a.*
3. *$a \in A$.*

Proof Since every state of A can be extended to a state of B, (1) and (2) are obviously equivalent. Furthermore, (3) obviously implies (1). Thus, we only need to show that (1) implies (3).

Let $m : A \to B$ be the inclusion of A in B, and let $s : S(B) \to S(A)$ be the corresponding surjection of states. We need to show that $C_a = s^{-1}(U)$ for some clopen subset U of $S(A)$.

By (1), for any $x \in S(A)$, either $s^{-1}\{x\} \subseteq C_a$ or $s^{-1}\{x\} \subseteq C_{\neg a}$. By Proposition 3.6.29, s is a closed map. Since C_a is open, Proposition 3.6.28 entails that the sets,

$$U = \{x \in S(B) \mid s^{-1}\{x\} \subseteq C_a\}, \quad \text{and} \quad V = \{x \in S(B) \mid s^{-1}\{x\} \subseteq C_{\neg a}\},$$

are open. Since $U = S(A) \backslash V$, it follows that U is clopen. Finally, it's clear that $s^{-1}(U) = C_a$. □

PROPOSITION 3.7.8 *In* **Bool**, *epimorphisms are surjective.*

Proof Suppose that $f : A \to B$ is not surjective. Then $f(A)$ is a proper subalgebra of B. By Proposition 3.7.7, there are states g, h of B such that $g \neq h$, but $g|_{f(A)} = h|_{f(A)}$. In other words, $g \circ f = h \circ f$, and f is not an epimorphism. □

Combining the previous two theorems, we have the following equivalences:

$$\mathbf{Th} \cong \mathbf{Bool} \cong \mathbf{Stone}^{op}.$$

We will now exploit these equivalences to explore the structure of the category of theories.

PROPOSITION 3.7.9 *Let T be a propositional theory in a countable signature. Then there is a conservative translation $f : T \to T_0$, where T_0 is an empty theory – i.e., a theory with no axioms.*

Proof After proving the above equivalences, we have several ways of seeing why this result is true. In terms of Boolean algebras, the proposition says that every countable Boolean algebra is embeddable into the free Boolean algebra on a countable number of generators (i.e., the Boolean algebra of clopen subsets of the Cantor space). That well-known result follows from the fact that Boolean algebras are always generated by their finite subalgebras. (In categorical terms, every Boolean algebra is a filtered colimit of finite Boolean algebras.)

In terms of Stone spaces, the proposition says that for every separable Stone space Y, there is a continuous surjection $p : X \to Y$, where X is the Cantor space. That fact is well known to topologists. One interesting proof uses the fact that a Stone space Y is *profinite* – i.e., Y is a limit of finite Hausdorff (hence discrete) spaces. One then shows that the Cantor space X has enough surjections onto discrete spaces and lifts these up to a surjection $p : X \to Y$. See, for example, Ribes and Zalesskii (2000). Or, for a more direct argument: each clopen subset U of Y corresponds to a continuous map $p_U : Y \to \{0, 1\}$. There are countably many such clopen subsets of Y. Since $X \simeq \prod_{i \in \mathbb{N}} \{0, 1\}$, these p_U induce a continuous function $p : X \to Y$. Moreover, since every point $y \in Y$ has a neighborhood basis of clopen sets, p is surjective. □

DISCUSSION 3.7.10 (Quine on eliminating posulates) It's no surprise that one can be charitable to a fault. Suppose that I am a theist, and you are an extremely charitable atheist. You are so charitable that you want to affirm the things I say. Here's how you can do it: when I say "God," assume that I really mean "kittens." Then when I say "God exists," you can interpret me to be saying "kittens exist." Then you can smile and say "I completely agree!"

Proposition 3.7.9 provides a general recipe for charitable interpretation. Imagine that I accept a theory T, which might be controversial. Imagine that you, on the other hand, like to play it safe: you only accept tautologies, viz. empty theory T_0. The previous proposition shows that there is a conservative translation $f : T \to T_0$. In other words, you can reinterpret my sentences in such a way that everything I say comes out as true by your lights – i.e., true by logic alone.

Since we're dealing merely with propositional logic, this result might not seem very provocative. However, a directly analogous result – proven by Quine and Goodman (1940); Quine (1964) – was thought to refute the analytic–synthetic distinction that was central to the logical positivist program. Quine's argument runs as follows: suppose that T is intended to represent a contingently true theory, such as (presumably) quantum mechanics or evolutionary biology. By making a series of clever definitions, the sentences of T can be reconstrued as tautologies. That is, any contingently true theory T can be reconstrued so that all of its claims come out as true by definition.

What we see here is an early instance of a strategy that Quine was to use again and again throughout his philosophical career. There is a supposedly important distinction in a theory T. Quine shows that this distinction doesn't survive translation of T into some other theory T_0. This result, Quine claims, shows that the distinction must be rejected.

Whether or not Quine's strategy is generally good, we should be a bit suspicious in the present case. The translation $f : T \to T_0$ is not an *equivalence* of theories – i.e., it does not show that T is equivalent to T_0. Since f is conservative, it does show a sense in which T is *embeddable in* or *reducible to* T_0. But we are left wondering: why should the existence of a formal relation $f : T \to T_0$ undercut the importance of the distinctions that are made within T?

If Proposition 3.7.9 was surprising, then the following result is even more surprising:

PROPOSITION 3.7.11 *Let T be a consistent propositional theory in a countably infinite signature. If T has a finite number of axioms, then T is equivalent to the empty theory T_0.*

Sketch of proof Suppose that T has a finite number of axioms. Without loss of generality, we assume that T has a single axiom ϕ. Let X be the Cantor space – i.e., the Stone space of the empty theory T_0. Let $U_\phi \subseteq X$ be the clopen subset of all models in which

ϕ is true. Then U_ϕ is homeomorphic to the Stone space of T. Assume for the moment any nonempty clopen subset of the Cantor space is homeomorphic to the Cantor space. In that case, U_ϕ is homeomorphic to the Cantor space X; and by Stone duality, T is equivalent to T_0.

We now argue that nonempty clopen subset of the Cantor space is homeomorphic to the Cantor space. (This result admits of several proofs, some more topologically illuminating than the one we give here.) We begin by arguing that if ϕ is a conjunction of literals (atomic or negated atomic sentences), then U_ϕ is homeomorphic to the Cantor space. Indeed, there is a direct proof that the theory $\{\phi\}$ is equivalent to the empty theory; hence, by Stone duality, U_ϕ is homeomorphic to X. Now, an arbitrary clopen subset U of X has the form U_ϕ for some sentence ϕ. We may rewrite ϕ in disjunctive normal form – i.e., as a finite disjunction of conjunctions of literals. Thus, U_ϕ is a disjoint union of $U_{\phi_1}, U_{\phi_2}, \dots, U_{\phi_n}$. By the previous argument, each U_{ϕ_i} is homeomorphic to X, and a disjoint union of copies of X is also homeomorphic to X. □

The previous proposition might suggest that the notion of equivalence we have adopted (Definition 1.4.6) is too *liberal* – i.e., that it counts too many theories as equivalent. If you think that's the case, we enjoin you to propose another criterion and explore its consequences.

DISCUSSION 3.7.12 The Stone duality theorem suggests that accepting a theory T involves accepting some claims about nearness/similarity relations among possible worlds. One theory T leads to a particular topological structure on the set of possible worlds, and another theory T' leads to a different topological structure on the set of possible worlds. That fact applies not just to propositional theories, but also to real-life scientific theories. For example, when one accepts the general theory of relativity, one doesn't simply believe that our universe is isomorphic to one of its models. Rather, one believes that the situtation we find ourselves in is one among many other situations that obey the laws of this theory. Moreover, some such situations are more similar than others. See Fletcher (2016) for an extended discussion of this example.

3.8 Notes

We have given only the most cursory introduction to the rich mathematical fields of Boolean algebras, topology, and the interactions between them. There is much more to be learned and many good books on these topics. Some of our favorites are the following:

- For more on Boolean algebras, see Sikorski (1969); Dwinger (1971); Koppelberg (1989); Givant and Halmos (2008); Monk (2014),
- There are many good books on topology. We learned originally from Munkres (2000), and our favorites include Engelking (1989) and Willard (1970). The latter is notable for its presentation of the ultrafilter approach to convergence.

- Stone spaces, being a particular kind of topological space, are sometimes mentioned in books about topology. But for a more systematic treatment of Stone spaces, you'll need to consult other resources. For a fully general and categorical treatment of Stone duality, see Johnstone (1986). For briefer and more pedestrian treatments, see Bell and Machover (1977); Halmos and Givant (1998); Cori and Lascar (2000). For a proof that Stone spaces are profinite, see Ribes and Zalesskii (2000).

4 Syntactic Metalogic

First-order logic plays a starring role in our best account of the structure of human knowledge. There is reason to believe that first-order logic is fully sufficient to encode *all* deductively valid reasoning. It was discovered in the early twentieth century that first-order logic is powerful enough to axiomatize many of the theories that mathematicians use, such as number theory, group theory, ring theory, field theory, etc. And although there are other mathematical theories that are overtly second-order (e.g., the theory of topological spaces quantifies over subsets, and not just individual points), nonetheless first-order logic can be used to axiomatize set theory, and any second-order theory can be formalized within set theory. Thus, first-order logic provides an expansive framework in which much, if not all, deductively valid human reasoning can be represented.

In this chapter, we will study the properties of first-order logic, the theories that can be formulated within it, and the relations that hold between them. Let's begin from the concrete – with examples of some theories that can be regimented in first-order logic.

4.1 Regimenting Theories

Example 4.1.1 (The theory of partial orders) We suppose that there is a relation, which we'll denote by $\leq$, and we then proceed to lay down some postulates for this relation. In particular:

- Postulate 1: The relation $\leq$ is reflexive in the sense that it holds between anything and itself. For example, if we were working with numbers, we could write $2 \leq 2$, or, more generally, we could write $n \leq n$ for any n. For this last phrase, we have a shorthand: we abbreviate it by $\forall n(n \leq n)$, which can be read out as "for all n, $n \leq n$." The symbol $\forall$ is called the **universal quantifier**.
- Postulate 2: The relation $\leq$ is transitive in the sense that if $x \leq y$ and $y \leq z$, then $x \leq z$. Again, we can abbreviate this last sentence as

  $$\forall x \forall y \forall z((x \leq y \wedge y \leq z) \rightarrow x \leq z),$$

  which can be read as, "for all x, for all y, and for all z, if . . . "
- Postulate 3: The relation $\leq$ is antisymmetric in the sense that if $x \leq y$ and $y \leq x$, then $x = y$. This postulate can be formalized as

  $$\forall x \forall y((x \leq y \wedge y \leq x) \rightarrow x = y).$$

In these previous postulates, we see the same logical connectives that we used in propositional logic, such as $\wedge$ and $\to$. But now these connectives might hold between things that are not themselves sentences. For example, $x \leq y$ is not itself a sentence, because x and y aren't names of things. We say that x and y are **variables**, that $\leq$ is a **relation symbol**, and that $x \leq y$ is a **formula**. Finally, the familiar symbol $=$ is also a relation symbol. ⌟

We've described just the barest of bones of the theory of a partial order. There are a couple of further things that we would definitely like to be able to do with this theory. First, we would like to be able to derive consequences from the postulates – i.e., we would like to derive theorems from the axioms. In order to do so, we will need to specify the **rules of derivation** for first-order logic. We will do that later in this chapter. We would also like to be able to identify mathematical structures that exemplify the axioms of partial order. To that end, we devote the following chapter to the **semantics**, or **model theory**, of first-order logic.

Example 4.1.2 (The theory of a linear order) Take the axioms of the theory of a partial order, and then add the following axiom:

$$\forall x \forall y((x \leq y) \vee (y \leq x)).$$

This axiom says that any two distinct things stand in the relation $\leq$. In other words, the elements from the domain form a total order. There are further specifications that we could then add to the theory of a linear order. For example, we could add an axiom saying that the linear order has endpoints. Alternatively, we could add an axiom saying that the linear order does not have endpoints. (Note, incidentally, that since either one of those axioms could be added, the original theory of linear orders is not complete – i.e., it leaves at least one sentence undecided.) We could also add an axiom saying that the linear order is dense, i.e., that between any two elements there is yet another element. ⌟

Example 4.1.3 (The theory of an equivalence relation) Let R be a binary relation symbol. The following axioms give the theory of an equivalence relation:

reflexive	$\vdash R(x,x)$
symmetric	$\vdash R(x,y) \to R(y,x)$
transitive	$\vdash (R(x,y) \wedge R(y,z)) \to R(x,z)$

Here when we write an open formula, such as $R(x,x)$, we mean to implicitly quantify universally over the free variables. That is, $\vdash R(x,x)$ is shorthand for $\vdash \forall x R(x,x)$. ⌟

Example 4.1.4 (The theory of abelian groups) We're all familiar with number systems such as the integers, the rational numbers, and the real numbers. What do these number systems have in common? One common structure between them is that they have a binary relation $+$ and a neutral element 0, and each number has a unique inverse. We also notice that the binary relation $+$ is associative in the sense that $x+(y+z)=(x+y)+z$, for all x, y, z. We can formalize this last statement as

$$\forall x \forall y \forall z(x+(y+z)=(x+y)+z).$$

In many familiar cases, the operation $+$ is also commutative; that is,

$$\forall x \forall y(x + y = y + x).$$

Bringing these postulates together, we have the theory of abelian groups. Notice that in this case, we've enlarged our vocabulary to include a symbol $+$ and a symbol 0. The symbol $+$ is not exactly a relation symbol, but instead is a function symbol. Intuitively speaking, given any names n and m of numbers, $n + m$ also names a number. Similarly, 0 is taken to be the name of some specific number, and in this sense it differs from a variable. ⌟

Example 4.1.5 (Boolean algebra) Suppose that $+$ and $\cdot$ are binary function symbols, and that 0 and 1 are constant symbols. If you look back at our discussion of Boolean algebras (Section 3.2), you'll see that each of the axioms amounts to a first-order sentence, where we use the $+$ symbol instead of the $\vee$ symbol, and the $\cdot$ symbol instead of the $\wedge$ symbol (since those symbols are already being used as our logical connectives). The theory of Boolean algebras is an example of an **algebraic theory**, which means that it can be axiomatized using only function symbols and equations. ⌟

Example 4.1.6 (Arithmetic) It's possible to formulate a first-order theory of arithmetic, e.g., Peano arithmetic. For this, we could use a signature Σ with constant symbols 0 and 1, and binary function symbols $+$ and $\cdot$. ⌟

Example 4.1.7 (Set theory) It's possible to formulate a first-order theory of sets, e.g., Zermelo–Fraenkel set theory. For this, we could use a signature Σ with a single relation symbol $\in$. However, for the elementary theory of the category of sets (ETCS), as we developed in Chapter 2, it would be more natural to use the framework of many-sorted logic, having one sort for sets and another sort for functions between sets. For more on many-sorted logic, see Chapter 5. ⌟

Example 4.1.8 (Mereology) There are various ways to formulate a first-order theory of mereology. Most presentations begin with a relation symbol $pt(x, y)$ to indicate that x is a part of y. Then we add some axioms that look a lot like the axioms for the less-than relation $<$ for a finite Boolean algebra. ⌟

4.2 Logical Grammar

Abstracting from the previous examples and many others like them throughout mathematics, we now define the language of first-order logic as follows.

DEFINITION 4.2.1 The **logical vocabulary** consists of the symbols:

$$\bot \; \forall \; \exists \; \wedge \; \vee \; \neg \; \rightarrow \; (\;)$$

The symbol $\bot$ will serve as a propositional constant. The final two symbols here, the parentheses, are simply punctuation symbols that will allow us to keep track of groupings of the other symbols.

Please note that we intentionally excluded the equality symbol $=$ from the list of logical vocabulary. Several philosophers in the twentieth century discussed the question of whether the axioms for equality were analytic truths or whether they should be considered to form a specific, contingent theory. We will not enter into the philosophical discussion at this point, but it will help us to separate out the theory of equality from the remaining content of our logical system. We will also take a more careful approach to variables by treating them as part of a theory's nonlogical vocabulary. Our reason for doing so will become clear when we discuss the notion of translations between theories.

DEFINITION 4.2.2 A **signature** Σ consists of

1. A countably infinite collection of **variables**.
2. A collection of **relation symbols**, each of which is assigned a natural number called its **arity**. A 0-ary relation symbol is called a **propositional constant**.
3. A collection of **function symbols**, each of which is assigned a natural number called its arity. A 0-ary function symbol is called a **constant symbol**.

DISCUSSION 4.2.3 Some logicians use the name **similarity type** as a synonym for **signature**. There is also a tendency among philosophers to think of a signature as the vocabulary for an **uninterpreted language**. The idea here is that the elements of the signature are symbols that receive meaning by means of a semantic interpretation. Nonetheless, we should be careful with this kind of usage, which might suggest that formal languages lie on the "mind side" of the mind–world divide, and that an interpretation relates a mental object to an object in the world. In fact, formal languages, sentences, and theories are all *mathematical objects* – of precisely the same ontological kind as the models that interpret them. We discuss this issue further in the next chapter.

Although a list of variables is technically part of a signature, we will frequently omit mention of the variables and defer to using the standard list $x, y, x_1, x_2, \ldots$ Only in cases where we are comparing two theories will we need to carefully distinguish their variables from each other.

Example 4.2.4 Every propositional signature is a special case of a signature in the sense just defined. ⌟

Example 4.2.5 For the theory of abelian groups, we used a signature Σ that has a binary function symbol $+$ and a constant symbol 0. Some other presentations of the theory of abelian groups use a signature Σ' that also has a unary function symbol "$-$" for the inverse of an element. Still other presentations of the theory use a signature that doesn't have the constant symbol 0. We will soon see that there is a sense in which these different theories all deserve to be called *the* theory of abelian groups. ⌟

DISCUSSION 4.2.6 Let Σ be the signature consisting of a binary relation symbol r, and let Σ' be the signature consisting of a binary relation symbol R. Are these signatures the same or different? That depends on what implicit background conventions that we adopt – in particular, whether our specification of a signature is case sensitive or not. In fact, we could adopt a convention that was even stricter in how it individuates signatures.

For example, let Σ'' be the signature consisting of a binary relation symbol r. One could say that Σ'' is a different signature from Σ because the r in Σ'' occurs at a different location on the page than the r that occurs in Σ. Of course, we would typically assume that $\Sigma'' = \Sigma$, but such a claim depends on an implicit background assumption that there is a single letterform of which the two occurences of r are instances.

We will generally leave these implicit background assumptions unmentioned. Indeed, to make these background assumptions explicit, we would have to rely on further implicit background assumptions, and we would never make progress in our study of first-order logic.

Let Σ be a fixed signature. We first define the sets of Σ-terms and Σ-formulas.

DEFINITION 4.2.7 We simultaneously define the set of **Σ-terms**, and the set $FV(t)$ of **free variables** of a Σ-term t as follows:

1. If x is a variable of Σ, then x is a Σ-term and $FV(x) = \{x\}$.
2. If f is a function symbol of Σ, and $t_1, \dots, t_n$ are Σ-terms, then $f(t_1, \dots, t_n)$ is a Σ-term and

$$FV(f(t_1, \dots, t_n)) := FV(t_1) \cup \dots \cup FV(t_n).$$

DEFINITION 4.2.8 We simultaneously define the set of **Σ-formulas** and the set $FV(\phi)$ of free variables of each Σ-formula ϕ as follows:

1. $\perp$ is a formula and $FV(\perp) = \emptyset$.
2. If r is an n-ary relation symbol in Σ, and $t_1, \dots, t_n$ are terms, then $r(t_1, \dots, t_n)$ is a formula and

$$FV(r(t_1, \dots, t_n)) := FV(t_1) \cup \dots \cup FV(t_n).$$

3. If ϕ and ψ are formulas, then $\phi \wedge \psi$ is a formula with $FV(\phi \wedge \psi) = FV(\phi) \cup FV(\psi)$. Similarly for the other Boolean connectives $\neg$, $\vee$, $\rightarrow$.
4. If ϕ is a formula, then so is $\exists x \phi$ and $FV(\exists x \phi) = FV(\phi) \backslash \{x\}$. Similarly for $\forall x \phi$.

A formula ϕ is called **closed**, or a **sentence**, if $FV(\phi) = \emptyset$.

A more fully precise definition of Σ-formulas would take into account the precise location of parentheses. For example, we would want to say that $(\phi \wedge \psi)$ is a Σ-formula when ϕ and ψ are Σ-formulas. Nonetheless, we will continue to allow ourselves to omit parentheses when no confusion is likely to result.

NOTE 4.2.9 Our definition of the set of formulas allows for redundant quantification. For example, the string $\exists x \forall x (x = x)$ is a well-formed formula according to our definition. This formula results from applying the quantifier $\exists x$ to the sentence $\forall x (x = x)$. We will have to be careful in our definition of derivation rules, and semantic rules, to take the case of empty quantification into account.

DEFINITION 4.2.10 The **elementary formulas** are those of the form $r(t_1, \dots, t_n)$, i.e., formulas that involve no Boolean connectives or quantifiers.

It is helpful to think of formulas in terms of their **parse trees**. For example, the formula $\forall x \exists y(r(x,x) \to r(x,y))$ has the following parse tree:

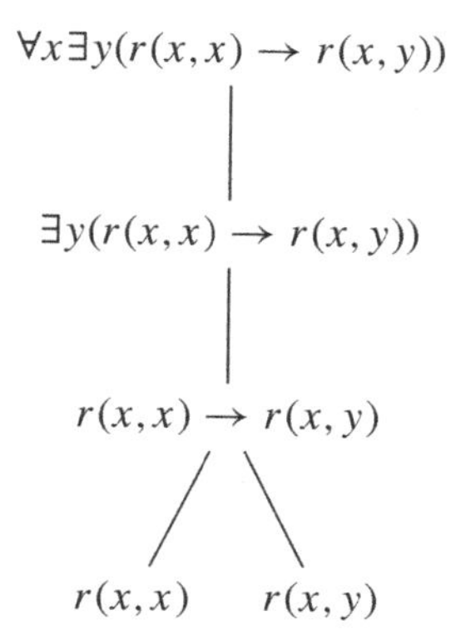

The bottom nodes must each be elementary formulas, i.e., either $\bot$ or a relation symbol followed by the appropriate number of terms. Each parent–child relationship in the tree corresponds to one of the Boolean connectives or to one of the quantifiers.

Formulas stand in one-to-one correspondence with parse trees: each well-formed tree ends with a specific formula, and no other tree yields the same formula. Using the identity of formulas and parse trees, we can easily define a few further helpful notions:

DEFINITION 4.2.11 Let ϕ be a Σ-formula. The family of **subformulas** of ϕ consists of all those formulas that occur at some node in its parse tree.

DEFINITION 4.2.12 If a quantifier $\exists x$ occurs in the formula ϕ, then the **scope** of that occurrence is the formula that occurs at the immediately previous node in the parse tree.

For example, in the formula $\forall x \exists y(r(x,x) \to r(x,y))$, the scope of $\exists y$ is the formula $r(x,x) \to r(x,y)$. In contrast, in the formula $\forall x(r(x,x) \to \exists y r(x,y))$, the scope of $\exists y$ is the formula $r(x,y)$.

We can now make the notion of free and bound variables even more precise. In particular, each individual occurrence of a variable in ϕ is either free or bound. For example, in the formula $p(x) \wedge \exists x p(x)$, x occurs freely in the first subformula and bound in the second subformula.

DEFINITION 4.2.13 (Free and bound occurrences) An occurence of a variable x in ϕ is **bound** just in case that occurrence is within the scope of either $\forall x$ or $\exists x$. Otherwise that occurence of x is free.

We could now perform a sanity check to make sure that our two notions of bound/free variables coincide with each other.

FACT 4.2.14 *A variable x is free in ϕ (in the sense of the definition of Σ-formulas) if and only if there is a free occurence of x in ϕ (in the sense that this occurence does not lie in the scope of any corresponding quantifier).*

It is also sometimes necessary to distinguish particular occurrences of a subformula of a formula and to define the **depth** at which such an instance occurs.

DEFINITION 4.2.15 Let ψ be a node in the parse tree of ϕ. The **depth** of ψ is the number of steps from ψ to the root node. We say that ψ is a **proper subformula** of ϕ if ψ occurs with depth greater than 0.

The parse trees of formulas are finite by definition. Therefore, the depth of every occurrence of a subformula of ϕ is some finite number.

There are a number of other properties of formulas that are definable in purely syntactic terms. For example, we could define the **length** of a formula. We could then note that the connectives take formulas of a certain length and combine them to create formulas of a certain greater length.

EXERCISE 4.2.16 Show that no Σ-formula can occur as a proper subformula of itself.

We now define a substitution operation $\phi \mapsto \phi[t/x]$ on formulas, where t is a fixed term and x is a fixed variable. The intention here is that $\phi[t/x]$ results from replacing all free occurrences of x in ϕ with t. We first define a corresponding operation on terms.

DEFINITION 4.2.17 Let t be a fixed term, and let x be a fixed variable. We define the operation $s \mapsto s[t/x]$, where s is an arbitrary term, as follows:

1. If s is a variable, then $s[t/x] \equiv s$ when $s \not\equiv x$, and $s[t/x] \equiv t$ when $s \equiv x$. (Here $\equiv$ means literal identity of strings of symbols.)
2. Suppose that $s \equiv f(t_1, \dots, t_n)$, where f is a function symbol and $t_1, \dots, t_n$ are terms. Then we define

$$s[t/x] \equiv f(t_1[t/x], \dots, t_n[t/x]).$$

This includes the special case where f is a 0-ary function symbol, where $f[t/x] \equiv f$.

DEFINITION 4.2.18 Let t be a fixed term, and let x be a fixed variable. We define the operation $\phi \mapsto \phi[t/x]$, for ϕ an arbitrary formula, as follows:

1. For the proposition $\bot$, let $\bot[t/x] := \bot$.
2. For an elementary formula $r(t_1, \dots, t_n)$, let

$$r(t_1, \dots, t_n)[t/x] := r(t_1[t/x], \dots, t_n[t/x]).$$

3. For a Boolean combination $\phi \wedge \psi$, let

$$(\phi \wedge \psi)[t/x] := \phi[t/x] \wedge \psi[t/x],$$

and similarly for the other Boolean connectives.
4. For an existentially quantified formula $\exists y\phi$, let

$$(\exists y\phi)[t/x] := \begin{cases} \exists y(\phi[t/x]) & \text{if } x \not\equiv y, \\ \exists y\phi & \text{if } x \equiv y. \end{cases}$$

5. For a universally quantifier formula $\forall y\phi$, let

$$(\forall y\phi)[t/x] := \begin{cases} \forall y(\phi[t/x]) & \text{if } x \not\equiv y, \\ \forall y\phi & \text{if } x \equiv y. \end{cases}$$

PROPOSITION 4.2.19 *For any formula ϕ, the variable x is not free in $\phi[y/x]$.*

Proof We first show that $x \notin FV(t[y/x])$ for any term t. That result follows by a simple induction on the construction of terms.

Now let ϕ be an elementary formula. That is, $\phi = r(t_1, \ldots, t_n)$. Then we have

$$\begin{aligned} FV(\phi[y/x]) &= FV(r(t_1, \ldots, t_n)[y/x]) \\ &= FV(r(t_1[y/x], \ldots, t_n[y/x])) \\ &= FV(t_1[y/x]) \cup \cdots \cup FV(t_n[y/x]). \end{aligned}$$

Since $x \notin FV(t_i[y/x])$, for $i = 1, \ldots, n$, it follows that $x \notin FV(\phi[y/x])$.

The argument for the Boolean connectives is trivial, so we turn to the argument for the quantifiers. Suppose that the result is true for ϕ. We need to show that it's also true for $\exists v\phi$. Suppose first that $v \equiv x$. In this case, we have

$$(\exists v\phi)[y/x] = (\exists x\phi)[y/x] = \exists x\phi.$$

Since $x \notin FV(\exists x\phi)$, it follows that $x \notin FV((\exists v\phi)[y/x])$. Suppose now that $v \not\equiv x$. In this case, we have

$$(\exists v\phi)[y/x] = \exists v(\phi[y/x]).$$

Since $x \notin FV(\phi[y/x])$, it follows then that $x \notin FV((\exists v\phi)[y/x])$. The argument is analogous for the quantifier $\forall v$. Therefore, for any formula ϕ, the variable x is not free in $\phi[y/x]$. □

4.3 Deduction Rules

We suppose again that Σ is a fixed signature. The goal now is to define a relation $\Gamma \vdash \phi$ of derivability, where Γ is a finite sequence of Σ-formulas and ϕ is a Σ-formula. Our derivation rules come in three groupings: rules for the Boolean connectives, rules for the $\bot$ symbol, and rules for the quantifiers.

Boolean Connectives

We carry over all of the rules for the Boolean connectives from propositional logic (see Section 1.2). These rules require no special handling of variables. For example, the following is a valid instance of $\wedge$-elim:

$$\frac{\Gamma \vdash \phi(x) \wedge \psi(y)}{\Gamma \vdash \phi(x)}.$$

Falsum

We intend for the propositional constant $\bot$ to serve as shorthand for "the false." To this end, we define its introduction and elimination rules as follows.

$\bot$ **intro** $\quad \dfrac{\Gamma \vdash \phi \wedge \neg\phi}{\Gamma \vdash \bot}$	$\bot$ **elim** $\quad \dfrac{\Gamma \vdash \bot}{\Gamma \vdash \phi}$

Quantifiers

In order to formulate good derivation rules for the quantifiers, we have to make a couple of strategic choices. In actual mathematical practice, mathematicians simply introduce new vocabulary whenever they need it. In some cases, new vocabulary is introduced by way of definition – for example, when a mathematician says something like, "we say that a number x is *prime* just in case ..." where the words following the dots refer to previously understood mathematical concepts. In other cases, the newly introduced vocabulary is really just newly introduced notation – for example, when a mathematician says something like, "let n be a natural number." In this latter case, the letter "n" wasn't a part of the original vocabulary of the theory of arithmetic, and was introduced as a matter of notational convenience.

Nonetheless, for our purposes it will be most convenient to have a fixed vocabulary Σ for a theory. But this means that if Σ has no constant symbols, then we might have trouble making use of the quantifier introduction and elimination rules. For example, imagine trying to derive a theorem in the theory of Boolean algebras if you weren't permitted to say, "let a be an arbitrary element of the Boolean algebra B." In order to simulate mathematics' free use of new notation, we'll simply be a bit more liberal in the way that we allow free variables to be used. To this end, we define the following notion.

DEFINITION 4.3.1 We say that t **is free for** x **in** ϕ just in case one of the following conditions holds:

1. ϕ is atomic, or
2. ϕ is a Boolean combination of formulas, in each of which t is free for x, or
3. $\phi := \exists y\psi$, and $y \notin FV(t)$, and t is free for x in ψ, where $x \neq y$.

Intuitively speaking, t is free for x in ϕ just in case substituting t in for x in ϕ does not result in any of the variables in t being captured by quantifiers. For example, in the formula $p(x)$, the variable y is free for x (since y is free in $p(y)$). In contrast, in the formula $\exists y p(x)$, the variable y is not free for x (since y is not free in $\exists y p(y)$). We will need this notion in order to coordinate our intro and elim rules for the quantifiers. For example, the rule of $\forall$-elim should say something like: $\forall x\phi(x) \vdash \phi(y)$. However, if this rule were not restricted in some way, then it would yield

$$\forall x \exists y(x \neq y) \vdash \exists y(y \neq y),$$

which is intuitively invalid.

$\forall$ **intro**	$\dfrac{\Gamma \vdash \phi}{\Gamma \vdash \forall x\phi}$	where x is not free in Γ.

$\forall$ **elim**	$\dfrac{\Gamma \vdash \forall x\phi}{\Gamma \vdash \phi[t/x]}$	where t is free for x.

The $\forall$-intro rule is easy to apply, for we only need to check that the variable x doesn't occur in the assumptions Γ from which ϕ is derived. Note that application of the $\forall$-intro rule can result in empty quantification; for example, $\forall x \forall x p(x)$ follows from $\forall x p(x)$.

To understand the restrictions on $\forall$-elim, note that it does license

$$\forall x\, r(x,x) \;\vdash\; r(y,y),$$

since $r(x,x)[y/x] \equiv r(y,y)$. In contrast, $\forall$-elim does not license

$$\forall x\, r(x,x) \;\vdash\; r(x,y),$$

since it is not the case that $r(x,x)[y/x] \equiv r(x,y)$. Similarly, $\forall$-elim does not license

$$\forall x \exists y\, r(x,y) \;\vdash\; \exists y\, r(y,y),$$

since y is not free for x in $\exists y\, r(x,y)$. Finally, $\forall$-elim permits universal quantifiers to be peeled off when they don't bind any variables. For example, $\forall x\, p \vdash p$ is licensed by $\forall$-elim.

Now we turn to the rules for the existential quantifier. First we state the rules in all their sequential glory:

$\exists$ intro	$\dfrac{\Gamma \vdash \phi[t/x]}{\Gamma \vdash \exists x \phi}$	provided t is free for x in ϕ.

$\exists$ elim	$\dfrac{\Gamma, \phi \vdash \psi}{\Gamma, \exists x \phi \vdash \psi}$	provided x is not free in ψ or Γ.

If we omit the use of auxiliary assumptions, we can rewrite the $\exists$ rules as follows:

$\exists$ intro	$\dfrac{\vdash \phi[t/x]}{\vdash \exists x \phi}$	provided t is free for x in ϕ.

$\exists$ elim	$\dfrac{\phi \vdash \psi}{\exists x \phi \vdash \psi}$	provided x is not free in ψ.

Again, let's look at some examples to illustrate the restrictions. First, in the case of the $\exists$-intro rule, suppose that there were no restriction on the term t. Let ϕ be the formula $\forall y\, r(x,y)$, and let t be the variable y, in which case $\phi[t/x] \equiv \forall y\, r(y,y)$. Then the $\exists$-in rule would yield

$$\forall y\, r(y,y) \;\vdash\; \exists x \forall y\, r(x,y),$$

which is intuitively invalid. (Consider, for example, the case where r is the relation $\leq$ on integers.) The problem, of course, is that the variable y is captured by the quantifier $\forall y$ when substituted into ϕ. Similarly, in the case of the $\exists$-elim rule, if there were no restriction on the variable x, then we could derive ϕ from $\exists x \phi$, and then, using $\forall$-intro, we could derive $\exists x \phi \vdash \forall x \phi$.

Structural Rules

In any proof system, there are some more or less tacit rules that arise from how the system is set up. For example, when someone learns natural deduction – e.g., via the system presented in Lemmon's *Beginning Logic* – then she will tacitly assume that she's allowed to absorb dependencies – e.g., if $\phi, \phi \vdash \psi$ then $\phi \vdash \psi$. These more or less tacit rules are called **structural rules** of the system – and there is a lot of interesting research on logical systems that drop one or more of these structural rules (see Restall, 2002). In this book, we stay within the confines of classical first-order logic; and we will not need to be explicit about the structural rules, except for the rule of **cut**, which allows sequents to be combined. Loosely speaking, cut says that if you have sequents $\Gamma \vdash \phi$ and $\Delta, \phi \vdash \psi$, then you may derive the sequent $\Gamma, \Delta \vdash \psi$.

As was the case with propositional logic, we will not specify a canonical way of writing predicate logic proofs. After all, our goal here is not to teach you the art of logical deduction; rather, our goal is to reflect on the relations between theories in formal logic.

Equality

As we mentioned before, there's something of a philosophical debate about whether the equality symbol $=$ should be considered as part of the logical or the nonlogical vocabulary of a theory. We don't want to get tangled up in that argument, but we do wish to point out how the axioms for equality compare to the axioms for a generic equivalence relation.

It is typical to write two axioms for equality, an introduction and an elimination rule. Equality introduction permits $\vdash t = t$ with any term t. Equality elimination permits

$$\frac{t = s \quad \phi[t/s]}{\phi}$$

so long as t is free for s in ϕ. Note that equality elimination allows us to replace single instances of a term. For example, if we let ϕ be the formula $r(s,t)$, then $\phi[t/s]$ is the formula $r(t,t)$. Hence, from $t = s$ and $r(t,t)$, equality elimination permits us to derive $r(s,t)$.

From the equality axioms, we can easily show that it's an equivalence relation. The introduction rule shows that it's reflexive. For symmetry, we let ϕ be the formula $y = x$, in which case $\phi[x/y]$ is the formula $x = x$. Thus, we have

$$\frac{x = y \quad x = x}{y = x}.$$

For transitivity, let ϕ be the formula $x = z$, in which case $\phi[y/x]$ is the formula $y = z$. Thus, we have

$$\frac{y = x \quad y = z}{x = z}.$$

This completes the list of the proof rules for our system of first-order logic, i.e., our definition of the relation $\vdash$. Before proceeding to investigate the properties of this relation, let's see a couple of examples of informal proofs.

Example 4.3.2 Let's show that $\exists x(\phi(x) \wedge \psi(x)) \vdash \exists x\phi(x)$. First note that $\phi(x) \wedge \psi(x) \vdash \phi(x)$ from $\wedge$-elim. Then $\phi(x) \wedge \psi(x) \vdash \exists x\phi(x)$ from $\exists$-intro. Finally, since $\exists x\phi(x)$ contains no free occurrences of x, we have $\exists x(\phi(x) \wedge \psi(x)) \vdash \exists x\phi(x)$. ⌟

Example 4.3.3 Of course, we should have $\forall x\phi \vdash \forall y(\phi[y/x])$, so long as y is free for x in ϕ. Using the rules we have, we can derive this result in two steps. First, we have $\forall x\phi(x) \vdash \phi[y/x]$ from $\forall$-elim, and then $\phi[y/x] \vdash \forall y\phi[y/x]$ by $\forall$-intro. We only need to verify that x is not free in $\phi[y/x]$. This can be shown by a simple inductive argument. ⌟

Recall that propositional logic is compositional in the following sense: Suppose that ϕ is a formula, and ψ is a subformula of ϕ. Let ϕ' denote the result of replacing ψ in ϕ with another formula ψ' where $\vdash \psi \leftrightarrow \psi'$. Then $\vdash \phi \leftrightarrow \phi'$. That result is fairly easy to prove by induction on the construction of proofs. It also follows from the truth-functionality of the Boolean connectives, by means of the completeness theorem. In this section, we are going to prove an analogous result for predicate logic. To simplify notation, we introduce the following.

DEFINITION 4.3.4 For formulas ϕ and ψ, we say that ϕ and ψ are **logically equivalent**, written $\phi \simeq \psi$, just in case both $\phi \vdash \psi$ and $\psi \vdash \phi$.

It is not hard to show that $\simeq$ is an equivalence relation on the set of formulas. Note that formulas ϕ and ψ can be equivalent in this sense even if they don't share all free variables in common – as long as the nonmatching variables occur vacuously. For example, $p(x)$ is equivalent to $p(x) \wedge (y = y)$, and it's also equivalent to $p(x) \vee (y \neq y)$. (The issue here has nothing in particular to do with the equality relation. The variable y also occurs vacuously in $p(y) \vee \neg p(y)$.) In contrast, the formulas $p(x)$ and $p(y)$ are not equivalent, since it's not universally valid that $\vdash \forall x\forall y(p(x) \leftrightarrow p(y))$.

LEMMA 4.3.5 *The relation $\simeq$ is compatible with the Boolean connectives in the following sense: if $\phi \simeq \phi'$ and $\psi \simeq \psi'$, then $(\phi \wedge \psi) \simeq (\phi' \wedge \psi')$, and similarly for the other Boolean connectives.*

The proof of this lemma is a fairly simple application of the introduction and elimination rules for the connectives. To complete the proof of the replacement theorem, we need one more lemma.

LEMMA 4.3.6 *If $\phi \simeq \psi$ then $\exists x\phi \simeq \exists x\psi$.*

Proof Suppose that $\phi \simeq \psi$, which means that $\phi \vdash \psi$ and $\psi \vdash \phi$. We're now going to show that $\exists x\phi \vdash \exists x\psi$. By $\exists$-in we have $\psi \vdash \exists x\psi$, hence by cut we have $\phi \vdash \exists x\psi$. Since x does not occur free in $\exists x\psi$, we have $\exists\phi \vdash \exists x\psi$ by $\exists$-out. □

THEOREM 4.3.7 (Replacement) *Suppose that ϕ is a formula in which ψ occurs as a subformula, and ϕ' is the result of replacing ψ with ψ'. If $\psi \simeq \psi'$ then $\phi \simeq \phi'$.*

In most presentations of the predicate calculus (i.e., the definition of the relation $\vdash$) the two central results are the soundness and completeness theorems. Intuitively speaking, the soundness theorem shows that the definition doesn't overgenerate, and the completeness theorem shows that it doesn't undergenerate. However, in fact, these results show something quite different – they show that the definition of $\vdash$ matches the definition of another relation $\vDash$. We will discuss this other relation $\vDash$ in Chapter 6, where we will also prove the traditional soundness and completeness theorems. In the remainder of this section, we show that the predicate calculus is consistent in the following purely syntactic sense.

DEFINITION 4.3.8 We say that the relation $\vdash$ is **consistent** just in case there is some formula ϕ that is not provable. Similarly, we say that a theory T is **consistent** just in case there is a formula ϕ such that $T \nvdash \phi$.

Note that the definition of consistency for $\vdash$ presupposes a fixed background signature Σ.

PROPOSITION 4.3.9 *A theory T is consistent iff $T \nvdash \bot$.*

Proof If T is inconsistent, then $T \vdash \phi$ for all formulas ϕ. In particular, $T \vdash \bot$. Conversely, if $T \vdash \bot$, then RA and DN yield $T \vdash \phi$ for any formula ϕ. □

THEOREM 4.3.10 *The predicate calculus is consistent.*

Proof Let Σ be a fixed predicate logic signature, and let Σ' be a propositional signature whose cardinality is greater than or equal to that of Σ. We will use the symbol $\vdash^*$ to denote derivability in the propositional calculus. Define a map $\phi \mapsto \phi^*$ from the formulas of Σ to the formulas of Σ' as follows:

- $\bot^* = \bot$.
- For any terms $t_1, \ldots, t_n$, $(p_i(t_1, \ldots, t_{n_i}))^* = q_i$.
- $(\phi \wedge \psi)^* = \phi^* \wedge \psi^*$, and similarly for the other Boolean connectives.
- $(\forall x \phi)^* = \phi^*$ and $(\exists x \phi)^* = \phi^*$.

We now use induction on the definition of $\vdash$ to show that if $\Gamma \vdash \phi$, then $\Gamma^* \vdash^* \phi^*$. We will provide a few representative steps, and leave it to the reader to supply the others.

- The base case, rule of assumptions, is trivial.
- Consider the case of $\wedge$-out. Suppose that $\Gamma \vdash \phi$ follows from $\Gamma \vdash \phi \wedge \psi$ by $\wedge$-out. By the inductive hypothesis, $\Gamma^* \vdash^* (\phi \wedge \psi)^*$. Using the definition of $(\phi \wedge \psi)^*$, it follows that $\Gamma^* \vdash^* \phi^* \wedge \psi^*$. Hence, by $\wedge$-out, we have $\Gamma^* \vdash^* \phi^*$.
- Consider the case of $\forall$-intro. That is, suppose that $\Gamma \vdash \forall x \phi$ is derived from $\Gamma \vdash \phi$ using $\forall$-in. In this case, the induction hypothesis tells us that $\Gamma^* \vdash^* \phi^*$. And since $(\forall x \phi)^* = \phi^*$, we have $\Gamma^* \vdash^* (\forall x \phi)^*$.

Completing the previous steps shows that if $\Gamma \vdash \phi$, then $\Gamma^* \vdash^* \phi^*$. Since the propositional calculus is consistent, $\nvdash^* \bot$ and, therefore, $\nvdash \bot$. □

DISCUSSION 4.3.11 Notice that the previous proof does not use the fact that our $\forall$-intro rule demands that x not occur free in Γ. Thus, this proof also shows the consistency of a proof system with an *unrestricted* $\forall$-intro rule.

But an unrestricted $\forall$-intro rule would nonetheless severely restrict the expressive power of our logic. Indeed, it would license

$$x \neq y \vdash \forall y(x \neq y) \vdash x \neq x,$$

the last of which contradicts the axioms for equality. Thus, an unrestricted $\forall$-intro would make $\forall x \forall y(x = y)$ a tautology.

4.4 Empirical Theories

Here we use the phrase "empirical theory" or "scientific theory," to mean a theory that one intends to describe the physical world. You know many examples of such theories: Newtonian mechanics, Einstein's general theory of relativity, quantum mechanics, evolutionary biology, the phlogiston theory of combustion, etc. You may also know many examples of theories from pure mathematics, such as set theory, group theory, ring theory, topology, and the theory of smooth manifolds. Intuitively, empirical theories differ in some important way from pure mathematical theories. We stress "intuitively" here because Quine brought into question the idea that there is a principled distinction between two types of theories. For the time being, we won't engage directly with Quine's more philosophical arguments against this distinction. Instead, we will turn back the clock to the time when Rudolf Carnap, among others, hoped that formal logic might illuminate the structure of scientific theories.

Rudolf Carnap was the primary advocate of the idea that philosophers ought to pursue a *syntactic* analysis of scientific theories. The story is typically told as follows: Carnap sought to construct a theory *of* scientific theories. Moreover, following in the footsteps of Bertrand Russell and Gottlob Frege, Carnap believed that philosophy had no business directly engaging in empirical questions. As Russell (1914b) had argued, philosophers ought to leave empirical questions to the empirical sciences. Thus, Carnap thought that a good philosophical theory of scientific theories ought to restrict itself to the purely formal aspects of those theories. In particular, the "metascientist" – i.e., the philosopher of science – ought to make use only of syntactic concepts.

Carnap begins his *Wissenschaftslogik* program in earnest in his first major book, *Logische Aufbau der Welt.* Already here we see the emphasis on "explication" – i.e., taking an intuitive concept and providing a precise formal counterpart. Carnap's paradigms of explication are those from nineteenth- and early-twentieth-century mathematics – explications of concepts such as "infinity" and "continuous function" and "open subset." Nonetheless, in the *Aufbau*, Carnap hasn't yet found his primary tool of analysis. That would only come from the development of logical metatheory in the 1920s. Carnap was working at the time in Vienna, among the other members of the infamous Vienna Circle. One of the youngest members of the circle was Kurt Gödel, whose 1929 PhD dissertation contained the first proof of the completeness of the predicate calculus. Thus, logical metatheory – or metamathematics – was in the air in Vienna, and Carnap was to try his hand at applying an analogous methodology to the empirical sciences. As the goal of metamathematics is to provide a rigorous theory *about* mathematics, Carnap wished to create a rigorous theory *about* the empirical sciences.

By the mid 1930s, Carnap had found his vision. In *Die Logische Syntax der Sprache*, Carnap states that his goal is to formalize scientific theories in the same way that Russell and Whitehead had formalized arithmetic – but with one important addition. With a theory of pure mathematics, the job is done once the relevant primitive concepts and axioms have been written down. However, empirical theories are, by their nature, "world directed" – i.e., they try to say something about concrete realities. Thus, an adequate analysis of a scientific theory cannot rest content with explaining that theory's formal structure. This analysis must also say something about how the theory gains its *empirical content.*

The task of explaining how a theory gains empirical content was to occupy Carnap for most of the remainder of his career. In fact, it became the stone on which the entire logical positivist movement stumbled. But we've gotten ahead of ourselves. We need first to see how Carnap proposed to analyze the structure of empirical theories.

What then is a theory? From the point of view of first-order logic, a theory T is specified by a signature Σ, and a set of axioms in that signature. Amazingly, many of the theories of pure mathematics can be described in terms of this simple schema. If, however, we intend for our theory T to describe concrete reality, what more do we need to add? Carnap's first proposal was a blunt instrument: he suggests identifying the empirical content of a theory by means of a division of that theory's vocabulary into two parts:

> The total language of science, L, is considered as consisting of two parts, the observation language L_O and the theoretical language L_T ... Let the observation vocabulary V_O be the class of the descriptive constants of L_O ... The terms of V_O are predicates designating observable properties of events or things (e.g., "blue," "hot," "large," etc.) or observable relations between them (e.g., "x is warmer than y," "x is contiguous to y," etc.). (Carnap, 1956, pp. 40-41)

Let's rewrite all of this in a better notation: the language of science consists of all the formulas built on some particular signature Σ, where Σ has a subset $O \subseteq \Sigma$ of observation vocabulary. The idea here is that terms in O have ostensive definitions – e.g., O might contain predicates such as "x is red" or "x is to the left of y." The elements of $\Sigma \backslash O$ are theoretical vocabulary, which need not have any direct empirical meaning. For example, $\Sigma \backslash O$ might contain predicates such as "x is a force." Thus, Carnap hopes to isolate empirical content by means of specifying a preferred subvocabulary of the language of science.

Before proceeding, note that Carnap – in this 1956 article – explicitly states that "for each language part the admitted types of variables are specified." That phrase was completely ignored by Carnap's subsequent critics, as we will soon see. And why did they ignore it? The reason, we suspect, is that they had been convinced by Quine that the notion of "types of variables" couldn't possibly make any difference in any philosophical debate. Well, Quine wasn't exactly right about that, as we discuss in

Section 5.3. However, at present, our goal is to see Carnap through the eyes of his critics, and according to these critics, Carnap's proposal amounts to saying the following.

DEFINITION 4.4.1 A formula ϕ of Σ is an **observation sentence** (alternatively, **protocol sentence**) just in case no symbol in ϕ comes from $\Sigma \backslash O$. If T is a theory in Σ, then we let $T|_O$ denote all the consequences of T in the sublanguage based on O.

In the light of these definitions, Carnap's proposal would amount to saying that the **empirical content** of a theory T is $T|_O$. Indeed, that's precisely what people took him to be saying – and they judged him accordingly. In fact, one of the standard "challenges for scientific realism" was to point out that the empirical subtheory $T|_O$ has the same empirical content as the original theory T. Thus, every nontrivial theory T has an empirically equivalent rival!

DEFINITION 4.4.2 Let T_1 and T_2 be theories in Σ. Then T_1 and T_2 have the same empirical content – i.e., are **empirically equivalent** – just in case $T_1|_O = T_2|_O$.

This definition fits right in with the picture that the logical positivists treat sentences as synonymous whenever those sentences have the same empirical content. Indeed, many people take the positivists to be saying that two scientific theories T_1 and T_2 should be considered equivalent *tout court* if they have the same observational consequences.

Before we go on to consider the criticisms that were brought against Carnap's picture of empirical content, let's ask ourselves what purpose the picture was supposed to serve. In other words, what questions was Carnap trying to answer by means of this proposal? In fact, it seems that Carnap was trying to answer several questions simultaneously. First, Carnap, along with many other logical positivists, was concerned with epistemological questions, such as, "am I justified in believing theory T?" Apropos of this question, the goal of isolating empirical content is to make some headway on understanding how it is that we can be warranted in believing a theory. To be clear, it's not only empiricists who should want to understand how we can use evidence to regulate our belief in a theory. That's a problem for anyone who thinks that we can learn from experience – and that's everybody besides the most extreme rationalists.

Nonetheless, there were some logical positivists – and perhaps sometimes Carnap himself – who thought that the empirical content of a theory provides the *only* route to justifying belief in that theory. For that kind of radical empiricist, isolating empirical content takes on an additional negative role: showing which parts of a theory do *not* contribute to our reasons for believing (or accepting) it.

It is sometimes forgotten, however, that epistemology was not the only reason that Carnap wanted to isolate empirical content. In fact, there are good reasons to think that epistemology wasn't even the primary reason that Carnap wanted to isolate empirical content. To the contrary, Carnap – who was, by training, a neo-Kantian – was concerned with how the abstract, highly mathematical theories of physics function in making assertions about the world. To understand this, we have to remember that Carnap was vividly aware of the upheaval caused by the discovery, in the mid-nineteenth century, of non-Euclidean geometries. One result of this upheaval was that mathematical formalism became *detached* from the empirical world, and the words that occur in it were *de-interpreted*. For example, in pre-nineteenth-century geometry, mathematicians were wont to think that a word such as "line" refers to those things in physical reality that are, in fact, lines. But insofar as the word "line" occurs in pure geometry, it has no reference at all – it is merely a symbol in a formal calculus.

Given the flight of pure mathematics away from empirical reality, the task of the mathematized empirical sciences is to tie mathematics back down. In other words, the task of the mathematical physicist is to take the uninterpreted symbols of pure mathematics and to endow them with empirical significance. It is precisely this methodological maneuver – peculiar to the new physics – that drives Carnap's desire to analyze the notion of the empirical content of a theory.

In the middle of the twentieth century, analytic philosophy moved west – from Vienna and Berlin to Oxford, Cambridge (in both old and New England), and then to Princeton, Pittsburgh, UCLA, etc. As analytic philosophy moved west, the focus on narrowly epistemological questions increased. It's no surprise, then, that Carnap's critics – first Quine, then Putnam, etc. – read him as attempting first and foremost to develop an empiricist epistemology. And their criticisms are directed almost exclusively at these aspects of his view. In fact, philosophers have been so focused with epistemological questions that they seem to have forgotten the puzzle that Carnap faced, and that we still face today: how do the sciences use abstract mathematical structures to represent concrete empirical reality?

In any case, we turn now to the criticisms of Carnap's account of the empirical content of a theory T as its restriction $T|_O$ to consequences in the observation subvocabulary O of Σ. Doubtless, all these criticisms descend, in one sense or other, from Quine's master criticism in "Two Dogmas of Empiricism" (Quine, 1951b). Here Quine's target is ostensibly statements, rather than theories. He argues that it makes no sense to talk about a statement's admitting of confirming or infirming (i.e., disconfirming) instances, at least when that statement is taken in isolation. While Quine doesn't apply his moral to the theories of the empirical sciences, it is only natural to transfer his conclusions to that case: it doesn't make sense to talk about the empirical content of a theory T.

To get an explicit statement of this criticism of Carnap's point of view, we have to wait a decade – for Putnam's paper "What Theories Are Not" (1962). Here Putnam claims that the attempt to select a subset $O \subseteq \Sigma$ of observation vocabulary is "completely broken-backed." His argument focuses on showing the incoherence of the notion of an observation term. To this end, he assumes that

> If $P(x)$ is an observation predicate, then it is never the case that $P(t)$, where t is a theoretical entity.

Putnam then simply enumerates examples where observation predicates have been applied to theoretical entities, e.g., Newton speaking of "red corpuscles."

For the sake of argument, let's assume that Putnam is correct that scientific theories sometimes use a single term in both observational and theoretical roles. Already that would pose a challenge to the adequacy of Carnap's account. Carnap assumes that among the terms of a mature scientific theory, there are some that are simply not used in observation reports – except possibly when a scientist is speaking loosely, e.g., if she says, "I saw an electron in the cloud chamber." Nonetheless, even if Putnam is right about that, his argument equivocates between formal and material modes of speech. On the one hand, Putnam speaks of observation predicates (formal mode); on the other hand, Putnam speaks of unobservable entities (material mode). Putnam's worry seems to be that some confusion might result if the philosopher of science classifies $P(x)$ as an observation predicate and then a scientist attributes $P(x)$ to a theoretical entity. Or perhaps the problem is that we cannot divide the vocabulary of Σ because we need to use predicates together with terms even when they would lie on opposite sides of the divide?

The anti-Carnap sentiment must have been in the air, for in the very same year, Maxwell (1962) also argued for the incoherence of the distinction between theoretical and observational terms. What's more, Maxwell explicitly claims that, in absence of this distinction, the only rational attitude toward a successful scientific theory is *full belief* – i.e., one must be a **scientific realist**.

Putnam and Maxwell seem to have convinced an entire generation of philosophers that Carnap's approach cannot be salvaged. In fact, the conclusion seems to have been that *nothing* of Carnap's approach could be salvaged, save the tendency to invoke results from mathematical logic. By the 1970s, there was no longer any serious debate about these issues. Instead, we find postmortem reflections on the "received view of scientific theories," as philosophers rushed headlong in the direction of Quinean holistic realism about everything (science, math, metaphysics).

4.5 Translation

Almost every discussion in twentieth-century philosophy of science has something or other to do with relations between theories. For example, philosophers of science have shown great interest in the notion that one theory is **reducible** to another. Similarly, several philosophical discussions pivot on the notion of a **conservative extension** of a theory. For example, Hartry Field (1980) aims to show that standard physical theories are conservative extensions over their "purely nominalistic parts" – hoping to undercut Quine's claim that belief in the existence of mathematical entities is demanded by belief in our best scientific theories.

We turn now to the task of explicating relations bewteen theories – i.e., giving a mathematically precise account of what these relations can be. One of the main questions considered in this book is

When are two theories T and T' the *same*, or *equivalent*?

Perhaps the answer seems clear: if a theory is a set of sentences, then two theories are the same if the corresponding sets of sentences are literally identical. However, there are numerous problems with that idea. First, that idea is not as clear as it might seem. When are two sets of sentences the same? What if the first set of sentences occurs in a book in the Princeton University library and the second set of sentences occurs on a chalkboard in Munich? Why would we say that those are the *same* sentences, when they occur in different spacetime locations?

Of course, the standard philosophical response to this worry is to shift focus from sentences to propositions – those abstract objects that are supposed to be expressed by concrete sentence tokens. Let's be completely clear: while we have no problems with abstract entities such as propositions, they won't help us make any progress deciding when sentences are synonymous, or when theories are equivalent. In other words, to say that sentences are synonymous if they express the same proposition may be *true*, but it is not an *explication* of synonymy. For the purposes of this book, we will set aside appeals to propositions or other such Platonic entities. We wish, instead, to provide clear and explicit definitions of equivalence (and other relations between theories) that could be applied to concrete cases (such as the debate whether Lagrangian and Hamiltonian mechanics are equivalent).

Let's suppose then that T and T' are first-order theories in a common signature Σ. Then we have the following obvious explication of equivalence:

DEFINITION 4.5.1 Let T and T' be theories in signature Σ. We say that T and T' are **logically equivalent** just in case $\text{Cn}(T) = \text{Cn}(T')$.

However, there are a couple of reasons why logical equivalence may not be a perfect explication of the notion of theoretical equivalence. First, there are cases of theories in the same signature that are the same "up to relabelling," but are not logically equivalent. For example, in the propositional signature $\Sigma = \{p,q\}$, let $T = \{p\}$ and let $T' = \{q\}$. Certainly there is one sense in which T and T' are different theories, since they disagree on which of the two propositional constants p and q should be affirmed. Nonetheless, there is another sense in which T' could be considered as a mere relabelling of T. At least structurally speaking, these two theories appear to be the same: they both have two propositional constants, and they assert precisely one of these two.

A second reason to worry about logical equivalence is that it cannot detect sameness of theories written in different signatures.

Example 4.5.2 Consider two signatures $\Sigma = \{f\}$ and $\Sigma' = \{\mathfrak{f}\}$, where both f and $\mathfrak{f}$ are one-place function symbols. (If you can't see the difference, the second f is written in Fraktur font. That raises an interesting question about whether f and $\mathfrak{f}$ are really the same letter or not.) Now let T be the theory with axiom $(f(x) = f(y)) \to (x = y)$, and let T' be the theory with axiom $(\mathfrak{f}(x) = \mathfrak{f}(y)) \to (x = y)$. Being written in different signatures, these two theories cannot be logically equivalent. But come now! Surely this is just a matter of different notations. Can't we write the *same* theory in different notation? ⌟

The problems with the previous example might be chalked up to needing a better criterion of sameness of signatures. However, that response won't help with the following sort of example.

Example 4.5.3 There are some theories that are intuitively equivalent, but not logically equivalent. In this example, we discuss two different formulations of the mathematical theory of groups.

Let $\Sigma_1 = \{\cdot, e\}$ be a signature where $\cdot$ is a binary function symbol and e is a constant symbol. Let T_1 be the following Σ_1-theory:

$$\big\{\forall x \forall y \forall z\big((x \cdot y) \cdot z = x \cdot (y \cdot z)\big), \forall x(x \cdot e = x \wedge e \cdot x = x), \\ \forall x \exists y(x \cdot y = e \wedge y \cdot x = e)\big\}$$

Now let $\Sigma_2 = \{\cdot, {}^{-1}\}$, where $\cdot$ is again a binary function symbol and ${}^{-1}$ is a unary function symbol. Let T_2 be the following Σ_2-theory:

$$\big\{\forall x \forall y \forall z\big((x \cdot y) \cdot z = x \cdot (y \cdot z)\big), \\ \exists x \forall y\big(y \cdot x = y \wedge x \cdot y = y \wedge y \cdot y^{-1} = x \wedge y^{-1} \cdot y = x\big)\big\}$$

If you open one textbook of group theory, you might find the axiomatization T_1. If you open another textbook of group theory, you might find the axiomatization T_2. And yet, the authors believe themselves to be talking about the same theory. How can this be so? The theories T_1 and T_2 are written in different signatures, and so are not even candidates for logical equivalence. ⌟

DEFINITION 4.5.4 If Σ_1 and Σ_2 are signatures, we call a map from elements of the signature Σ_1 to Σ_2-formulas a **reconstrual** $F : \Sigma_1 \to \Sigma_2$ if it satisfies the following three conditions.

- For every n-ary predicate symbol $p \in \Sigma_1$, $Fp(\vec{x})$ is a Σ_2-formula with n free variables.
- For every n-ary function symbol $f \in \Sigma_1$, $Ff(\vec{x}, y)$ is a Σ_2-formula with $n+1$ free variables.
- For every constant symbol $c \in \Sigma_1$, $Fc(y)$ is a Σ_2-formula with one free variable.

One can think of the Σ_2-formula $Fp(\vec{x})$ as a "translation" of the Σ_1-formula $p(\vec{x})$ into the signature Σ_2. Similarly, $Ff(\vec{x}, y)$ and $Fc(y)$ can be thought of as translations of the Σ_1-formulas $f(\vec{x}) = y$ and $c = y$, respectively.

A reconstrual $F : \Sigma_1 \to \Sigma_2$ naturally induces a map from Σ_1-formulas to Σ_2-formulas. In order to describe this map, we first need to describe the map that F induces from Σ_1-terms to Σ_2-formulas. Let $t(\vec{x})$ be a Σ_1-term. We define the Σ_2-formula $Ft(\vec{x}, y)$ recursively as follows.

- If t is the variable x_i then $Ft(x_i, y)$ is the Σ_2-formula $x_i = y$.
- If t is the constant symbol $c \in \Sigma_1$ then $Ft(y)$ is the Σ_2-formula $Fc(y)$.
- Suppose that t is the term $f(t_1(\vec{x}), \dots, t_k(\vec{x}))$ and that each of the Σ_2-formulas $Ft_i(\vec{x}, y)$ have been defined. Then we define $Ft(x_1, \dots x_n, y)$ to be the Σ_2-formula

$$\exists z_1 \dots \exists z_k \big(Ft_1(\vec{x}, z_1) \wedge \dots \wedge Ft_k(\vec{x}, z_k) \wedge Ff(z_1, \dots, z_k, y)\big).$$

We use this map from Σ_1-terms to Σ_2-formulas to describe how F maps Σ_1-formulas to Σ_2-formulas. Let $\phi(\vec{x})$ be a Σ_1-formula. We define the Σ_2-formula $F\phi(\vec{x})$ recursively as follows.

- If $\phi(\vec{x})$ is the Σ_1-atom $s(\vec{x}) = t(\vec{x})$, where s and t are Σ_1-terms, then $F\phi(\vec{x})$ is the Σ_2-formula
$$\exists z\big(Ft(\vec{x}, z) \wedge Fs(\vec{x}, z)\big).$$
- If $\phi(\vec{x})$ is the Σ_1-atom $p(t_1(\vec{x}), \ldots, t_k(\vec{x}))$, with $p \in \Sigma_1$ a k-ary predicate symbol, then $F\phi(\vec{x})$ is the Σ_2-formula
$$\exists z_1 \ldots \exists z_k\big(Ft_1(\vec{x}, z_1) \wedge \ldots \wedge Ft_k(\vec{x}, z_k) \wedge Fp(z_1, \ldots, z_k)\big).$$
- The definition of the Σ_2-formula $F\phi$ extends to all Σ_1-formulas ϕ in the now familiar manner.

In this way, a reconstrual $F : \Sigma_1 \to \Sigma_2$ gives rise to a map between Σ_1-formulas and Σ_2-formulas.

DEFINITION 4.5.5 We call a reconstrual $F : \Sigma_1 \to \Sigma_2$ a **translation** of a Σ_1-theory T_1 into a Σ_2-theory T_2 if $T_1 \vdash \phi$ implies that $T_2 \vdash F\phi$ for all Σ_1-sentences ϕ. We will use the notation $F : T_1 \to T_2$ to denote a translation of T_1 into T_2.

Example 4.5.6 Let Σ be the empty signature, let T_1 be the theory in Σ that says "there are at least n things," and let T_2 be the theory in Σ that says "there are exactly n things." Since Σ is empty, there is precisely one reconstrual $F : \Sigma \to \Sigma$, namely the identity reconstrual. This reconstrual is a translation from T_1 to T_2, but it is *not* a translation from T_2 to T_1. ⌟

DISCUSSION 4.5.7 A translation $F : T_1 \to T_2$ is in some ways quite rigid – e.g., it must preserve all numerical claims. To see this, observe first that for any atomic formula $x = y$, we have $F(x = y) \equiv (x = y)$. Moreover, since F preserves the Boolean connectives and quantifiers, it follows that F preserves all statements of the form, "there are at least n things," and "there are at most n things," and "there are exactly n things."

The notion of a translation between signatures gives us a particularly nice way to understand the notion of **substitution**. Informally speaking, we perform a substitution on a formula ϕ by replacing a predicate symbol p (or a function symbol f, or a constant symbol c) in ϕ uniformly with some other formula θ that has the same free variables. Intuitively speaking, since the validity of an argument depends only on form, such a substitution should map valid arguments to valid arguments. In short, if $\phi \vdash \psi$, and if ϕ^* and ψ^* are the result of a uniform substitution, then we should also have $\phi^* \vdash \psi^*$.

The notion of substitution is, in fact, a special case of the notion of a reconstrual. In the working example, we define a reconstrual $F : \Sigma \to \Sigma$ by setting $Fp = \theta$, and $Fs = s$ for every other symbol. Then to substitute θ for p in ϕ is simply to apply the function F to ϕ, yielding the formula $F\phi$. We might hope then to show that
$$\phi \vdash \psi \quad \Longrightarrow \quad F\phi \vdash F\psi.$$
But this isn't quite right yet. For example, suppose that F is a reconstrual that maps the constant symbol c to the formula $\theta(y)$. In this case, $\vdash \exists! y(y = c)$, but it is not

necessarily the case that $\vdash \exists !\, y\psi(y)$. To deal with this sort of case, we need to introduce the notion of admissibility conditions for function and constant symbols.

Suppose then that $F : \Sigma \to \Sigma'$ is a reconstrual. If $f \in \Sigma$ is an n-ary function symbol, then $Ff(\vec{x}, y)$ is a $(n+1)$-ary formula in Σ'. The **admissibility condition** for $Ff(\vec{x}, y)$ is simply the sentence

$$\forall \vec{x} \exists !\, y\, Ff(\vec{x}, y),$$

which says that Ff is a functional relation. In the case that f is a constant symbol (i.e., a 0-ary function symbol), Ff is a formula $\psi(y)$, and its admissibility condition is the formula $\exists ! y\psi(y)$.

DEFINITION 4.5.8 If $* : \Sigma \to \Sigma'$ is a reconstrual, then we let Δ be the set of Σ'-formulas giving the admissibility conditions for all function symbols in Σ.

Thus, the correct version of the substitution theorem can be stated as follows:

$$\phi \vdash \psi \implies \Delta, \phi^* \vdash \psi^*,$$

where Δ are the admissibility conditions for the reconstrual $*$. To prove this, we show first that for any term t of Σ, Δ implies that t^* is a functional relation.

LEMMA 4.5.9 *Let $* : \Sigma \to \Sigma'$ be a reconstrual, and let Δ be the admissibility conditions for the function symbols in Σ. Then for any term t of Σ, $\Delta \vdash \exists ! y\, t^*(\vec{x}, y)$.*

Proof We prove it by induction on the construction of t. The case where t is a variable is trivial. Suppose then that $t \equiv f(t_1, \ldots, t_n)$, where the result already holds for $t_1, \ldots, t_n$. In this case, $f(t_1, \ldots, t_n)^*$ is the relation

$$\exists z_1 \cdots \exists z_n (t_1^*(\vec{x}, z_1) \wedge \cdots \wedge t_n^*(\vec{x}, z_n) \wedge f^*(z_1, \ldots, z_n, y)).$$

Fix the n-tuple $\vec{x}$. We need to show that there is at least one y that stands in this relation, and that there is only one such y. For the former, since t_i^* is functional, there is a z_i such that $t_i^*(\vec{x}, z_i)$. Moroever, $\Delta \vdash \exists y f(z_1, \ldots, z_n, y)$. Thus, we've established the existence of at least one such y. For uniqueness, we first use the fact that if $t_i^*(\vec{x}, z_i)$ and $t^*(\vec{x}, z_i')$, then $z_i = z_i'$. Then we use the fact that

$$\Delta \vdash (f^*(z_1, \ldots, z_n, y) \wedge f^*(z_1, \ldots, z_n, y')) \to y = y'.$$

This establishes that $f(t_1, \ldots, t_n)^*$ is a functional relation. □

The next two lemmas show that, modulo these admissibility conditions, reconstruals preserve the validity of the intro and elim rules for equality.

LEMMA 4.5.10 *Let $* : \Sigma \to \Sigma'$ be a reconstrual, and let Δ be the admissibility conditions for the function symbols in Σ. Then for any term t of Σ, $\Delta \vdash (t = t)^*$.*

Proof Here $(t = t)^*$ is the formula $\exists y(t^*(\vec{x}, y) \wedge t^*(\vec{x}, y))$, which is equivalent to $\exists y\, t^*(\vec{x}, y)$. Thus, the result follows immediately from the fact that Δ entails that t^* is a functional relation. □

LEMMA 4.5.11 *Let $* : \Sigma \to \Sigma'$ be a reconstrual, and let Δ be the admissibility conditions for the function symbols in Σ. Then $\Delta, \phi(s)^*, (s = t)^* \vdash \phi(t)^*$.*

Proof Recall that $\phi(s)^*$ is the formula $\exists y(s^*(\vec{x}, y) \wedge \phi^*(y))$, and similarly for $\phi(t)^*$. Thus, we need to show that

$$\Delta, \exists y(s^*(\vec{x}, y) \wedge \phi^*(y)), \exists z(s^*(\vec{x}, z) \wedge t^*(\vec{x}, z)) \vdash \exists w(t^*(\vec{x}, w) \wedge \phi^*(w)).$$

The key fact, again, is that Δ implies that t^* and s^* are functional relations. We can then argue intuitively: holding $\vec{x}$ fixed, from $\exists y(s^*(\vec{x}, y) \wedge \phi^*(y))$ and $\exists z(s^*(\vec{x}, z) \wedge t^*(\vec{x}, z))$, we are able to conclude this y and z are the same and, thus, that $\exists w(t^*(\vec{x}, w) \wedge \phi^*(w))$. □

Recall that a reconstrual $* : \Sigma \to \Sigma'$ maps a formula such as $t(\vec{x}) = y$ to a formula $t^*(\vec{x}, y)$. Here the variable y is chosen arbitrarily, but in such a manner as not to conflict with any variables already in use. Intuitively, this y is the only new free variable in t^*. We now validate that intuition.

LEMMA 4.5.12 *Let $* : \Sigma \to \Sigma'$ be a reconstrual. Then for each term t of Σ, $FV(t^*) = FV(t) \cup \{y\}$.*

Proof Base cases: If t is a variable x, then $t^*(x, y) \equiv (x = y)$. In this case, $FV(t^*(x, y)) = FV(t) \cup \{y\}$. If t is a constant symbol c, then t^* is the formula $c = y$, and $FV(t^*) = FV(c) \cup \{y\}$.

Inductive case: Suppose that the result holds for $t_1, \ldots, t_n$. Then

$$\begin{aligned} FV(f(t_1, \ldots, t_n)) &= FV(t_1) \cup \cdots \cup FV(t_n) \\ &= FV(t_1^*) \cup \cdots FV(t_n^*). \end{aligned}$$

Recall that $f(t_1, \ldots, t_n)^*$ is defined as the formula

$$\exists z_1 \cdots \exists z_n \, (t_1(\vec{x}, z_1) \wedge \cdots t_n(\vec{x}, z_n) \wedge f(z_1, \ldots, z_n, y)),$$

from which it can easily be seen that

$$FV(f(t_1, \ldots t_n)^*) = FV(t_1^*) \cup \cdots \cup FV(t_n^*) \cup \{y\}.$$

□

LEMMA 4.5.13 *Let $* : \Sigma \to \Sigma'$ be a reconstrual. Then for each formula ϕ of Σ, $FV(\phi^*) = FV(\phi)$.*

Proof We prove this by induction on the construction of ϕ. Base case: Suppose that ϕ is the formula $s = t$, where s and t are terms. Then ϕ^* is the formula

$$\exists y(s^*(\vec{x}, y) \wedge t^*(\vec{x}, y)).$$

By the previous lemma, y is the only free variable in s^* that doesn't occur in s, and similarly for t^* and t. Therefore, $FV(\phi^*) = FV(\phi)$.

Base case: Suppose that ϕ is the formula $p(t_1, \ldots, t_n)$, where p is a relation symbol, and $t_1, \ldots, t_n$ are terms. Here the free variables in ϕ are just all those free in the terms t_i. Moreover, $p(t_1, \ldots, t_n)^*$ is the formula that says: there are $z_1, \ldots, z_n$ such that $t_i^*(\vec{x}, z_i)$

and $p^*(z_1, \dots, z_n)$. By the previous lemma, $FV(t_i^*(\vec{x}, z_i)) = FV(t_i) \cup \{z_i\}$. Therefore, $p(t_1, \dots, t_n)^*$ has the same free variables as $p(t_1, \dots, t_n)$.

Inductive cases: The cases for the Boolean connectives are easy, and are left to the reader. Let's just check the case of the universal quantifier. Suppose that the result is true for ϕ. Then

$$FV((\forall x\phi)^*) = FV(\forall x\phi^*) = FV(\phi^*)\backslash\{x\} = FV(\phi)\backslash\{x\} = FV(\forall x\phi).$$

□

THEOREM 4.5.14 (Substitution) *Let $*: \Sigma \to \Sigma'$ be a reconstrual, and let Δ be the admissibility conditions for function symbols in Σ. Then for any formulas ϕ and ψ of Σ, if $\phi \vdash \psi$, then $\Delta, \phi^* \vdash \psi^*$.*

Proof We prove this by induction on the definition of the relation $\vdash$. The base case is the rule of assumptions: show that when $\phi \vdash \phi$, then also $\Delta, \phi^* \vdash \phi^*$. However, the latter follows immediately by the rule of assumptions, plus monotonicity of $\vdash$.

The clauses for the Boolean connectives follow immediately from the fact that $*$ is compositional. We will now look at the clause for $\forall$-intro. Suppose that $\Gamma \vdash \forall x\phi$ results from $\Gamma \vdash \phi$, where x is not free in Γ. Assume that $\Delta, \Gamma^* \vdash \phi^*$. By Lemma 4.5.13, x is not free in Γ^*. Moreover, since the admissibility conditions are sentences, x is not free in Δ. Therefore, $\Delta, \Gamma^* \vdash \forall x\phi^*$, and since $(\forall x\phi)^* \equiv \forall x\phi^*$, it follows that $\Delta, \Gamma^* \vdash (\forall x\phi)^*$. □

We began this section with a discussion of various relations bewteen theories that have been interesting to philosophers of science – e.g., equivalence, reducibility, conservative extension. We now look at how such relations might be represented as certain kinds of translations between theories. We begin by considering the proposal that theories are **equivalent** just in case they are **intertranslatable**. The key here is in specifying what is meant by "intertranslatable." Do we only require a pair of translations $F : T \to T'$ and $G : T' \to T$, with no particular relation between F and G? Or do we require more? The following condition requires that F and G are inverses, relative to the notion of "sameness" of formulas internal to the theories T and T'. To be more specific, two formulas ϕ and ϕ' are the "same" relative to theory T just in case $T \vdash \phi \leftrightarrow \phi'$.

DEFINITION 4.5.15 Let T be a Σ-theory and T' a Σ'-theory. Then T and T' are said to be **strongly intertranslatable** or **homotopy equivalent** if there are translations $F : T \to T'$ and $G : T' \to T$ such that

$$T \vdash \phi \leftrightarrow GF\phi \qquad \text{and} \qquad T' \vdash \psi \leftrightarrow FG\psi, \tag{4.1}$$

for every Σ-formula ϕ and every Σ'-formula ψ.

The conditions (4.1) can be thought of as requiring the translations $F : T \to T'$ and $G : T' \to T$ to be "almost inverse" to one another. Note, however, that F and G need not be literal inverses. The Σ-formula $GF\phi$ is not required to be *equal* to the Σ-formula ϕ. Rather, these two formulas are merely required to be *equivalent* according to the theory T.

DISCUSSION 4.5.16 Let T and T' be theories in a common signature Σ. It should be fairly obvious that if T and T' are logically equivalent, then they are homotopy equivalent. Indeed, it suffices to let both F and G be the identity reconstrual on the signature Σ.

It should also be obvious that not all homotopy equivalent theories are logically equivalent – not even theories in the same signature. For example, let $\Sigma = \{p, q\}$ be a propositional signature, let T be the theory with axiom p, and let T' be the theory with axiom q. Obviously T and T' are not logically equivalent. However, the reconstrual $Fp = q$ shows that T and T' are homotopy equivalent.

DISCUSSION 4.5.17 One might legitimately wonder: what's the motivation for the definition of homotopy equivalence, with a word "homotopy" that is not in most philosophers' active vocabulary? One might also wonder, more generally: what is the right method for deciding on an account of equivalence? What is at stake, and how do we choose between various proposed explications? Are we supposed to have strong intuitions about what "equivalent" really means? Or, at the opposite extreme, is the definition of "equivalent" merely a convention to be judged by its utility?

These are difficult philosophical questions that we won't try to answer here (but see Chapter 8). However, there is both a historical and a mathematical motivation for the definition of homotopy equivalence. The historical motivation for this definition is its appearance in various works of logic and philosophy of science beginning in the 1950s. As for mathematical motivation, the phrase "homotopy equivalence" originally comes from topology, where it denotes a kind of "sameness" that is weaker than the notion of a homeomorphism. Interestingly, the idea of weakening isomorphism is also particularly helpful in category theory. In category theory, the natural notion of "sameness" of categories is not isomorphism, but categorical equivalence. Recall that two categories $\mathbf{C}$ and $\mathbf{D}$ are equivalent just in case there is a pair of functors $F : \mathbf{C} \to \mathbf{D}$ and $G : \mathbf{D} \to \mathbf{C}$ such that both FG and GF are naturally isomorphic to the respective identity functors (see 3.3.6). In the case of Makkai and Reyes' logical categories, the natural notion of equivalence is simply categorical equivalence (see Makkai and Reyes, 1977).

At this point, we will set aside further discussion of equivalence until Section 4.6. We will now turn to some other relations between theories.

Suppose that T is a theory in signature Σ, and T' is a theory in signature Σ', where $\Sigma \subseteq \Sigma'$. We say that T' is an **extension** of T just in case: if $T \vdash \phi$, then $T' \vdash \phi$, for all sentences ϕ of Σ. An extension of of a theory amounts to the addition of new concepts – i.e., new vocabulary – to a theory. (Here we are using the word "concepts" in a nontechnical sense. To be technically precise, a conservative extension of a theory results when new symbols are added to the signature Σ.) In the development of the sciences, there can be a variety of reasons for adding new concepts; e.g., we might use them as a convenient shorthand for old concepts, or we might feel the new to expand our conceptual repertoire. For example, many physicists would say that the concept of a "quantum state" is a genuinely novel addition to the stock of concepts used in classical physics.

One very interesting question in philosophy of science is whether there are any sorts of "rules" or "guidelines" for the expansion of our conceptual repertoire. Must the new concepts be connected to the old ones? And if so, what sorts of connections should we hope for them to have?

A **conservative extension** of a theory T adds new concepts, but without in any way changing the logical relations between old concepts. In the case of formal theories, there are two extreme cases of a conservative extension: (1) the new vocabulary is shorthand for old vocabulary, and (2) the new vocabulary is unrelated to the old vocabulary. The paradigm example of the former (new vocabulary as shorthand) is a definitional extension of a theory, as discussed in the preceding paragraphs. As an example of the latter (new vocabulary unrelated), consider first the theory T in the empty signature Σ which says that "there are exactly two things." Now let T' have the same axiom as T, but let it be formulated in a signature $\Sigma' = \{p\}$, where p is a unary predicate symbol. Let $F : T \to T'$ be the translation given by the inclusion $\Sigma \subseteq \Sigma'$. It's clear, then, that F is a conservative translation.

The example we have just given is somewhat atypical, since the new theory T' says nothing about the new vocabulary $\Sigma' \backslash \Sigma$. However, the point would be unchanged if, for example, we equipped T' with the axiom $\forall x p(x)$.

DEFINITION 4.5.18 A translation $F : T \to T'$ is said to be **conservative** just in case: if $T' \vdash F\phi$, then $T \vdash \phi$.

The idea behind this definition of a conservative extension is that the target theory T' adds no new claims that can be formulated in the language Σ of the original theory. In other words, if T' says that some relation holds between sentences $F\phi_1$ and $F\phi_2$, then T already asserts that this relation holds. Of course, T' might say nontrivial things in its new vocabulary, i.e., in those Σ'-sentences that are not in the image of the mapping F.

Example 4.5.19 There are many intuitive examples of conservative extensions in mathematics. For example, the theory of the integers is a conservative extension of the theory of natural numbers, and the theory of complex numbers is a conservative extension of the theory of real numbers. ⌟

EXERCISE 4.5.20 Suppose that $F: T \to T'$ is conservative. Show that if T is consistent, then T' is consistent.

EXERCISE 4.5.21 Suppose that T is a consistent and complete theory in signature Σ. Let $\Sigma \subseteq \Sigma'$, and let T' be a consistent theory in Σ'. Show that if T' is an extension of T, then T' is a conservative extension of T.

EXERCISE 4.5.22 Let T be the theory from the previous example, and let $\Sigma' = \{p\}$, where p is a unary predicate symbol. Which theories in Σ' are extensions of T? Which of these extensions is conservative? More difficult: classify all extensions of T in the language Σ', up to homotopy equivalence. (In other words, consider two extensions to be the same if they are homotopy equivalent. Hint: consider the question, "how many p are there?")

A conservative translation $F : T \to T'$ is like a monomorphism from T to T'. Thus, we might also be interested in a dual sort of notion – something like an epimorphism from T to T'. As with propositional theories, it works well to consider a notion of surjectivity up to logical equivalence. Borrowing terminology from category theory, we call this notion "essential surjectivity."

DEFINITION 4.5.23 Let $F : T \to T'$ be a translation between theories. We say that F is **essentially surjective** (abbreviated **eso**) just in case for each Σ'-formula ψ, there is a Σ-formula ϕ such that $T \vdash \psi \leftrightarrow F\phi$.

The idea behind an essentially surjective translation is that the domain T is ideologically as rich as the codomain theory T'. We are using "ideology" here in the sense of Quine, i.e., the language in which a theory is formulated. If $F : T \to T'$ is essentially surjective, then (up to logical equivalence), T can express all of the concepts that T' can express.

A paradigm example of an essentially surjective translation is a "specialization" of a theory, i.e., where we add some new axioms, but without adding any new vocabulary. Indeed, suppose that T is a theory in Σ, and that T' results from adding some axioms to T. Then the identity reconstrual $I : \Sigma \to \Sigma$ yields an essentially surjective translation $I : T \to T'$. Of course, there are other sorts of essentially surjective translations.

EXERCISE 4.5.24 Let $\Sigma = \{p\}$, where p is a unary predicate, and let $\Sigma' = \{r\}$, where r is a binary relation. Let T be the empty theory in Σ, and let T' be the theory in Σ' that says that r is symmetric, i.e., $r(x, y) \to r(y, x)$. Let $F : \Sigma \to \Sigma'$ be the reconstrual that takes p to $\exists z\, r(x, z)$. Is F essentially surjective? (This exercise will be a lot easier to answer after Chapter 6.)

EXERCISE 4.5.25 Let Σ be the signature with a single unary predicate symbol p, and let Σ' be the empty signature. Let T' be the theory in Σ' that says "there are exactly two things," and let T be the extension of T' in Σ that also says "there is a unique p." Is there an essentially surjective translation $F : T \to T'$? (This exercise will be a lot easier after Chapter 6.)

In the case of propositional theories, we saw that a translation is an equivalence iff it is conservative and essentially surjective. We now show the same for first-order theories.

PROPOSITION 4.5.26 *Suppose that $F : T \to T'$ is one-half of a homotopy equivalence. Then F is conservative and essentially surjective.*

Proof The proof here is structurally identical to the one for propositional theories. Suppose that $G : T' \to T$ is the other half of a homotopy equivalence so that $T \vdash \phi \leftrightarrow GF\phi$ and $T' \vdash \psi \leftrightarrow FG\psi$. To see that F is conservative, suppose that $T' \vdash F\phi$. Then $T \vdash GF\phi$, and since $T \vdash \phi \leftrightarrow GF\phi$, it follows that $T \vdash \phi$. Therefore, F is conservative. To see that F is essentially surjective, let ψ be a Σ'-formula. Then $G\psi$ is a Σ-formula, and we have $T' \vdash \psi \leftrightarrow FG\psi$. Therefore, F is essentially surjective. □

PROPOSITION 4.5.27 *Suppose that $F : T \to T'$ is conservative and essentially surjective. Then F is one-half of a homotopy equivalence.*

Proof Again, the proof here is structurally identical to the proof in the propositional case.

Fix a relation symbol p of Σ'. Since F is essentially surjective, there is a formula ϕ_p of Σ such that $T' \vdash p \leftrightarrow F\phi_p$. Define $Gp = \phi_p$. Thus, for each relation symbol p, we have $T' \vdash p \leftrightarrow FGp$ by definition. And since FG is (by definition) compositional, $T' \vdash \psi \leftrightarrow FG\psi$ for all formulas ψ of Σ'.

Now we claim that G is a translation from T' into T. Indeed, if $T' \vdash \psi$, then $T' \vdash FG\psi$, and since F is conservative, $T \vdash G\psi$. Therefore, $G : T' \to T$ is a translation.

Finally, given an arbitrary formula ϕ of Σ, we have $T' \vdash F\phi \leftrightarrow FGF\phi$, and hence $T' \vdash F(\phi \leftrightarrow GF\phi)$. Since F is conservative, it follows that $T \vdash \phi \leftrightarrow GF\phi$. Therefore, F and G form a homotopy equivalence. □

DISCUSSION 4.5.28 Using translations between theories as arrows, we could now define a category **Th** of first-order theories, and we could explore the features of this category. However, we will resist this impulse – because it turns out that this category isn't very interesting.

DISCUSSION 4.5.29 Does the notion of translation capture every philosophically interesting relation between theories? There are few questions here: First, can every interesting relation between theories be explicated syntactically? And if the answer to the first question is yes, then can every such relation be described as a translation? And if the answer to that question is yes, then have we given an adequate account of translation?

Recall that Carnap (1934) seeks a theory of science that uses only syntactic concepts. If we were to follow Carnap's lead, then we would have to answer yes to the first question. But of course, many philosophers of science have convinced themselves that the first question must receive a negative answer. Indeed, some philosophers of science claimed that the interesting relations between theories (e.g., equivalence, reducibility) cannot be explicated syntactically. We will return later to this claim.

4.6 Definitional Extension and Equivalence

Mathematicians frequently define new concepts out of old ones, and logicians have only begun to understanding the varieties of ways that mathematicians do so. We do have a sense that not all definitions are created equal. On the one hand, definitions can seem quite trivial, e.g., when we come up with a new name for an old concept. On the other hand, some definitions are overtly inconsistent. For example, if we said "let n be the largest prime number," then we could prove both that there is a largest prime number and that there is not. The goal of a logical theory of definition is to steer a course between these two extremes – i.e., to account for those definitions that are both fruitful and safe.

In this section, we'll look at some of the simplest kinds of definitions. Our general setup will consist of a pair of signatures Σ and Σ^+ with $\Sigma \subseteq \Sigma^+$. Here we think of Σ as "old concepts" and we think of $\Sigma^+ \backslash \Sigma$ as "new concepts."

DEFINITION 4.6.1 If p is a relation symbol in Σ^+, then an **explicit definition of p in terms of Σ** is a Σ^+-sentence of the form

$$\forall \vec{x}(p(\vec{x}) \leftrightarrow \phi(\vec{x})),$$

where $\phi(\vec{x})$ is a Σ-formula.

Here p can be thought of as "convenient shorthand" for the formula ϕ, which might itself be quite complex. For example, from the predicates "is a parent" and "is a male," we could explicitly define a predicate "is a father." Of course, the definition itself is a sentence in the larger signature Σ^+ and not in the smaller signature Σ.

DISCUSSION 4.6.2 What are we doing when we define new concepts out of old ones? Some philosophers might worry that defining a new concept amounts to making a theoretical commitment – to the existence of some worldly structure corresponding to that concept. For example, suppose that we initially have a theory with the concepts $\mathsf{male}(x)$ and $\mathsf{parent}(x)$. If we define $\mathsf{father}(x)$ in terms of these two original concepts, then are we committing to the existence of some further worldly structure, viz. the property of fatherhood?

The default answer in first-order logic is no. Using a predicate symbol r does not amount to any kind of postulating of worldly structure corresponding to r. Accordingly, adding definitions to a theory does not change the content of that theory – it only changes the resources we have for expressing that content. (We don't mean to say that these views are uncontroversial or mandatory.)

Not only can we define new relations; we can also define new functions and constants. Certainly a function can be defined in terms of other functions. For example, if we begin with functions g and f, then we can define a composite function $g \circ f$. In fact, this composite $g \circ f$ can be defined explicitly by the formula

$$((g \circ f)(x) = y) \leftrightarrow \exists z((f(x) = z) \wedge (g(z) = y)).$$

Similarly, a constant symbol c can be defined in terms of a function symbol f and other constants $d_1, \ldots, d_n$, namely

$$c := f(d_1, \ldots, d_n).$$

However, these ways of defining functions in terms of functions, and defining constants in terms of functions and constants, can be subsumed into a more general way of defining functions and constants in terms of relations.

DEFINITION 4.6.3 An explicit definition of an n-ary function symbol $f \in \Sigma^+$ in terms of Σ is a Σ^+-sentence of the form

$$\forall \vec{x} \forall y(f(\vec{x}) = y \leftrightarrow \phi(\vec{x}, y)), \tag{4.2}$$

where $\phi(\vec{x}, y)$ is a Σ formula.

DEFINITION 4.6.4 An explicit definition of a constant symbol $c \in \Sigma^+$ is a Σ^+-sentence of the form

$$\forall y(y = c \leftrightarrow \psi(y)), \tag{4.3}$$

where $\psi(y)$ is a Σ-formula.

Although they are Σ^+-sentences, (4.2) and (4.3) have consequences in the signature Σ. In particular, (4.2) implies $\forall \vec{x} \exists ! y \phi(\vec{x}, y)$ and (4.3) implies $\exists ! y \psi(y)$. These two sentences are called the **admissibility conditions** for the explicit definitions (4.2) and (4.3).

DEFINITION 4.6.5 A **definitional extension** of a Σ-theory T to the signature Σ^+ is a Σ^+-theory

$$T^+ = T \cup \{\delta_s : s \in \Sigma^+ \backslash \Sigma\},$$

that satisfies the following two conditions. First, for each symbol $s \in \Sigma^+ \backslash \Sigma$, the sentence δ_s is an explicit definition of s in terms of Σ. And second, if s is a constant symbol or a function symbol and α_s is the admissibility condition for δ_s, then $T \vdash \alpha_s$.

Example 4.6.6 Let $\Sigma = \{p\}$, where p is a unary predicate symbol, and let T be *any* theory in Σ. We can then define a relation r by means of the formula

$$r(x, y) \leftrightarrow (p(x) \leftrightarrow p(y)).$$

It's easy to see that r is an equivalence relation. Thus, every unary predicate symbol defines a corresponding equivalence relation. In fact, this equivalence relation r has precisely two equivalence classes.

The converse is not exactly true. In fact, suppose that a theory T entails that r is an equivalence relation with exactly two equivalence classes. One might try to define a predicate p so that the first equivalence class consists of elements satisfying p, and the second equivalence class consists of elements not satisfying p. However, this won't work, because the relation r itself does not provide the resources to name the individual classes. Intuitively speaking, r can't tell the difference between the two equivalence classes, but p can, and therefore p cannot be defined from r. We'll be able to see this fact more clearly after Chapter 6. ⌟

Example 4.6.7 The following example is from Quine and Goodman (1940). Let $\Sigma = \{r\}$, where r is a binary relation symbol. Now define a new relation symbol s by setting

$$s(x, y) \leftrightarrow \forall w(r(x, w) \rightarrow r(y, w)).$$

Then it follows that s is a transitive relation, i.e.,

$$\vdash (s(x, y) \wedge s(y, z)) \rightarrow s(x, z).$$

⌟

Example 4.6.8 Let T be the theory of Boolean algebras. Then one can define a relation symbol $\leq$ by setting

$$x \leq y \leftrightarrow x \wedge y = y.$$

It follows that $\leq$ is a partial order. ⌟

Let T^+ be a definitional extension of T. We now define two translations $I : T \to T^+$ and $R : T^+ \to T$. The translation $I : T \to T^+$ is simply the inclusion: it acts as the identity on elements of the signature Σ. The latter we define as follows: for each symbol r in $\Sigma^+ \backslash \Sigma$, let $Rr = \theta_r$, where θ_r is the Σ-formula in the explicit definition

$$T^+ \vdash \forall \vec{x}(r(\vec{x}) \leftrightarrow \theta_r(\vec{x})).$$

For $r \in \Sigma$, let $Rr = r$.

DISCUSSION 4.6.9 The translation $R : T^+ \to T$ is an example of a **reduction** of the theory T^+ to the theory T. Here we simply replace the definiendum r with its definiens θ_r. However, this particular R has another feature that not all reductions have – namely, it's an equivalence between the theories T^+ and T (as we show in the subsequent lemmas).

This kind of strict reduction is similar to what Carnap hoped to achieve in the *Aufbau* – and for which he was so severely critized by Quine. However, it should be noted that Carnap's permissible constructions were stronger than explicit definitions in the sense we've explained here. At the very least, Carnap permitted maneuvers such as "extension by abstraction," which are akin to the "extensions by sorts" that we consider in the following chapter.

Similarly, advocates of the old-fashioned mind–brain identity theory presumably believed that folk psychology (to the extent that it is accurate) could be reduced, in this strict sense, to neuroscience, and perhaps ultimately to fundamental physics.

Scientists do often talk about one theory T^+ being reducible to another T, e.g., thermodynamics being reducible to statistical mechanics. However, it's beyond doubtful that all such cases of successful reduction could be faithfully modelled as a simple expansion of definitions. We don't think, however, that the moral is that philosophers of science should resort to vague and imprecise accounts of reduction. Instead, they should find more sophisticated tools for their explications.

We now show that the pair I, R form a homotopy equivalence between T and T^+. For this we need a few auxiliary lemmas.

LEMMA 4.6.10 *For any term $t(\vec{x})$ of Σ^+, we have $T^+ \vdash (t(\vec{x}) = y) \leftrightarrow Rt(\vec{x}, y)$.*

Proof We prove this by induction on the construction of t. In the case that t is a variable x, the claim is

$$T^+ \vdash (x = y) \leftrightarrow (x = y),$$

which obviously holds. Now suppose that t is the term $f(t_1, \ldots, t_n)$ and that the result holds for the terms $t_1, \ldots, t_n$. That is,

$$T^+ \vdash (t_i(\vec{x}) = z_i) \leftrightarrow Rt_i(\vec{x}, z_i). \tag{4.4}$$

Since T^+ defines f in terms of Rf, we also have

$$T^+ \vdash (f(z_1, \ldots, z_n) = y) \leftrightarrow Rf(z_1, \ldots, z_n, y). \tag{4.5}$$

By the definition of R on terms, $R(f(t_1,\ldots,t_n))(\vec{x},y)$ is the formula

$$\exists z_1\cdots\exists z_n(Rt_1(\vec{x},z_1)\wedge Rt_n(\vec{x},z_n)\wedge Rf(z_1,\ldots,z_n,y)).$$

Thus, (4.4) and (4.5) imply that

$$T^+ \ \vdash\ (f(t_1(\vec{x}),\ldots,t_n(\vec{x}))=y)\leftrightarrow R(f(t_1,\ldots,t_n))(\vec{x},y).$$

□

LEMMA 4.6.11 *For any Σ^+-formula ϕ, we have $T^+\vdash\phi\leftrightarrow R\phi$.*

Proof We prove this by induction on the construction of ϕ. Since R is defined compositionally on formulas, it will suffice to establish the two base cases.

1. Suppose first that ϕ is the formula $s(\vec{x})=t(\vec{x})$, in which case $R\phi$ is the formula $\exists y(Rs(\vec{x},y)\wedge Rt(\vec{x},y))$. By the previous result,

$$T^+ \ \vdash\ (t(\vec{x})=y)\leftrightarrow Rt(\vec{x},y),$$

and

$$T^+ \ \vdash\ (s(\vec{x})=y)\leftrightarrow Rs(\vec{x},y).$$

By assumption, we have

$$T^+ \ \vdash\ \exists y(s(\vec{x})=y\wedge t(\vec{x})=y),$$

and the result immediately follows.

2. Suppose now that ϕ is the formula $p(t_1,\ldots,t_n)$, in which case $R\phi$ is the formula

$$\exists y_1\cdots\exists y_n(Rt_1(\vec{x},y_1)\wedge\cdots\wedge Rt_n(\vec{x},y_n)\wedge Rp(y_1,\ldots,y_n)).$$

By the previous result again,

$$T^+ \ \vdash\ (t_i(\vec{x})=y_i)\leftrightarrow Rt_i(\vec{x},y_i).$$

Moreover, since T^+ explicitly defines p in terms of Rp, we have

$$T^+ \ \vdash\ p(\vec{y})\leftrightarrow Rp(\vec{y}).$$

The result follows immediately. □

LEMMA 4.6.12 *For any Σ^+-formula ϕ, if $T^+\vdash R\phi$ then $T\vdash R\phi$.*

Proof To say that $T^+\vdash R\phi$ means that there is a finite family $\theta_1,\ldots,\theta_n$ of axioms of T^+ such that $\theta_1,\ldots,\theta_n\vdash R\phi$. By the substitution theorem (4.5.14),

$$\Delta, R\theta_1,\ldots,R\theta_n \ \vdash\ RR\phi,$$

where Δ consists of the admissibility conditions for function symbols in Σ^+. Since T^+ is a definitional extension, T implies the admissibility conditions in Δ. Moreover, since $RR\phi=\phi$, we have

$$T, R\theta_1,\ldots,R\theta_n \ \vdash\ R\phi.$$

Thus, it will suffice to show that $T\vdash R\theta_i$, for each i.

Fix i and let $\theta \equiv \theta_i$. Now, θ is either an axiom of T or an explicit definition of a symbol $s \in \Sigma^+ \backslash \Sigma$. If θ is an axiom of T, then it's also a formula in signature Σ, in which case $R\theta = \theta$ and $T \vdash R\theta$. If θ is an explicit definition of a symbol s, then $R\theta$ is the tautology $Rs \leftrightarrow Rs$, and $T \vdash R\theta$. □

PROPOSITION 4.6.13 *If T^+ is a definitional extension of T, then $I : T \to T^+$ and $R : T^+ \to T$ form a homotopy equivalence.*

Proof Since the axioms of T are a subset of the axioms of T^+, it follows that I is a translation from T to T^+. Next we show that R is a translation from T^+ to T – i.e., that if $T^+ \vdash \phi$, then $T \vdash R\phi$. By the previous two lemmas, we have $T^+ \vdash \phi \leftrightarrow R\phi$, and if $T^+ \vdash \phi$, then $T \vdash \phi$. Thus, if $T^+ \vdash \phi$, then $T^+ \vdash R\phi$, and $T \vdash R\phi$.

Next we show that $T \vdash \phi \leftrightarrow RI\phi$ and $T^+ \vdash \psi \leftrightarrow IR\psi$. For this, recall that both $I : \Sigma \to \Sigma^+$ and $R : \Sigma^+ \to \Sigma$ act as the identity on Σ-formulas. Thus, we immediately get $T \vdash \phi \leftrightarrow IR\phi$ for any Σ-formula ϕ. Furthermore, Lemma 4.6.11 entails that $T^+ \vdash \psi \leftrightarrow R\psi$. Since $R\psi$ is a Σ-formula, it follows that $T^+ \vdash \psi \leftrightarrow IR\psi$. □

COROLLARY 4.6.14 *If T^+ is a definitional extension of T, then T^+ is a conservative extension of T.*

The previous results show, first, that a definitional extension is conservative: it adds no new results in the old vocabulary. In fact, Proposition 4.6.13 shows that a definitional extension is, in one important sense, equivalent to the original theory. You may want to keep that fact in mind as we turn to a proposal that some logicians made in the 1950s and 1960s, and that was applied to philosophy of science by Glymour (1971). According to Glymour, two scientific theories should be considered equivalent only if they have a common definitional extension.

DEFINITION 4.6.15 Let T_1 be a Σ_1-theory and T_2 be a Σ_2-theory. Then T_1 and T_2 are said to be **definitionally equivalent** if there is a definitional extension T_1^+ of T_1 to the signature $\Sigma_1 \cup \Sigma_2$ and a definitional extension T_2^+ of T_2 to the signature $\Sigma_1 \cup \Sigma_2$ such that T_1^+ and T_2^+ are logically equivalent.

If T_1 and T_2 are definitionally equivalent, then they in fact have a **common definitional extension**, namely the theory $T^+ := \mathrm{Cn}(T_1^+) = \mathrm{Cn}(T_2^+)$. These three theories then form a span:

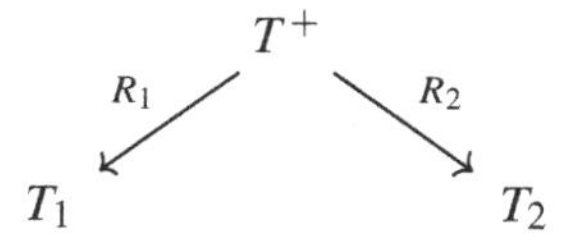

Here $R_i : T^+ \to T_i$ is the translation that results from replacing definienda in the signature $\Sigma_1 \cup \Sigma_2$ with their definiens in signature Σ_i. Note that if T_1 and T_2 are both Σ-theories (i.e., if they are formulated in the same signature), then T_1 and T_2 are definitionally equivalent if and only if they are logically equivalent.

Definitional equivalence captures a sense in which theories formulated in different signatures might nonetheless be theoretically equivalent. For example, although they are not logically equivalent, the theory of groups$_1$ and the theory of groups$_2$ are definitionally equivalent.

Example 4.6.16 Recall the two formulations of group theory from Example 4.5.3. Consider the following two $\Sigma_1 \cup \Sigma_2$-sentences.

$$\delta_{-1} := \forall x \forall y \big(x^{-1} = y \leftrightarrow (x \cdot y = e \wedge y \cdot x = e)\big)$$
$$\delta_e := \forall x \big(x = e \leftrightarrow \forall z (z \cdot x = z \wedge x \cdot z = z)\big).$$

The theory T_1 defines the unary function symbol $^{-1}$ with the sentence δ_{-1}, and the theory T_2 defines the constant symbol e with the sentence δ_e. One can verify that T_1 satisfies the admissibility condition for δ_{-1} and that T_2 satisfies the admissibility condition for δ_e. The theory of groups$_1$ $\cup \{\delta_{-1}\}$ and the theory of groups$_2$ $\cup \{\delta_e\}$ are logically equivalent. This implies that these two formulations of group theory are definitionally equivalent. ⌟

We're now ready for the first big result relating different notions of equivalence.

THEOREM 4.6.17 (Barrett) *Let T_1 and T_2 be theories with a common definitional extension. Then there are translations $F : T_1 \to T_2$ and $G : T_2 \to T_1$ that form a homotopy equivalence.*

Proof Let T^+ be a common definitional extension of T_1 and T_2. By Prop. 4.6.13, there are homotopy equivalences $I_1 : T_1 \to T^+$ and $R_2 : T^+ \to T_2$. Thus, $R_2 I_1 : T_1 \to T_2$ is a homotopy equivalence. □

We prove the converse of this theorem in 6.6.21.

DISCUSSION 4.6.18 In this section, we've discussed methods for defining relation, function, and constant symbols. It's commonly assumed, however, that other sorts of definitions are also possible. For example, we might define an exclusive "or" connective $\oplus$ by means of the recipe

$$\phi \oplus \psi \;\leftrightarrow\; (\phi \vee \psi) \wedge \neg(\phi \wedge \psi).$$

For more on the notion of defining new connectives, see Dewar (2018a).

The same might be said for quantifiers. Given existential quantifiers $\exists x$ and $\exists y$, we might introduce a new quantifier $\exists x \exists y$ over pairs. But does this new syntactic entity, $\exists x \exists y$, deserve to be called a "quantifier"?

Supposing that $\exists x \exists y$ does deserve to be called a quantifier, then we need to rethink the notion of the "ontological commitments" of a theory – and along with that, a whole slew of attitudes toward ontology that come along with it. It's common for philosophers of science to raise the question: "What are the ontological commitments of this theory?" The idea here is that if the scientific community accepts a theory, then we should accept that theory's ontological commitments. For example, some philosophers argue that we should believe in the existence of mathematical objects since our best scientific theories (such as general relativity and quantum mechanics) quantify over them. Others, such as

Field (1980) attempt to "nominalize" these theories – i.e., to reformulate them in such a way that they don't quantify over mathematical objects.

Both parties to this dispute about mathematical objects share a common presupposition: Once a theory is regimented in first-order logic, then its ontological commitments can be read off from the formalism. But this presupposition is brought into question by the fact that first-order theories can implicitly define new quantifiers. Thus, a theory might have *more* ontological commitments than are shown in its original quantifiers. Conversely, a theory isn't necessarily committed to the ontology encoded in its initial quantifiers. Those quantifiers might capture some derivative ontology, and the actual ontology might be captured by quantifiers that are defined in terms of those original quantifiers. In short, regimenting a theory in first-order logic does *not* settle all ontological disputes.

4.7 Notes

- Carnap (1935) gives a readable, nontechnical overview of his *Wissenschaftslogik* program. The amount of high-quality historical research on Carnap is on the steady rise. See, e.g., Friedman (1982); Awodey and Klein (2004); Andreas (2007); Creath and Friedman (2007); Hudson (2010); Friedman (2011). For the relevance of Carnap's views to contemporary issues, see, e.g., Price (2009); Blatti and Lapointe (2016).
- The substitution theorem is rarely proven in detail. One notable exception is Kleene (1952). We prove another, more general, version of the theorem in the following chapter.
- The word "reconstrual" comes from Quine (1975), where he uses it to propose a notion of theoretical equivalence. We find his notion to be far too liberal, as discussed in Barrett and Halvorson (2016a).
- **Definitional equivalence** and **common definition extension** have been part of the logical folklore since the 1960s, and many results about them have been proven – see, e.g., Hodges (1993, §2.6), de Bouvére (1965), Kanger (1968), Pinter (1978), Pelletier and Urquhart (2003), Andréka et al. (2005), and Friedman and Visser (2014) for some results. To our knowledge, Glymour was the first philosopher of science to recognize the significance of these notions for discussions of theoretical equivalence. For an application of definitional equivalence in recent metaphysical debate, see McSweeney (2016a).
- For overviews of recent work on scientific reduction, see Scheibe (2013); Van Riel and van Gulick (2014); Love and Hüttemann (2016); Hudetz (2018b). Nagel's pioneering work on the topic can be found in Nagel (1935, 1961). For recent discussions of Nagel's view, see Dizadji-Bahmani et al. (2010); Sarkar (2015). We discuss semantic accounts of reduction in Chapter 6.

5 Syntactic Metalogic Redux

5.1 Many-Sorted Logic

We now turn to a generalization of first-order logic – a generalization that has proven to be surprisingly controversial. This generalization proceeds by noting that in ordinary first-order logic, it is implicitly assumed that all syntactic objects are compatible. For example, for any two terms s, t, it makes sense to write $s = t$; and for any relation symbol r, and terms $t_1, \ldots, t_n$, it makes sense to write $r(t_1, \ldots, t_n)$. However, that assumption is not obviously warranted. Instead, one might insist that syntactic objects, such as terms, come with a **type** or **sort**, and that there are sort-based rules about how these objects can be combined.

This generalization can provoke two responses that pull in completely opposite directions. On the one hand, one might think that many-sorted logic is stronger than single-sorted logic, and hence that its theoretical commitments outrun those of single-sorted logic. (The obvious analogy here is with second-order logic.) On the other hand, some philosophers, such as Quine (1963, 267–268), argue that many-sorted logic is reducible to single-sorted logic, and hence is dispensable. If we give pride of place to classical (single-sorted) first-order logic, then both of these responses would undermine our motivation to study many-sorted logic. However, the presuppositions of these two responses cannot both be correct – i.e., many-sorted logic cannot both exceed the resources of single-sorted logic and also be reducible to it. So which view is the right one?

The view we will advance here is that many-sorted logic is, in one clear sense, reducible to single-sorted logic, but that this reduction does not mean that many-sorted logic is dispensible. Before we take up this argument, we need to explain how many-sorted logic works.

DEFINITION 5.1.1 A many-sorted **signature** Σ is a set of sort symbols, predicate symbols, function symbols, and constant symbols. Σ must have at least one sort symbol. Each predicate symbol $p \in \Sigma$ has an **arity** $\sigma_1 \times \ldots \times \sigma_n$, where $\sigma_1, \ldots, \sigma_n \in \Sigma$ are (not necessarily distinct) sort symbols. Likewise, each function symbol $f \in \Sigma$ has an **arity** $\sigma_1 \times \ldots \times \sigma_n \to \sigma$, where $\sigma_1, \ldots, \sigma_n, \sigma \in \Sigma$ are again (not necessarily distinct) sort symbols. Lastly, each constant symbol $c \in \Sigma$ is assigned a sort $\sigma \in \Sigma$. In addition to the elements of Σ, we also have a stock of variables. We use the letters x, y, and z to denote these variables, adding subscripts when necessary. Each variable has a sort $\sigma \in \Sigma$.

NOTE 5.1.2 The symbol $\sigma_1 \times \cdots \times \sigma_n$ has no intrinsic meaning. To say that "p has arity $\sigma_1 \times \cdots \times \sigma_n$" is simply an abbreviated way of saying that p can be combined with n terms, whose sorts must respectively be $\sigma_1, \ldots, \sigma_n$.

A Σ-term can be thought of as a "naming expression" in the signature Σ. Each Σ-term has a sort $\sigma \in \Sigma$.

DEFINITION 5.1.3 The **Σ-terms** of sort σ are recursively defined as follows. Every variable of sort σ is a Σ-term of sort σ, and every constant symbol $c \in \Sigma$ of sort σ is also a Σ-term of sort σ. Furthermore, if $f \in \Sigma$ is a function symbol with arity $\sigma_1 \times \ldots \times \sigma_n \to \sigma$ and $t_1, \ldots, t_n$ are Σ-terms of sorts $\sigma_1, \ldots, \sigma_n$, then $f(t_1, \ldots, t_n)$ is a Σ-term of sort σ. We will use the notation $t(\vec{x})$ to denote a Σ-term in which all of the variables that appear in t are in the sequence $\vec{x} \equiv x_1, \ldots, x_n$, but we leave open the possibility that some of the x_i do not appear in the term t.

A **Σ-atom** is an expression either of the form $s(x_1, \ldots, x_n) = t(x_1, \ldots, x_n)$, where s and t are Σ-terms of the same sort $\sigma \in \Sigma$, or of the form $p(t_1, \ldots, t_n)$, where $t_1, \ldots, t_n$ are Σ-terms of sorts $\sigma_1, \ldots, \sigma_n$ and $p \in \Sigma$ is a predicate of arity $\sigma_1 \times \ldots \times \sigma_n$.

DEFINITION 5.1.4 The **Σ-formulas** are defined recursively as follows.

- Every Σ-atom is a Σ-formula.
- If ϕ is a Σ-formula, then $\neg\phi$ is a Σ-formula.
- If ϕ and ψ are Σ-formulas, then $\phi \to \psi$, $\phi \wedge \psi$, $\phi \vee \psi$, and $\phi \leftrightarrow \psi$ are Σ-formulas.
- If ϕ is a Σ-formula and x is a variable of sort $\sigma \in \Sigma$, then $\forall_\sigma x\phi$ and $\exists_\sigma x\phi$ are Σ-formulas.

In addition to the preceding formulas, we will use the notation $\exists_{\sigma=1} y\phi(x_1, \ldots, x_n, y)$ to abbreviate the formula

$$\exists_\sigma y(\phi(x_1, \ldots, x_n, y) \wedge \forall_\sigma z(\phi(x_1, \ldots, x_n, z) \to y = z)).$$

As before, the notation $\phi(\vec{x})$ will denote a Σ-formula ϕ in which all of the free variables appearing in ϕ are in the sequence $\vec{x} \equiv x_1, \ldots, x_n$, but we again leave open the possibility that some of the x_i do not appear as free variables in ϕ.

DEFINITION 5.1.5 A **Σ-sentence** is a Σ-formula that has no free variables.

We will not give an explicit listing of the derivation rules for many-sorted logic. Suffice it to say that they are direct generalizations of the derivation rules for single-sorted logic, provided that one observe all restrictions on syntactic compatibility. For example, in many sorted logic, we can infer $\forall x\phi(x)$ from $\phi(y)$ only if the variables x and y are of the same type. If they were not of the same type, then one of these two formulas would fail to be well-formed.

As a result of these restrictions, we need to exercise some caution about carrying over intuitions that we might have developed in using single-sorted logic. For example, in single-sorted logic, for any two terms s and t, we have a tautology

$$\vdash (s = t) \vee (s \neq t).$$

However, in many sorted logic, the expressions $s = t$ and $s \neq t$ are well-formed only when s and t are terms of the same sort. Thus, to the question "do s and t denote the same object?" many-sorted logic sometimes offers no answer.

One might be tempted, nonetheless, to think that if s and t are terms of different sorts, then we can just add $t \neq s$ as axiom. However, that suggestion can lead to disaster. For example, suppose that s denotes the number 0 and that t denotes the renowned actor David Hasselhoff. Because I accept Peano arithmetic, I assume that every natural number besides 0 is greater than 0. In other words, I assume that

$$\forall x(x \neq s \rightarrow (x > 0)),$$

where x is a variable ranging over natural numbers. If I now added $t \neq s$ to my total theory, then I would be committed to the claim that David Hasselhoff is greater than 0. These considerations show that we need to exercise caution when moving between many- and single-sorted frameworks.

Example 5.1.6 Let $\Sigma = \{\sigma_1, \sigma_2\}$, and let T be the empty theory in Σ. Note that both $\exists_{\sigma_1} x(x = x)$ and $\exists_{\sigma_2} y(y = y)$. This might seem like a strange consequence: T is the empty theory, and you might think that the empty theory should have no nontrivial consequences. But the combination of $\exists_{\sigma_1} x(x = x)$ and $\exists_{\sigma_2} y(y = y)$ seems like a nontrivial consequence, viz. that there are at least two things!

However, there is a mistake in our reasoning. Those two sentences together do not imply that there are at least two things. For there is no third quantifier $\exists$ such that $\exists v \exists w(v \neq w)$ is guaranteed to hold.

These considerations show that distinct sort symbols do not necessarily represent different kinds of things. Indeed, it is not generally valid to infer that there are $n + m$ objects from the fact that there are n objects of sort σ_1 and m objects of sort σ_2. ⌟

Example 5.1.7 Let $\Sigma = \{\sigma_1, \sigma_2, i\}$, where $i : \sigma_1 \rightarrow \sigma_2$. Let T be the theory that says that i is bijective; that is, i is injective:

$$(i(x) = i(y)) \rightarrow x = y,$$

and i is surjective:

$$\exists x(i(x) = z).$$

Then T defines a functional relation ϕ of sort $\sigma_2 \times \sigma_1$ by means of

$$\phi(z, x) \leftrightarrow (i(x) = z).$$

The function $j : \sigma_2 \rightarrow \sigma_1$ corresponding to ϕ is the inverse of i. ⌟

Example 5.1.8 The theory of categories can conveniently be formulated as a many-sorted theory. Let $\Sigma = \{O, A, d_0, d_1, i, \circ\}$, where O and A are sorts, $d_0 : A \rightarrow O$, $d_1 : A \rightarrow O$, $i : O \rightarrow A$, and $\circ$ is a relation of sort $A \times A \times A$. (The relation $\circ$ is used as the composition function on arrows – i.e., a partial function defined for compatible arrows.) We will leave it as an exercise for the reader to write down the axioms corresponding to the following ideas:

1. For each arrow f, $d_0 f$ is the domain object, and $d_1 f$ is the codomain object. Thus, we may write $f : d_0 f \to d_1 f$. More generally, we write $f : x \to y$ to indicate that $x = d_0 f$ and $y = d_1 f$. The function $\circ$ is defined on pairs of arrows where the first arrow's domain matches the second arrow's codomain.
2. The function $\circ$ is associative.
3. For each object x, $i(x) : x \to x$. Moreover, for any arrow f such that $d_1 f = x$, we have $i(x) \circ f = f$. And for any arrow g such that $d_0 g = x$, we have $g \circ i(x) = g$.

⌟

What can many-sorted logic do for us? In pure mathematics, it can certainly have pragmatic advantages to introduce sorts. For example, in axiomatizing category theory, it seems more intuitive to think about objects and arrows as different sorts of things, rather than introducing some predicate that is satisfied by objects but not by arrows. Similarly, in axiomatizing the theory of vector spaces, it is convenient to think of vectors and scalars as different sorts of things. Indeed, in this latter case, it's hard to imagine a mathematician investigating the question: "is c a scalar or a vector?" Instead, it seems that general words like "vector" and "scalar" function more like labels than they do as names of properties that mathematicians are interested in investigating.

But what about empirical theories? Could a many-sorted formulation of an empirical theory provide a more perspicuous representation of the structure of reality? Let's focus on a more specific question, that was central to twentieth-century philosophy of science: can the distinction between observable and unobservable be encoded into the syntax of a theory?

Suppose then that in formulating a theory T, we begin by introducing a sort symbol O for observable objects, and a sort symbol P for theoretical objects. Then, any relation symbol R must be explicitly sorted – i.e., each slot after R can be occupied only by terms of one particular sort. Similarly, formulas such as $t = t'$ and $t \neq t'$ are well-formed only if t and t' are terms of the same type. It should be clear now that this language does not have a predicate "is unobservable," nor does it have any well-formed expression corresponding to the sentence:

($*$) No theoretical entity (i.e., entity of type P) is an observable entity (i.e., entity of type O).

The grammatical malformity of ($*$) is sometimes brushed right over in criticisms of the syntactic view of theories (e.g., van Fraassen, 1980), and in criticisms of the observation–theory distinction (e.g., Dicken and Lipton, 2006).

5.2 Morita Extension and Equivalence

Glymour (1971) claims that definitional equivalence (see 4.6.15) is a necessary condition on the equivalence of scientific theories. However, there are several reasons to believe that this criterion is too strict.

First, it is frequently argued that many-sorted logic is reducible to single-sorted logic (see Schmidt, 1951; Manzano, 1996). What is actually shown in these arguments is that for any many-sorted theory T, a corresponding single-sorted theory T' can be constructed. But what is the relation between T and T'? Obviously, the two theories T and T' cannot be definitionally equivalent, since that criterion applies only to single-sorted theories. Therefore, to make sense of the claim that many-sorted logic can be reduced to single-sorted logic, we need a generalization of definitional equivalence.

Second, there are well-known examples of theories that could naturally be formulated either within a single-sorted framework or within a many-sorted framework – and we need a generalization of definitional equivalence to explain in what sense these two formulations are equivalent. For example, category theory can be formulated as a many-sorted theory, using both a sort of "objects" and a sort of "arrows" (Eilenberg and Mac Lane, 1942, 1945); and category theory can also be formulated as a single-sorted theory using only "arrows" (Mac Lane, 1948). (Freyd [1964, p. 5] and Mac Lane [1971 p. 9] also describe this alternate formulation.) These two formulations of category theory are in some sense equivalent, and we would like an account of this more general notion of equivalence.

Third, definitional equivalence is too restrictive even for single-sorted theories. For example, affine geometry can be formalized in a way that quantifies over points, or it can be formalized in a way that quantifies over lines (see Schwabhäuser et al., 1983). But saying that the point theory (T_p) and the line theory (T_ℓ) both are formulations of the same theory indicates again that T_p and T_ℓ are in some sense equivalent – although T_p and T_ℓ are *not* definitionally equivalent. Indeed, the smallest model of T_p has five elements, which we can think of as the four corners of a square and its center point. On the other hand, the smallest model of T_ℓ has six elements. But if T_p and T_ℓ were definitionally equivalent, then every model M of T_ℓ would be the reduct of an expansion of a model M' of T_p (de Bouvére, 1965). In particular, we would have $|M| = |M'|$, which entails that T_ℓ has a model of cardinality five – a contradiction. Therefore, T_p and T_ℓ are not definitionally equivalent.

Finally, even if we ignore the complications mentioned previously, and even if we assume that each many-sorted theory T can be replaced by a single-sorted variant T' (by the standard procedure of unifying sorts), definitional equivalence is still inadequate. Consider the following example.

Example 5.2.1 Let T_1 be the objects-and-arrows formulation of category theory, and let T_2 be the arrows-only formulation of category theory. Intuitively, T_1 and T_2 are equivalent theories; but their single-sorted versions T_1' and T_2' are not definitionally equivalent. Indeed, $T_2' = T_2$, since T_2 is single sorted. However, T_1' has a single sort that includes both objects and arrows. Thus, while T_2' has a model with one element (i.e., the category with a single arrow), T_1' has no models with one element (since every model of T_1' has at least one object and at least one arrow). Therefore, T_1' and T_2' are not definitionally equivalent. ⌟

These examples all show that definitional equivalence does not capture the sense in which some theories are equivalent. If one wants to capture this sense, one needs a more general criterion for theoretical equivalence than definitional equivalence. Our aim here is to introduce one such criterion. We will call it **Morita equivalence**. This criterion is a natural generalization of definitional equivalence. In fact, Morita equivalence is essentially the same as definitional equivalence, except that it allows one to define new sort symbols in addition to new predicate symbols, function symbols, and constant symbols. In order to state the criterion precisely, we again need to do some work. We begin by defining the concept of a Morita extension. In Chapter 7, we will show the sense in which Morita equivalence is a natural generalization of definitional equivalence.

As we did for predicates, functions, and constants, we need to say how to define new sorts. Let $\Sigma \subseteq \Sigma^+$ be signatures and consider a sort symbol $\sigma \in \Sigma^+ \backslash \Sigma$. One can define the sort σ as a product sort, a coproduct sort, a subsort, or a quotient sort. In each case, one defines σ using old sorts in Σ and new function symbols in $\Sigma^+ \backslash \Sigma$. These new function symbols specify how the new sort σ is related to the old sorts in Σ. We describe these four cases in detail.

product sort In order to define σ as a product sort, one needs two function symbols $\pi_1, \pi_2 \in \Sigma^+ \backslash \Sigma$ with π_1 of arity $\sigma \to \sigma_1$, π_2 of arity $\sigma \to \sigma_2$, and $\sigma_1, \sigma_2 \in \Sigma$. The function symbols π_1 and π_2 serve as the "canonical projections" associated with the product sort σ. A sort definition of the symbols σ, π_1, and π_2 as a **product sort** in terms of Σ is a Σ^+-sentence of the form

$$\forall_{\sigma_1} x \forall_{\sigma_2} y \exists_{\sigma=1} z (\pi_1(z) = x \wedge \pi_2(z) = y).$$

One should think of a product sort σ as the sort whose elements are ordered pairs, where the first element of each pair is of sort σ_1 and the second is of sort σ_2.

coproduct sort One can also define σ as a coproduct sort. One again needs two function symbols $\rho_1, \rho_2 \in \Sigma^+ \backslash \Sigma$ with ρ_1 of arity $\sigma_1 \to \sigma$, ρ_2 of arity $\sigma_2 \to \sigma$, and $\sigma_1, \sigma_2 \in \Sigma$. The function symbols ρ_1 and ρ_2 are the "canonical injections" associated with the coproduct sort σ. A sort definition of the symbols σ, ρ_1, and ρ_2 as a **coproduct sort** in terms of Σ is a Σ^+-sentence of the form

$$\forall_{\sigma} z \big(\exists_{\sigma_1=1} x (\rho_1(x) = z) \vee \exists_{\sigma_2=1} y (\rho_2(y) = z)\big) \wedge \forall_{\sigma_1} x \forall_{\sigma_2} y \neg \big(\rho_1(x) = \rho_2(y)\big)$$

One should think of a coproduct sort σ as the disjoint union of the elements of sorts σ_1 and σ_2.

When defining a new sort σ as a product sort or a coproduct sort, one uses two sort symbols in Σ and two function symbols in $\Sigma^+ \backslash \Sigma$. The next two ways of defining a new sort σ only require one sort symbol in Σ and one function symbol in $\Sigma^+ \backslash \Sigma$.

subsort In order to define σ as a subsort, one needs a function symbol $i \in \Sigma^+ \backslash \Sigma$ of arity $\sigma \to \sigma_1$ with $\sigma_1 \in \Sigma$. The function symbol i is the "canonical inclusion" associated with the subsort σ. A sort definition of the symbols σ and i as a **subsort** in terms of Σ is a Σ^+-sentence of the form

$$\forall_{\sigma_1} x\big(\phi(x) \leftrightarrow \exists_\sigma z(i(z) = x)\big) \wedge \forall_\sigma z_1 \forall_\sigma z_2\big(i(z_1) = i(z_2) \rightarrow z_1 = z_2\big), \quad (5.1)$$

where $\phi(x)$ is a Σ-formula. One can think of the subsort σ as consisting of "the elements of sort σ_1 that are ϕ." The sentence (5.1) entails the Σ-sentence $\exists_{\sigma_1} x\phi(x)$. As before, we will call this Σ-sentence the **admissibility condition** for the definition (5.1).

quotient sort Lastly, in order to define σ as a quotient sort, one needs a function symbol $\epsilon \in \Sigma^+\backslash\Sigma$ of arity $\sigma_1 \rightarrow \sigma$ with $\sigma_1 \in \Sigma$. A sort definition of the symbols σ and ϵ as a **quotient sort** in terms of Σ is a Σ^+-sentence of the form

$$\forall_{\sigma_1} x_1 \forall_{\sigma_1} x_2\big(\epsilon(x_1) = \epsilon(x_2) \leftrightarrow \phi(x_1, x_2)\big) \wedge \forall_\sigma z \exists_{\sigma_1} x(\epsilon(x) = z), \quad (5.2)$$

where $\phi(x_1, x_2)$ is a Σ-formula. This sentence defines σ as a quotient sort that is obtained by "quotienting out" the sort σ_1 with respect to the formula $\phi(x_1, x_2)$. The sort σ should be thought of as the set of "equivalence classes of elements of σ_1 with respect to the relation $\phi(x_1, x_2)$." The function symbol ϵ is the "canonical projection" that maps an element to its equivalence class. One can verify that the sentence (5.2) implies that $\phi(x_1, x_2)$ is an equivalence relation. In particular, it entails the following Σ-sentences:

$$\forall_{\sigma_1} x(\phi(x, x))$$
$$\forall_{\sigma_1} x_1 \forall_{\sigma_1} x_2(\phi(x_1, x_2) \rightarrow \phi(x_2, x_1))$$
$$\forall_{\sigma_1} x_1 \forall_{\sigma_1} x_2 \forall_{\sigma_1} x_3\big((\phi(x_1, x_2) \wedge \phi(x_2, x_3)) \rightarrow \phi(x_1, x_3)\big).$$

These Σ-sentences are the **admissibility conditions** for the definition (5.2).

Now that we have presented the four ways of defining new sort symbols, we can define the concept of a Morita extension. A Morita extension is a natural generalization of a definitional extension. The only difference is that now one is allowed to define new sort symbols.

DEFINITION 5.2.2 Let $\Sigma \subset \Sigma^+$ be signatures and T a Σ-theory. A **Morita extension** of T to the signature Σ^+ is a Σ^+-theory

$$T^+ = T \cup \{\delta_s : s \in \Sigma^+\backslash\Sigma\}$$

that satisfies the following conditions. First, for each symbol $s \in \Sigma^+\backslash\Sigma$, the sentence δ_s is an explicit definition of s in terms of Σ. Second, if $\sigma \in \Sigma^+\backslash\Sigma$ is a sort symbol and $f \in \Sigma^+\backslash\Sigma$ is a function symbol that is used in the sort definition of σ, then $\delta_f = \delta_\sigma$. (For example, if σ is defined as a product sort with projections π_1 and π_2, then $\delta_\sigma = \delta_{\pi_1} = \delta_{\pi_2}$.) And third, if α_s is an admissibility condition for a definition δ_s, then $T \vdash \alpha_s$.

Note that unlike a definitional extension of a theory, a Morita extension can have more sort symbols than the original theory.[1] The following is a particularly simple example of a Morita extension.

[1] Also note that if T^+ is a Morita extension of T to Σ^+, then there are restrictions on the arities of predicates, functions, and constants in $\Sigma^+\backslash\Sigma$. If $p \in \Sigma^+\backslash\Sigma$ is a predicate symbol of arity $\sigma_1 \times \ldots \times \sigma_n$, we immediately see that $\sigma_1, \ldots, \sigma_n \in \Sigma$. Taking a single Morita extension does not allow one to define predicate symbols that apply to sorts that are not in Σ. One must take multiple Morita extensions to do

---

Example 5.2.3 Let $\Sigma = \{\sigma, p\}$ and $\Sigma^+ = \{\sigma, \sigma^+, p, i\}$ be a signatures with σ and σ^+ sort symbols, p a predicate symbol of arity σ, and i a function symbol of arity $\sigma^+ \to \sigma$. Consider the Σ-theory $T = \{\exists_\sigma x p(x)\}$. The following Σ^+-sentence defines the sort symbol σ^+ as the subsort consisting of "the elements that are p."

$$\forall_\sigma x\big(p(x) \leftrightarrow \exists_{\sigma^+} z(i(z) = x)\big) \wedge \forall_{\sigma^+} z_1 \forall_{\sigma^+} z_2\big(i(z_1) = i(z_2) \to z_1 = z_2\big). \qquad (\delta_{\sigma^+})$$

The Σ^+-theory $T^+ = T \cup \{\delta_{\sigma^+}\}$ is a Morita extension of T to the signature Σ^+. The theory T^+ adds to the theory T the ability to quantify over the set of "things that are p." ⌟

DEFINITION 5.2.4 Let T_1 be a Σ_1-theory and T_2 a Σ_2-theory. T_1 and T_2 are **Morita equivalent** if there are theories $T_1^1, \ldots, T_1^n$ and $T_2^1, \ldots, T_2^m$ that satisfy the following three conditions:

- Each theory T_1^{i+1} is a Morita extension of T_1^i,
- Each theory T_2^{i+1} is a Morita extension of T_2^i,
- T_1^n and T_2^m are logically equivalent Σ-theories with $\Sigma_1 \cup \Sigma_2 \subseteq \Sigma$.

Two theories are Morita equivalent if they have a "common Morita extension." The situation can be pictured as follows, where each arrow in the figure indicates a Morita extension.

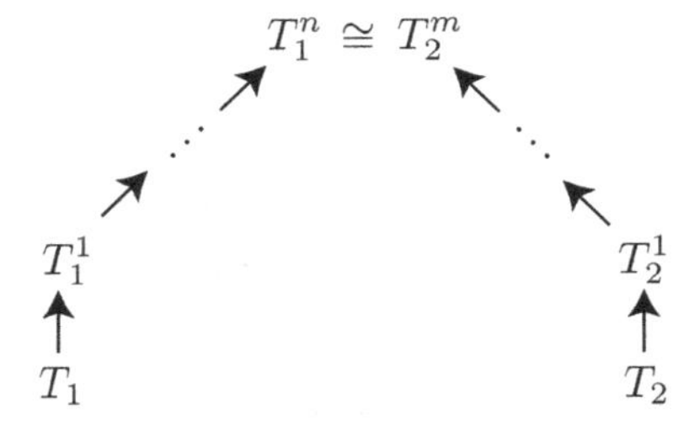

At first glance, Morita equivalence might strike one as different from definitional equivalence in an important way. To show that theories are Morita equivalent, one is allowed to take any finite number of Morita extensions of the theories. On the other hand, to show that two theories are definitionally equivalent, it appears that one is only allowed to take *one* definitional extension of each theory. One might worry that Morita equivalence is therefore not perfectly analogous to definitional equivalence.

Fortunately, this is not the case. By Theorem 4.6.17, if T' is a definitional extension of T, then T and T' are intertranslatable. Clearly intertranslatability is a transitive relation, and in Theorem 6.6.21, we will see that if two theories are intertranslatable, then they are definitionally equivalent. Therefore, if theories $T_1, \ldots, T_n$ are such that each T_{i+1} is a definitional extension of T_i, then T_n is in fact a definitional extension of T_1. (One can easily verify that this is not true of Morita extensions.) To show that two theories are

this. Likewise, any constant symbol $c \in \Sigma^+ \backslash \Sigma$ must be of sort $\sigma \in \Sigma$. And a function symbol $f \in \Sigma^+ \backslash \Sigma$ must either have arity $\sigma_1 \times \ldots \times \sigma_n \to \sigma$ with $\sigma_1, \ldots, \sigma_n, \sigma \in \Sigma$, or f must be one of the function symbols that appears in the definition of a new sort symbol $\sigma \in \Sigma^+ \backslash \Sigma$.

definitionally equivalent, therefore, one actually *is* allowed to take any finite number of definitional extensions of each theory.

If two theories are definitionally equivalent, then they are trivially Morita equivalent. Unlike definitional equivalence, however, Morita equivalence is capable of capturing a sense in which theories with different sort symbols are equivalent. The following example demonstrates that Morita equivalence is a more liberal criterion for theoretical equivalence.

Example 5.2.5 Let $\Sigma_1 = \{\sigma_1, p, q\}$ and $\Sigma_2 = \{\sigma_2, \sigma_3\}$ be signatures with σ_i sort symbols, and p and q predicate symbols of arity σ_1. Let T_1 be the Σ_1-theory that says p and q are nonempty, mutually exclusive, and exhaustive. Let T_2 be the empty theory in Σ_2. Since the signatures Σ_1 and Σ_2 have different sort symbols, T_1 and T_2 can't possibly be definitionally equivalent. Nonetheless, it's easy to see that T_1 and T_2 are Morita equivalent. Let $\Sigma = \Sigma_1 \cup \Sigma_2 \cup \{i_2, i_3\}$ be a signature with i_2 and i_3 function symbols of arity $\sigma_2 \to \sigma_1$ and $\sigma_3 \to \sigma_1$. Consider the following Σ-sentences.

$$\forall_{\sigma_1} x\big(p(x) \leftrightarrow \exists_{\sigma_2} y(i_2(y) = x)\big) \wedge \forall_{\sigma_2} y_1 \forall_{\sigma_2} y_2 \big(i_2(y_1) = i_2(y_2) \to y_1 = y_2\big) \qquad (\delta_{\sigma_2})$$

$$\forall_{\sigma_1} x\big(q(x) \leftrightarrow \exists_{\sigma_3} z(i_3(z) = x)\big) \wedge \forall_{\sigma_3} z_1 \forall_{\sigma_3} z_2 \big(i_3(z_1) = i_3(z_2) \to z_1 = z_2\big) \qquad (\delta_{\sigma_3})$$

$$\forall_{\sigma_1} x\big(\exists_{\sigma_2 = 1} y(i_2(y) = x) \vee \exists_{\sigma_3 = 1} z(i_3(z) = x)\big) \wedge \forall_{\sigma_2} y \forall_{\sigma_3} z \neg\big(i_2(y) = i_3(z)\big) \qquad (\delta_{\sigma_1})$$

$$\forall_{\sigma_1} x\big(p(x) \leftrightarrow \exists_{\sigma_2} y(i_2(y) = x)\big) \qquad (\delta_p)$$

$$\forall_{\sigma_1} x\big(q(x) \leftrightarrow \exists_{\sigma_3} z(i_3(z) = x)\big) \qquad (\delta_q)$$

The Σ-theory $T_1^1 = T_1 \cup \{\delta_{\sigma_2}, \delta_{\sigma_3}\}$ is a Morita extension of T_1 to the signature Σ. It defines σ_2 to be the subsort of "elements that are p" and σ_3 to be the subsort of "elements that are q." The theory $T_2^1 = T_2 \cup \{\delta_{\sigma_1}\}$ is a Morita extension of T_2 to the signature $\Sigma_2 \cup \{\sigma_1, i_2, i_3\}$. It defines σ_1 to be the coproduct sort of σ_2 and σ_3. Lastly, the Σ-theory $T_2^2 = T_2^1 \cup \{\delta_p, \delta_q\}$ is a Morita extension of T_2^1 to the signature Σ. It defines the predicates p and q to apply to elements in the "images" of i_2 and i_3, respectively. One can verify that T_1^1 and T_2^2 are logically equivalent, so T_1 and T_2 are Morita equivalent. ⌟

5.3 Quine on the Dispensability of Many-Sorted Logic

The notion of Morita equivalence bears directly on several central disputes in twentieth-century analytic philosophy. For example, in his debate with Carnap and the logical positivists, Quine claims that many-sorted logic is dispensable. Morita equivalence shows a precise sense in which Quine is right about that. Similarly, to motivate the rejection of metaphysical realism, Putnam claims that a geometric theory with points as primitives is equivalent to a theory with lines as primitives. (See, for example, Putnam, 1977,

489–491; Putnam, 1992, 109, 115–120; and Putnam, 2001.) Morita equivalence also shows a precise sense in which Putnam is right about that. We take up Quine's argument in the remainder of this section. We take up Putnam's argument in Section 7.4, after we have developed some semantic tools.

One proves Quine's claim by explicitly constructing a "corresponding" single-sorted theory $\widehat{T}$ for every many-sorted theory T. The basic idea behind the construction is intuitive. The theory $\widehat{T}$ simply replaces the sort symbols that the theory T uses with predicate symbols. This construction recalls the proof that every theory is definitionally equivalent to a theory that uses only predicate symbols (Barrett and Halvorson, 2016a, Prop. 2). Quine (1937, 1938, 1956, 1963) suggests the basic idea behind our proof, as do Burgess (2005, 12) and Manzano (1996, 221–222). However, the theorem that we prove here is more general than Quine's results because we make no assumption about what the theory T is, whereas Quine only considers Russell's theory of types and NBG set theory.

Let Σ be a signature with finitely many sort symbols $\sigma_1, \ldots, \sigma_n$. We begin by constructing a corresponding signature $\widehat{\Sigma}$ that contains one sort symbol σ. The symbols in $\widehat{\Sigma}$ are defined as follows. For every sort symbol $\sigma_j \in \Sigma$, we let q_{σ_j} be a predicate symbol of sort σ. For every predicate symbol $p \in \Sigma$ of arity $\sigma_{j_1} \times \ldots \times \sigma_{j_m}$, we let q_p be a predicate symbol of arity σ^m (the m-fold product of σ). Likewise, for every function symbol $f \in \Sigma$ of arity $\sigma_{j_1} \times \ldots \times \sigma_{j_m} \to \sigma_j$, we let q_f be a predicate symbol of arity σ^{m+1}. And, lastly, for every constant symbol $c \in \Sigma$ we let d_c be a constant symbol of sort σ. The single-sorted signature $\widehat{\Sigma}$ corresponding to Σ is then defined to be

$$\widehat{\Sigma} = \{\sigma\} \cup \{q_{\sigma_1}, \ldots, q_{\sigma_n}\} \cup \{q_p : p \in \Sigma\} \cup \{q_f : f \in \Sigma\} \cup \{d_c : c \in \Sigma\}.$$

We can now describe a method of "translating" Σ-theories into $\widehat{\Sigma}$-theories. Let T be an arbitrary Σ-theory. We define a corresponding $\widehat{\Sigma}$-theory $\widehat{T}$ and then show that $\widehat{T}$ is Morita equivalent to T.

We begin by translating the axioms of T into the signature $\widehat{\Sigma}$. This will take two steps. First, we describe a way to translate the Σ-terms into $\widehat{\Sigma}$-formulas. Given a Σ-term $t(x_1, \ldots, x_n)$, we define the $\widehat{\Sigma}$-formula $\widehat{\psi_t}(y_1, \ldots, y_n, y)$ recursively as follows.

- If $t(x_1, \ldots, x_n)$ is the variable x_i, then $\widehat{\psi_t}$ is the $\widehat{\Sigma}$-formula $y_i = y$.
- If $t(x_1, \ldots, x_n)$ is the constant c, then $\widehat{\psi_t}$ is the $\widehat{\Sigma}$-formula $d_c = y$.
- Suppose that $t(x_1, \ldots, x_n)$ is the term $f(t_1(x_1, \ldots, x_n), \ldots, t_k(x_1, \ldots, x_n))$ and that each of the $\widehat{\Sigma}$-formulas $\widehat{\psi_{t_i}}(y_1, \ldots, y_n, y)$ have been defined. Then $\widehat{\psi_t}(y_1, \ldots, y_n, y)$ is the $\widehat{\Sigma}$-formula

$$\exists_\sigma z_1 \ldots \exists_\sigma z_k \big(\widehat{\psi_{t_1}}(y_1, \ldots, y_n, z_1) \wedge \ldots \wedge \widehat{\psi_{t_k}}(y_1, \ldots, y_n, z_k) \wedge q_f(z_1, \ldots, z_k, y)\big).$$

One can think of the formula $\psi_t(y_1, \ldots, y_n, y)$ as the translation of the expression "$t(x_1, \ldots, x_n) = x$" into the signature $\widehat{\Sigma}$.

Second, we use this map from Σ-terms to $\widehat{\Sigma}$-formulas to describe a map from Σ-formulas to $\widehat{\Sigma}$-formulas. Given a Σ-formula $\psi(x_1, \ldots, x_n)$, we define the $\widehat{\Sigma}$-formula $\widehat{\psi}(y_1, \ldots, y_n)$ recursively as follows.

- If $\psi(x_1,\dots,x_n)$ is $t(x_1,\dots,x_n) = s(x_1,\dots,x_n)$, where s and t are Σ-terms of sort σ_i, then $\widehat{\psi}(y_1,\dots,y_n)$ is the $\widehat{\Sigma}$-formula

$$\exists_\sigma z\big(\widehat{\psi_t}(y_1,\dots,y_n,z) \wedge \widehat{\psi_s}(y_1,\dots,y_n,z) \wedge q_{\sigma_i}(z)\big).$$

- If $\psi(x_1,\dots,x_n)$ is $p(t_1(x_1,\dots,x_n),\dots,t_k(x_1,\dots,x_n))$, where $p \in \Sigma$ is a predicate symbol, then $\widehat{\psi}(y_1,\dots,y_n)$ is the $\widehat{\Sigma}$-formula

$$\exists_\sigma z_1 \dots \exists_\sigma z_k\big(\widehat{\psi_{t_1}}(y_1,\dots,y_n,z_1) \wedge \dots \wedge \widehat{\psi_{t_k}}(y_1,\dots,y_n,z_k) \wedge q_p(z_1,\dots,z_k)\big).$$

- This definition extends to all Σ-formulas in the standard way. We define the $\widehat{\Sigma}$-formulas $\widehat{\neg\psi} := \neg\widehat{\psi}$, $\widehat{\psi_1 \wedge \psi_2} := \widehat{\psi_1} \wedge \widehat{\psi_2}$, $\widehat{\psi_1 \vee \psi_2} := \widehat{\psi_1} \vee \widehat{\psi_2}$, and $\widehat{\psi_1 \to \psi_2} := \widehat{\psi_1} \to \widehat{\psi_2}$. Furthermore, if $\psi(x_1,\dots,x_n,x)$ is a Σ-formula, then we define both of the following:

$$\widehat{\forall_{\sigma_i} x\psi} := \forall_\sigma y(q_{\sigma_i}(y) \to \widehat{\psi}(y_1,\dots,y_n,y))$$
$$\widehat{\exists_{\sigma_i} x\psi} := \exists_\sigma y(q_{\sigma_i}(y) \wedge \widehat{\psi}(y_1,\dots,y_n,y)).$$

One should think of the formula $\widehat{\psi}$ as the translation of the Σ-formula ψ into the signature $\widehat{\Sigma}$.

This allows us to consider the translations $\widehat{\alpha}$ of the axioms $\alpha \in T$. The single-sorted theory $\widehat{T}$ will have the $\widehat{\Sigma}$-sentences $\widehat{\alpha}$ as some of its axioms. But $\widehat{T}$ will have more axioms than just the sentences $\widehat{\alpha}$. It will also have some **auxiliary axioms**. These auxiliary axioms will guarantee that the symbols in $\widehat{\Sigma}$ "behave like" their counterparts in Σ. We define auxiliary axioms for the predicate symbols $q_{\sigma_1},\dots,q_{\sigma_n} \in \widehat{\Sigma}$, $q_p \in \widehat{\Sigma}$, and $q_f \in \widehat{\Sigma}$, and for the constant symbols $d_c \in \widehat{\Sigma}$. We discuss each of these four cases in detail.

We first define auxiliary axioms to guarantee that the symbols $q_{\sigma_1},\dots,q_{\sigma_n}$ behave like sort symbols. The $\widehat{\Sigma}$-sentence ϕ is defined to be $\forall_\sigma y(q_{\sigma_1}(y) \vee \dots \vee q_{\sigma_n}(y))$.[2] Furthermore, for each sort symbol $\sigma_j \in \Sigma$ we define the $\widehat{\Sigma}$-sentence ϕ_{σ_j} to be

$$\exists_\sigma y(q_{\sigma_j}(y)) \wedge \forall_\sigma y\big(q_{\sigma_j}(y) \to (\neg q_{\sigma_1}(y) \wedge \dots \wedge \neg q_{\sigma_{j-1}}(y) \\ \wedge \neg q_{\sigma_{j+1}}(y) \wedge \dots \wedge \neg q_{\sigma_n}(y))\big).$$

One can think of the sentences $\phi_{\sigma_1},\dots,\phi_{\sigma_n}$, and ϕ as saying that "everything is of some sort, nothing is of more than one sort, and every sort is nonempty."

Next we define auxiliary axioms to guarantee that the symbols q_p, q_f, and d_c behave like their counterparts p, f, and c in Σ. For each predicate symbol $p \in \Sigma$ of arity $\sigma_{j_1} \times \dots \times \sigma_{j_m}$, we define the $\widehat{\Sigma}$-sentence ϕ_p to be

$$\forall_\sigma y_1 \dots \forall_\sigma y_m \Big(q_p(y_1,\dots,y_m) \to \Big(q_{\sigma_{j_1}}(y_1) \wedge \dots \wedge q_{\sigma_{jm}}(y_m)\Big)\Big).$$

This sentence restricts the extension of q_p to the subdomain of n-tuples satisfying $q_{\sigma_{j_1}},\dots,q_{\sigma_{jm}}$, guaranteeing that the predicate q_p has "the appropriate arity." Consider, for example, the case of a unary predicate p of sort σ_i. In that case, ϕ_p says that

[2] Note that if there were infinitely many sort symbols in Σ, then we could not define the $\widehat{\Sigma}$-sentence ϕ in this way.

$$\forall_\sigma y(q_p(y) \to q_{\sigma_i}(y)),$$

which means that nothing outside the subdomain q_{σ_i} satisfies q_p. Note, however, that here we have made a conventional choice. We could just as well have stipulated that q_p applies to *everything* outside of the subdomain q_{σ_i}. All that matters here is that q_p is trivial (either trivially true or trivially false) except on the subdomain of objects satisfying q_{σ_i}.

For each function symbol $f \in \Sigma$ of arity $\sigma_{j_1} \times \ldots \times \sigma_{j_m} \to \sigma_j$, we define the $\widehat{\Sigma}$-sentence ϕ_f to be the conjunction

$$\begin{aligned}&\forall_\sigma y_1 \ldots \forall_\sigma y_m \forall_\sigma y\big(q_f(y_1, \ldots, y_m, y) \to (q_{\sigma_{j_1}}(y_1) \wedge \ldots \wedge q_{\sigma_{jm}}(y_m) \wedge q_{\sigma_j}(y))\big)\\ &\wedge \forall_\sigma y_1 \ldots \forall_\sigma y_m \big((q_{\sigma_{j_1}}(y_1) \wedge \ldots \wedge q_{\sigma_{jm}}(y_m)) \to \exists_{\sigma=1} y(q_f(y_1, \ldots, y_m, y))\big).\end{aligned}$$

The first conjunct guarantees that the symbol q_f has "the appropriate arity," and the second conjunct guarantees that q_f behaves like a function. Lastly, if $c \in \Sigma$ is a constant symbol of arity σ_j, then we define the $\widehat{\Sigma}$-sentence ϕ_c to be $q_{\sigma_j}(d_c)$. This sentence guarantees that the constant symbol d_c also has "the appropriate arity."

We now have the resources to define a $\widehat{\Sigma}$-theory $\widehat{T}$ that is Morita equivalent to T:

$$\begin{aligned}\widehat{T} = \{\widehat{\alpha} : \alpha \in T\} &\cup \{\phi, \phi_{\sigma_1}, \ldots, \phi_{\sigma_n}\} \cup \{\phi_p : p \in \Sigma\}\\ &\cup \{\phi_f : f \in \Sigma\} \cup \{\phi_c : c \in \Sigma\}.\end{aligned}$$

The theory $\widehat{T}$ has two kinds of axioms, the translated axioms of T and the auxiliary axioms. These axioms allow $\widehat{T}$ to imitate the theory T in the signature $\widehat{\Sigma}$. Indeed, one can prove the following result.

THEOREM 5.3.1 (Barrett) *The theories T and $\widehat{T}$ are Morita equivalent.*

The proof of Theorem 5.3.1 requires some work, but the idea behind it is simple. The theory T needs to define symbols in $\widehat{\Sigma}$. It defines the sort symbol σ as a "universal sort" by taking the coproduct of the sorts $\sigma_1, \ldots, \sigma_n \in \Sigma$. The theory T then defines the symbols q_p, q_f, and d_c in $\widehat{\Sigma}$ simply by using the corresponding symbols p, f, and c in Σ. Likewise, the theory $\widehat{T}$ needs to define the symbols in Σ. It defines the sort symbol σ_j as the subsort of "things that are q_{σ_j}" for each $j = 1, \ldots, n$. And $\widehat{T}$ defines the symbols p, f, and c again by using the corresponding symbols q_p, q_f, and d_c.

We now proceed to the gory details. We prove a special case of the result for convenience. We will assume that Σ has only three sort symbols $\sigma_1, \sigma_2, \sigma_3$ and that Σ does not contain function or constant symbols. A perfectly analogous (though more tedious) proof goes through in the general case.

We prove the result by explicitly constructing a "common Morita extension" $T_4 \cong \widehat{T}_4$ of T and $\widehat{T}$ to the following signature:

$$\Sigma^+ = \Sigma \cup \widehat{\Sigma} \cup \{\sigma_{12}\} \cup \{\rho_1, \rho_2, \rho_{12}, \rho_3\} \cup \{i_1, i_2, i_3\}.$$

The symbol $\sigma_{12} \in \Sigma^+$ is a sort symbol. The symbols denoted by subscripted ρ are function symbols. Their arities are expressed in the following figure.

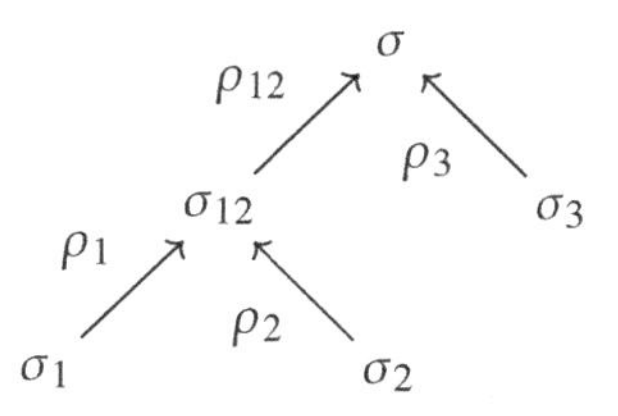

The symbols i_1, i_2, and i_3 are function symbols with arity $\sigma_1 \to \sigma$, $\sigma_2 \to \sigma$, and $\sigma_3 \to \sigma$, respectively.

We now turn to the proof.

Proof of Theorem 5.3.1 The following figure illustrates how our proof will be organized.

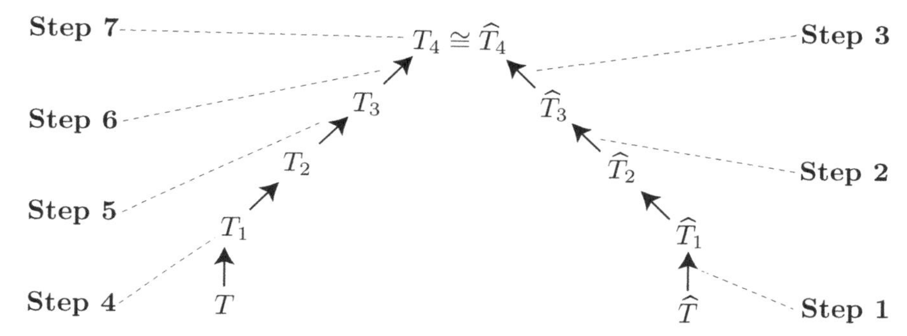

Steps 1–3 define the theories $\widehat{T}_1, \ldots, \widehat{T}_4$, Steps 4–6 define $T_1, \ldots, T_4$, and Step 7 shows that T_4 and $\widehat{T}_4$ are logically equivalent.

Step 1. We begin by defining the theory $\widehat{T}_1$. For each sort $\sigma_j \in \Sigma$ we consider the following sentence.

$$\begin{aligned} &\forall_\sigma y\big(q_{\sigma_j}(y) \leftrightarrow \exists_{\sigma_j} x(i_j(x) = y)\big) \\ &\quad \wedge \forall_{\sigma_j} x_1 \forall_{\sigma_j} x_2 (i_j(x_1) = i_j(x_2) \to x_1 = x_2) \end{aligned} \qquad (\theta_{\sigma_j})$$

The sentence θ_{σ_j} defines the symbols σ_j and i_j as the subsort of "things that are q_{σ_j}." The auxiliary axioms ϕ_{σ_j} of $\widehat{T}$ guarantee that the admissibility conditions for these definitions are satisfied. The theory $\widehat{T}_1 = \widehat{T} \cup \{\theta_{\sigma_1}, \theta_{\sigma_2}, \theta_{\sigma_3}\}$ is therefore a Morita extension of $\widehat{T}$ to the signature $\widehat{\Sigma} \cup \{\sigma_1, \sigma_2, \sigma_3, i_1, i_2, i_3\}$.

Step 2. We now define the theories $\widehat{T}_2$ and $\widehat{T}_3$. Let $\theta_{\sigma_{12}}$ be a sentence that defines the symbols $\sigma_{12}, \rho_1, \rho_2$ as a coproduct sort. The theory $\widehat{T}_2 = \widehat{T}_1 \cup \{\theta_{\sigma_{12}}\}$ is clearly a Morita extension of $\widehat{T}_1$. We have yet to define the function symbols ρ_{12} and ρ_3. The following two sentences define these symbols.

$$\forall_{\sigma_3} x \forall_\sigma y(\rho_3(x) = y \leftrightarrow i_3(x) = y) \qquad (\theta_{\rho_3})$$
$$\forall_{\sigma_{12}} x \forall_\sigma y(\rho_{12}(x) = y \leftrightarrow \psi(x, y)) \qquad (\theta_{\rho_{12}})$$

The sentence θ_{ρ_3} simply defines ρ_3 to be equal to the function i_3. For the sentence $\theta_{\rho_{12}}$, we define the formula $\psi(x, y)$ to be

$$\exists_{\sigma_1} z_1\big(\rho_1(z_1) = x \wedge i_1(z_1) = y\big) \vee \exists_{\sigma_2} z_2\big(\rho_2(z_2) = x \wedge i_2(z_2) = y\big).$$

We should take a moment here to understand the definition $\theta_{\rho_{12}}$. We want to define what the function ρ_{12} does to an element a of sort σ_{12}. Since the sort σ_{12} is the coproduct of the sorts σ_1 and σ_2, the element a must "actually be" of one of the sorts σ_1 or σ_2. (The disjuncts in the formula $\psi(x, y)$ correspond to these possibilities.) The definition $\theta_{\rho_{12}}$ stipulates that if a is "actually" of sort σ_j, then the value of ρ_{12} at a is the same as the value of i_j at a. One can verify that $\widehat{T_2}$ satisfies the admissibility conditions for θ_{ρ_3} and $\theta_{\rho_{12}}$, so the theory $\widehat{T_3} = \widehat{T_2} \cup \{\theta_{\rho_3}, \theta_{\rho_{12}}\}$ is a Morita extension of $\widehat{T_2}$ to the signature

$$\widehat{\Sigma} \cup \{\sigma_1, \sigma_2, \sigma_3, \sigma_{12}, i_1, i_2, i_3, \rho_1, \rho_2, \rho_3, \rho_{12}\}.$$

Step 3. We now describe the Σ^+-theory $\widehat{T_4}$. This theory defines the predicates in the signature Σ. Let $p \in \Sigma$ be a predicate symbol of arity $\sigma_{j_1} \times \ldots \times \sigma_{j_m}$. We consider the following sentence.

$$\forall_{\sigma_{j_1}} x_1 \ldots \forall_{\sigma_{jm}} x_m \left(p(x_1, \ldots, x_m) \leftrightarrow q_p(i_{j_1}(x_1), \ldots, i_{jm}(x_m))\right). \qquad (\theta_p)$$

The theory $\widehat{T_4} = \widehat{T_3} \cup \{\theta_p : p \in \Sigma\}$ is therefore a Morita extension of $\widehat{T_3}$ to the signature Σ^+.

Step 4. We turn to the left-hand side of our organizational figure and define the theories T_1 and T_2. We proceed in an analogous manner to the first part of Step 2. The theory $T_1 = T \cup \{\theta_{\sigma_{12}}\}$ is a Morita extension of T to the signature $\Sigma \cup \{\sigma_{12}, \rho_1, \rho_2\}$. Now let θ_σ be the sentence that defines the symbols $\sigma, \rho_{12}, \rho_3$ as a coproduct sort. The theory $T_2 = T_1 \cup \{\theta_\sigma\}$ is a Morita extension of T_1 to the signature $\Sigma \cup \{\sigma_{12}, \sigma, \rho_1, \rho_2, \rho_3, \rho_{12}\}$.

Step 5. This step defines the function symbols i_1, i_2, and i_3. We consider the following sentences.

$$\forall_{\sigma_3} x_3 \forall_\sigma y (i_3(x_3) = y \leftrightarrow \rho_3(x_3) = y) \qquad (\theta_{i_3})$$

$$\forall_{\sigma_2} x_2 \forall_\sigma y \big(i_2(x_2) = y \leftrightarrow \exists_{\sigma_{12}} z(\rho_2(x_2) = z \wedge \rho_{12}(z) = y)\big) \qquad (\theta_{i_2})$$

$$\forall_{\sigma_1} x_1 \forall_\sigma y \big(i_1(x_1) = y \leftrightarrow \exists_{\sigma_{12}} z(\rho_1(x_1) = z \wedge \rho_{12}(z) = y)\big) \qquad (\theta_{i_1})$$

The sentence θ_{i_3} defines the function symbol i_3 to be equal to ρ_3. The sentence θ_{i_2} defines the function symbol i_2 to be equal to the composition "$\rho_{12} \circ \rho_2$." Likewise, the sentence θ_{i_1} defines the function symbol i_1 to be "$\rho_{12} \circ \rho_1$." The theory $T_3 = T_2 \cup \{\theta_{i_1}, \theta_{i_2}, \theta_{i_3}\}$ is a Morita extension of T_2 to the signature $\Sigma \cup \{\sigma_{12}, \sigma, \rho_1, \rho_2, \rho_3, \rho_{12}, i_1, i_2, i_3\}$.

Step 6. We still need to define the predicate symbols in $\widehat{\Sigma}$. Let $\sigma_j \in \Sigma$ be a sort symbol and $p \in \Sigma$ a predicate symbol of arity $\sigma_{j_1} \times \ldots \times \sigma_{jm}$. We consider the following sentences.

$$\forall_\sigma y (q_{\sigma_j}(y) \leftrightarrow \exists_{\sigma_j} x(i_j(x) = y)) \qquad (\theta_{q_{\sigma_j}})$$

$$\forall_\sigma y_1 \ldots \forall_\sigma y_m \big(q_p(y_1, \ldots, y_m) \leftrightarrow \exists_{\sigma_{j_1}} x_1 \ldots \exists_{\sigma_{jm}} x_m (i_{j_1}(x_1) = y_1 \wedge \ldots \wedge i_{jm}(x_m) = y_m \wedge p(x_1, \ldots, x_m))\big) \qquad (\theta_{q_p})$$

These sentences define the predicates $q_{\sigma_j} \in \widehat{\Sigma}$ and $q_p \in \widehat{\Sigma}$. One can verify that T_3 satisfies the admissibility conditions for the definitions $\theta_{q_{\sigma_j}}$. And, therefore, the theory

$T_4 = T_3 \cup \{\theta_{q_{\sigma_1}}, \theta_{q_{\sigma_2}}, \theta_{q_{\sigma_3}}\} \cup \{\theta_{q_p} : p \in \Sigma\}$ is a Morita extension of T_3 to the signature Σ^+.

Step 7. It only remains to show that the Σ^+-theories T_4 and $\widehat{T_4}$ are logically equivalent. One can verify by induction on the complexity of ψ that

$$T_4 \vdash \psi \leftrightarrow \widehat{\psi} \quad \text{and} \quad \widehat{T_4} \vdash \psi \leftrightarrow \widehat{\psi}. \tag{5.3}$$

for every Σ-sentence ψ. One then uses (5.3) to show that T_4 and $\widehat{T_4}$ are logically equivalent. The argument involves a number of cases, but since each case is straightforward, we leave them to the reader to verify. The theories T_4 and $\widehat{T_4}$ are logically equivalent, which implies that T and $\widehat{T}$ are Morita equivalent. □

Theorem 5.3.1 validates Quine's claim that every many-sorted theory can be converted to a single-sorted theory. He concluded that many-sorted logic is dispensable. Whether Quine was right or wrong, his claims in this regard are probably the reason why many-sorted logic hasn't been part of the standard curriculum for analytic philosophers. We hope that our efforts here go some way toward remedying this unfortunate situation.

5.4 Translation Generalized

In the previous chapters, we've talked about various notions of a "translation" between theories. Of course, we did not find the definition of translation written on tablets of stone; nor did we have a Platonic vision of the one true form of a translation. No, we found Quine's definition in the literature, and it works quite well for some purposes, but it's also quite restrictive. In particular, Quine's notions of reconstrual and translation are not general enough to capture some well-known cases of translations between the theories of pure mathematics.

1. In the nineteenth century, the German mathematician Leopold Kronecker is reported to have said, "God made the integers, all else is the work of man." In more prosaic terms, talk about higher number systems – such as rational, real, and complex numbers – can be *reduced* to talk about integers. However, to effect such a reduction, one must treat each rational number as a pair of integers – or, more accurately, as an equivalence class of pairs of integers. Similarly, to reduce the complex numbers to the real numbers, one must treat a complex number as a pair of real numbers, viz. the real and imaginary parts of the complex number.
2. Now for a more controversial example, which we will take up at greater length in Section 7.4. There are different ways that one can write down axioms for Euclidean geometry. In one axiomatization, the basic objects are points; and in another axiomatization, the basic objects are lines. Is there a sense in which these two axiomatized theories could both be Euclidean geometry – in particular, that they could be equivalent? The answer is yes, but only if one allows translations that take a single variable of the first theory to a pair of variables of the second theory. In particular, a line needs to be treated as an equivalence class of pairs of points, and a point needs to be treated as a pair of intersecting lines.

In the previous chapter, we required that a formula $p(x)$ of Σ be translated to a formula $\phi(x)$ of Σ'. There's one particular part of this recipe that seems questionable: why would the same variable x occur in both formulas? In general, why suppose that two signatures Σ and Σ' should share the same variables in common? It's not like variables have some "trans-theoretical" meaning that must be preserved by any reasonable translation.

But how then can variables be reconstrued in moving from one theory to another? One natural proposal would be to include in a reconstrual a mapping from variables of Σ to variables of Σ' – i.e., a function that assigns a variable of Σ' to each variable of Σ. Even so, it's a nontrivial question whether there is an in-principle reason that a single variable in Σ must be reconstrued as a single variable in Σ'. Perhaps one theorist uses several variables to do the work that the other theorist manages to do with a single variable. Such cases are not hard to find in the sciences – for example, when the objects of one mathematical theory are reconstrued as "logical constructions" of objects in another mathematical theory.

Let's proceed then under the assumption that a single variable in one language could be reconstrued in terms of multiple variables in another language. Thus, a reconstrual, in the formal sense, should include a function that matches variables of the signature Σ to n-tuples of variables of the signature Σ'.

Consider again the case of reconstruing rational numbers (i.e., fractions) as pairs of integers. Of course, not *every* pair of integers gives a well-defined fraction. For example, there is no fraction of the form $\frac{1}{0}$. In that case, the "integer theorist" doesn't think of the domain of fractions as consisting of all pairs of integers; rather, she thinks of that domain as consisting of pairs of integers where the second entry is nonzero. To capture this nuance – the restriction of the domain of quantification – we stipulate that a reconstrual F includes a formula D of the target language Σ'. In the running example, the formula D could be given by

$$D(x, y) \equiv (x = x) \wedge (y \neq 0).$$

The integer theorist can then use the formula D to restrict her quantifiers to the domain of well-defined fractions.

Finally, and most controversially, let's consider how we might reconstrue the equality relation $=$ of the domain theory T as a relation of the target theory T'. (Our choice here will prove to be controversial when we show that it yields a positive verdict in favor of quantifier variance. See Example 5.4.16.) Recall that the single variables x and y will typically be reconstrued as n-tuples of variables $\vec{x}$ and $\vec{y}$. In that case, how should we reconstrue the formula $x = y$? One might naturally propose that $x = y$ be reconstrued as the formula

$$(x_1 = y_1) \wedge (x_2 = y_2) \wedge \cdots \wedge (x_n = y_n). \tag{5.4}$$

But here we need to think a bit harder about how and why variables of Σ are encoded as variables of Σ'. For this, let's consider again the example of rational numbers being reduced to integers.

Consider a formula $x = y$ in the theory of rational numbers. To the "integer theorist," the variables x and y really represent complex entities, namely fractions. What's more, to say that two fractions $\frac{x_1}{x_2}$ and $\frac{y_1}{y_2}$ are equal does not mean that $x_1 = y_1$ and $x_2 = y_2$. Rather, $\frac{x_1}{x_2} = \frac{y_1}{y_2}$ means that $x_1 \times y_2 = y_1 \times x_2$. In other words, the formula $x = y$ of the language of the rational numbers is reconstrued as the formula

$$x_1 \times y_2 = y_1 \times x_2, \tag{5.5}$$

in the language of the integers, where $\times$ is the multiplication operation.

This example suggests that we might not always want the formula $x = y$ to be reconstrued as Eqn. 5.4. Instead, we might prefer to reconstrue $x = y$ as some other Σ' formula $E(x_1, \ldots, x_n; y_1, \ldots, y_n)$. Of course, not everything goes: E will need to perform the same functions in the theory T' that the formula $x = y$ performs in the theory T. In particular, we will require that E be an equivalence relation relative to the theory T'.

We're now ready to consider ways in which the elements of one signature Σ can be reconstrued as syntactic structures built from a second signature Σ'. (We include here the case where $\Sigma' = \Sigma$. In that case, we will be considering substitutions and permutations of notation.) The case of relation symbols is relatively easy: an m-ary relation symbol r of Σ should correspond to a formula $F(r)$ of Σ' with mn free variables. To be even more precise, it's the relation symbol r and an n-tuple of variables $x_1, \ldots, x_n$ that corresponds to some particular formula $F(r)$ of Σ', and we require that $FV(F(r)) = \{\vec{x}_1, \ldots, \vec{x}_n\}$.

We will need to proceed with more caution for the function symbols in the signature Σ. The question at issue is: which syntactic structures over Σ' are the proper targets for a reconstrual of the function symbols in Σ? To say that the target must be another function symbol is too restrictive. Indeed, there's a well-known "theorem" that says that every first-order theory is equivalent to a theory that uses only relation symbols. (The reason that "theorem" is placed in quotes here is because the result cannot be proven with mathematical rigor until the word "equivalent" is defined with mathematical rigor.) The trick to proving that theorem is to reconstrue each function symbol f as a relation

$$p_f(x_1, \ldots, x_m, y) \equiv (f(x_1, \ldots, x_m) = y)$$

and then to add axioms saying that p_f relates each m-tuple $x_1, \ldots, x_m$ to a unique output y. If we are to be able to validate such a result (which is intuitively correct), then we ought to permit function symbols of Σ to be reconstrued as formulas of Σ'. We will deal with this issue by analogy with the way we dealt with relation symbols earlier: a function symbol f of Σ and $m + 1$ variables $x_1, \ldots, x_n, y$ of Σ ought to correspond to a formula $(Ff)(\vec{x}_1, \ldots, \vec{x}_n, \vec{y})$ of Σ'.

In order to define a more general notion of a translation, the key is to allow a single sort σ of Σ to be mapped to a sequence of sorts of Σ', including the case of repetitions of a single sort. The idea, in short, is to encode a single variable (or quantifier) in Σ by means of several variables (or quantifiers) in Σ'. In order to make this idea clearer, it will help to give a precise definition of the monoid of finite sequences from a set S.

DEFINITION 5.4.1 For a set S, we let S^* denote the **free monoid** on S, which is uniquely defined by the following universal property: there is a function $\eta_S : S \to S^*$, and for any monoid A, and function $f : S \to A$, there is a unique monoid morphism $f^* : S^* \to A$ such that $f^* \circ \eta_S = f$. Concretely speaking, S^* can be constructed as the set

$$S \amalg (S \times S) \amalg (S \times S \times S) \amalg \cdots ,$$

where $\eta_S : S \to S^*$ is the first coprojection. In this case, given $f : S \to A$, $f^* : S^* \to A$ is the function

$$f^*(s_1, \ldots, s_n) = f(s_1) \circ \cdots \circ f(s_n),$$

where $\circ$ is the monoid operation on A.

DEFINITION 5.4.2 Let Σ and Σ' be many-sorted signatures with sets of sorts S and S' respectively. A generalized **reconstrual** $F : \Sigma \to \Sigma'$ consists of the following:

1. A function $F : S \to (S')^*$. That is, F maps the sorts of Σ to nonempty sequences of sorts of S'. For each $\sigma \in S$, let $d(\sigma)$ be the length of the sequence $F(\sigma)$. We call $d : S \to \mathbb{N}$ the **dimension function** of F.
2. A corresponding function $x \mapsto \vec{x} = x_1, \ldots, x_{d(\sigma)}$ from Σ-variables to sequences of Σ'-variables, such that $x_i : F(\sigma)_i$. We require that if $x \not\equiv y$, then the sequences $\vec{x}$ and $\vec{y}$ have no overlap.
3. A function D from Σ-variables to Σ'-formulas. We call D_x a **domain formula**. We require the map $x \mapsto D_x$ to be natural in the following sense: if y is of the same sort as x, then $D_y = D_x[\vec{y}/\vec{x}]$.
4. A function F that takes a relation symbol p of Σ, and a suitable context $x_1, \ldots, x_n$ of variables from Σ, and yields a formula $(Fp)(\vec{x}_1, \ldots, \vec{x}_n)$ of Σ'. We again require this map to be natural in the sense that

$$(Fp)(\vec{y}_1, \ldots, \vec{y}_n) = (Fp)(\vec{x}_1, \ldots, \vec{x}_n)[\vec{y}_1, \ldots, \vec{y}_n/\vec{x}_1, \ldots, \vec{x}_n].$$

A reconstrual F naturally extends to a map from Σ-formulas to Σ'-formulas. We define this extension, also called F, so that for any Σ-formula ϕ, with x free in ϕ, the following two constraints are satisfied:

$$F(\phi) \vdash D(\vec{x}), \qquad F(\phi[y/x]) = F(\phi)[\vec{y}/\vec{x}].$$

The first restriction is not technically necessary – it is simply a convenient way to ignore whatever the formula $F(\phi)$ says about things outside of the domain $D(\vec{x})$. (This apparently minor issue plays a significant role in Quine's argument for the dispensability of many-sorted logic. See 5.4.17.) Accordingly, for a relation symbol p of Σ, we first redefine $(Fp)(\vec{x}_1, \ldots, \vec{x}_n)$ by conjoining with $D(\vec{x}_1) \wedge \cdots \wedge D(\vec{x}_n)$. (We could have also have included this condition in the very definition of a reconstrual.) The extension of F proceeds as follows:

- Let $F(\phi \wedge \psi) = F(\phi) \wedge F(\psi)$, and let $F(\phi \vee \psi) = F(\phi) \vee F(\psi)$.
- Let $F(\neg\phi) = \neg F(\phi) \wedge D(\vec{x}_1) \wedge \cdots \wedge D(\vec{x}_n)$, where $x_1, \ldots, x_n$ are all the free variables that occur in ϕ.
- Let $F(\phi \to \psi) = F(\neg\phi) \vee F(\psi)$.
- Let $F(\exists x\phi) = \exists\vec{x}(D(\vec{x}) \wedge F(\phi))$.
- Let $F(\forall x\phi) = \forall\vec{x}(D(\vec{x}) \to F(\phi))$.

DEFINITION 5.4.3 Let $F : \Sigma \to \Sigma'$ be a reconstrual. We say that F is a **translation** of T into T' just in case, for every Σ-sentence ϕ, if $T \vdash \phi$ then $T' \vdash F(\phi)$. In this case, we write $F : T \to T'$. In the case that Σ has a single sort σ, we say that F is a $d(\sigma)$-dimensional translation.

The definition of a translation allows us to handle the case where the domain signature Σ has equality relations and function symbols. In particular, for each theory T in Σ, we explicitly include the following axioms:

- The equality introduction axioms: $\vdash x =_\sigma x$.
- The equality elimination axioms: $\phi(x), (x =_\sigma y) \vdash \phi(y)$, for each atomic or negated atomic formula ϕ of Σ.

As usual, these axioms together entail that $=_\sigma$ is an equivalence relation. Thus, if $F : T \to T'$ is a translation, then $F(=_\sigma)(\vec{x}, \vec{y})$ is an equivalence relation on domain $D(\vec{x})$. We abbreviate this relation by $E_\sigma(\vec{x}, \vec{y})$ or, when no confusion can result, simply as $E(\vec{x}, \vec{y})$. In this case, for each relation symbol p of Σ,

$$T', (Fp)(\vec{x}), E(\vec{x}, \vec{y}) \vdash (Fp)(\vec{y}).$$

Roughly speaking, the predicate Fp has to be compatible with the equivalence relation E: it holds of something iff it holds of everything E-equivalent to that thing. Equivalently, the extension of Fp is a union of E-equivalence classes.

Now suppose that Σ contains a constant symbol c. Then, choosing a variable x of the same sort, $c = x$ is a unary formula, and $F(c = x)$ is a formula $\phi(\vec{x})$. The theory T entails that the formula $c = x$ is uniquely satisfied. Hence, if $F : T \to T'$ is a translation, then T' entails that $\phi(\vec{x})$ is uniquely satisfied – relative to the equivalence relation E. In short, T' implies both $\exists\vec{x}(D_x \wedge \phi(\vec{x}))$ and $\phi(\vec{x}) \wedge \phi(\vec{y}) \to E(\vec{x}, \vec{y})$. Intuitively speaking, this means that the extension of $\phi(\vec{x})$ is a single E-equivalence class.

Similar reasoning applies to the case of any function symbol f of Σ. The Σ-formula $f(x_1, \ldots, x_n) = y$ is reconstrued as some Σ'-formula $\phi(\vec{x}_1, \ldots, \vec{x}_n, \vec{y})$. If $F : T \to T'$ is a translation, then T' entails that ϕ is a functional relation relative to E-equivalence. What this means intuitively is that ϕ is a function from E-equivalence classes to E-equivalence classes.

Example 5.4.4 (Quantifier variance) We now undertake an extended discussion of an example that is near and dear to metaphysicians: the debate between mereological universalism and nihilism. To keep the technicalities to a bare minimum, we will consider a dispute over whether the composite of two things exists. Suppose that the parties to the dispute are are named Niels the Nihilist and Mette the Mereological Universalist. Niels says that there are exactly two things, whereas Mette says that there are exactly three things, one of which is composed of the other two.

Now, we press Niels and Mette to regiment their theories, and here's what they come up with. Niels has a signature Σ, which is empty, very much in line with his predilection for desert landscapes. Niels' theory has a single axiom, "there are exactly two things." Mette has a signature Σ' with a binary relation symbol p that she'll use to express the parthood relation. Mette's theory T' says that p is a strict partial order, that there are exactly two atoms, and exactly one thing above those two atoms. Note that Mette can define an open formula in Σ'

$$a(x) \equiv \neg\exists y\, p(y,x),$$

which intuitively expresses the claim that x is an atom.

At the turn of the twenty-first century, metaphysicians were engaged in a fierce debate about whether Niels or Mette has a better theory. Then some other philosophers, such as Eli Hirsch, said, "stop arguing – it's merely a verbal dispute, like an argument about whether there are six roses or half a dozen roses" (see Chalmers et al., 2009; Hirsch, 2011). These other philosophers espouse a position known as **quantifier variance**. One clear explication of quantifier variance would be to say that Niels and Mette's theories are **equivalent**. So are they equivalent or not? The answer to this question depends (unsurprisingly) on the standard of equivalence that we adopt. For example, it is easy to see that Niels and Mette's theories are not strictly intertranslatable in the sense of Defn. 4.5.15. However, we will now see that Niels and Mette's theories are intertranslatable in the weaker sense described in Defn. 5.4.14.

It seems clear that Mette can make sense of Niels' theory – in particular, that she can identify Niels' quantifier as a restriction of her own. The idea that Mette can "make sense of Niels' theory" can be cashed out formally as saying that Niels' theory can be translated into Mette's theory. Intuitively speaking, for any sentence ϕ asserted by Niels, there is a corresponding sentence ϕ^* asserted by Mette. For example, when Niels says,

There are exactly two things,

Mette can charitably interpret him as saying,

There are exactly two atoms.

Now we show that there is indeed a translation $F : T \to T'$, where T is Niels' theory, and T' is Mette's theory. Here Niels and Mette's theories are single-sorted, and we define F to be a one-dimensional reconstrual. We define the domain formula as $D_F(x) = a(x)$, and we translate Niels' equality relation as Mette's equality relation restricted to D_F.

Let's just check that F is indeed a translation. While a general argument is not difficult, let's focus on Niels' controversial claim ϕ: that there are *at most two* objects in the domain:

$$\phi \equiv \forall x\forall y\forall z((x = y) \vee (x = z) \vee (y = z)).$$

The reconstrual F takes $x = y$ to the formula $a(x') \wedge a(y') \wedge (x' = y')$, and hence $F(\phi)$ is the uncontroversially true statement that there are at most two atoms. Of course, Mette agrees with that claim, and so $F : T \to T'$ is a translation of Niels' theory into Mette's.

Indeed, F is a particularly nice translation: it's conservative, in the sense that if $T' \vdash F(\phi)$, then $T \vdash \phi$. Thus, not only does Mette affirm everything that Niels says about atoms; Niels also affirms everything that Mette says about atoms. Thus, there is a precise sense in which Niels' theory is simply a "sub-theory" of Mette's theory. They are in complete agreement relative to their shared language, and Mette simply has a larger vocabulary than Niels.

The existence of the translation $F : T \to T'$ comes as no surprise. But what about the other way around? Can Niels be as charitable to Mette as she has been to him? Can he find a way to affirm *everything* that she says? The answer to that question is far from clear. For example, Mette says things like, "x is a composite of y and z." How in the world could Niels make sense of that claim? How in the world could Niels say, "what Mette says here is perfectly correct, if only understood in the proper way"? Similarly, Mette says that "there are more than two things." How in the world could Niels validate such a claim?

We will now see that Niels can indeed charitably interpret, and endorse, all of Mette's assertions. Indeed, Niels needs only think of Mette's notion of "a thing" as corresponding to what he means by "a pair of things" – as long as two pairs are considered to be "the same" when they are permutations of each other.

More precisely, consider a two-dimensional reconstrual $G : \Sigma' \to \Sigma$ that encodes a Σ'-variable x as a pair x_1, x_2 of Σ-variables. Define $D_G(x_1, x_2)$ to be the formula $(x_1 = x_1) \wedge (x_2 = x_2)$ that holds for all pairs $\langle x_1, x_2 \rangle$. Define $E_G(x_1, x_2, y_1, y_2)$ to be the relation that holds between $\langle x_1, x_2 \rangle$ and $\langle y_1, y_2 \rangle$ just in case one is a permutation of the other. That is,

$$E_G(x_1, x_2, y_1, y_2) \equiv (x_1 = y_1 \wedge x_2 = y_2) \vee (x_1 = y_2 \wedge x_2 = y_1).$$

Clearly, T entails that E_G is an equivalence relation.

The signature Σ' consists of a single binary relation symbol p. Since G is two-dimensional, Gp must be defined to be a four-place relation in Σ. Here is the intuitive idea behind our definition of Gp: we will simulate atoms of Mette's theory by means of diagonal pairs, i.e., pairs of the form $\langle x, x \rangle$. We then say that Gp holds precisely between pairs when the first is diagonal, the second is not, and the first has a term in common with the second. More precisely,

$$(Gp)(x_1, x_2, y_1, y_2) \equiv (x_1 = x_2) \wedge (y_1 \neq y_2) \wedge (x_1 = y_1 \vee x_1 = y_2).$$

Recall that $a(x)$ is the formula of Σ' that says that x is an atom. We claim now that the translation $G(a(x))$ of $a(x)$ holds precisely for the pairs on the diagonal. That is,

$$T \vdash G(a)(x_1, x_2) \leftrightarrow (x_1 = x_2).$$

We argue by reductio ad absurdum. (Here we use the notion of a model, which will first be introduced in the next chapter. Hopefully, the intuition will be clear.) First, if

$$T \nvdash G(a)(x_1, x_2) \to (x_1 = x_2),$$

then there is a model M of T, and two distinct objects c,d of M such that $M \models G(a)(c,d)$. That means that

$$M \models \neg\exists y \exists z\, G(p)(y,z,c,d).$$

But, clearly, $M \models G(p)(c,c,c,d)$, a contradiction. To prove the other direction, it will suffice to show that for any model M of T, and for any $c \in M$, we have $M \models G(a)(c,c)$. Recalling that T only has one model, namely a model with two objects, the result easily follows.

When Mette the Mereologist says that there are more than two things, Niels the Nihilist understands her as saying that there are more than two pairs of things. Of course, Niels agrees with that claim. In fact, it's not hard to see that, under this interpretation, Niels affirms everything that Mette says. ⌟

DISCUSSION 5.4.5 We've shown that Niels' theory can be translated into Mette's, and vice versa. Granting that this is a good notion of "translation," does it follow that these two theories are equivalent? In short, no. Recall the simpler case of propositional theories. For example, let $\Sigma = \{p_0, p_1, \ldots\}$, let T be the empty theory in Σ, and let T' be the theory with axioms $p_0 \vdash p_1, p_0 \vdash p_2, \ldots$ Then there are translations $f : T \to T'$ and $g : T' \to T$, but T and T' are not equivalent theories. In general, mutual interpretability is not sufficient for equivalence. Nonetheless, we will soon see (Example 5.4.16) that there is a precise sense in which Niels' and Mette's theories are indeed equivalent. □

We are now ready to prove a generalized version of the **substitution theorem**. In its simplest form, the substitution theorem says a valid derivation $\phi_1, \ldots, \phi_n \vdash \psi$ is preserved under uniform substitution of the non-logical symbols in $\phi_1, \ldots, \phi_n$ and ψ. For example, from a valid derivation of $\exists x(p(x) \wedge q(x)) \vdash \exists x p(x)$, substitution of $\forall y r(y,z)$ for $p(x)$ yields a valid derivation of

$$\exists z(\forall y r(y,z) \wedge q(z)) \vdash \exists z \forall y r(y,z).$$

However, we need to be careful in describing what counts as a legitimate "substitution instance" of a formula. Let's test our intuitions against an example.

Example 5.4.6 Let Σ be a single-sorted signature with equality, but no other symbols. Let Σ' be a single-sorted signature with equality, and one other monadic predicate $D(x)$. We define a one-dimensional reconstrual $F : \Sigma \to \Sigma'$ by taking $D(x)$ to be the domain formula, and by taking $E(x,y)$ to be equality in Σ'. We will see now that the substitution theorem does *not* hold in the form: if $\phi \vdash \psi$ then $F(\phi) \vdash F(\psi)$.

In Σ, we have $x \neq y \vdash \exists z(x \neq z)$. Since F translates equality in Σ to equality in Σ', we have $F(x \neq y) \equiv (x \neq y)$. Furthermore, $F(\exists z(x \neq z))$ is the relativized formula $\exists z(D(z) \wedge x \neq z)$. But $x \neq y$ does not imply that there is a z such that $D(z)$ and $x \neq z$. For example, in the domain $\{a,b\}$, if the extension of D is $\{a\}$, then $a \neq b$, but not $\exists z(D(z) \wedge a \neq z)$. Thus, the substitution theorem does *not* hold in the form: if $\phi \vdash \psi$, then $F(\phi) \vdash F(\psi)$. So what's the problem here?

To speak figuratively, a reconstrual F maps a variable x of Σ to variables $\vec{x}$ that are relativized to the domain $D(\vec{x})$. However, the turnstile $\vdash$ for Σ' is not relativized in this fashion: a sequent $F(\phi) \vdash F(\psi)$ corresponds to a tautology $\vdash \forall\vec{x}(F(\phi) \rightarrow F(\psi))$. It wouldn't make sense to expect this last statement to hold, since the intention is for the variables in $F(\phi)$ and $F(\psi)$ to range over $D(\vec{x})$. Thus, the relevant question is whether $F(\phi) \vdash_{D(\vec{x})} F(\psi)$, where the latter is shorthand for

$$\vdash \forall\vec{x}(D(\vec{x}) \rightarrow (F(\phi) \rightarrow F(\psi))).$$

In the current example, then, the question is whether the following holds:

$$F(x \neq y), D(x), D(y) \vdash F(\exists y(x \neq y)).$$

And it obviously does. This example shows us how to formulate a substitution theorem for generalized reconstruals such as F. ⌟

THEOREM 5.4.7 (Substitution) *Let Σ be a signature without function symbols, and suppose that F is a reconstrual from Σ to Σ'. Then, for any formulas ϕ and ψ with free variables $x_1, \ldots, x_n$, if $\phi \vdash \psi$, then $F(\phi) \vdash_{D(\vec{x}_1,\ldots,\vec{x}_n)} F(\psi)$. In particular, if ϕ and ψ are Σ-sentences, then $F(\phi) \vdash F(\psi)$.*

Proof We will prove this result by induction on the construction of proofs. For the base case, the rule of assumptions justifies not only $\phi \vdash \phi$, but also $F(\phi) \vdash F(\phi)$, and hence $D(\vec{x}), F(\phi) \vdash F(\phi)$. The inductive cases for the Boolean connectives involve no special complications, and so we leave them to the reader.

Consider now the case of $\exists$-elim. Suppose that $\exists y\phi \vdash \psi$ results from application of $\exists$-elim to $\phi \vdash \psi$, in which case y is not free in ψ. We rewrite ϕ and ψ in the suggestive notation $\phi(x, y)$ and $\psi(x)$, indicating that x may be free in both ϕ and ψ, and that $y \not\equiv x$. (Note, however, that our argument doesn't depend on ϕ and ψ sharing exactly one free variable in common.) We want to show that $F(\exists y\phi(x, y)) \vdash_{D(\vec{x})} F(\psi(x))$, which expands to

$$D(\vec{x}), \exists\vec{y}(D(\vec{y}) \wedge F(\phi(x, y))) \vdash F(\psi(x)). \tag{5.6}$$

The inductive hypothesis here says that

$$D(\vec{x}), D(\vec{y}), F(\phi(x, y)) \vdash F(\psi(x)).$$

Since x and y are distinct variables, the sequences $\vec{x}$ and $\vec{y}$ have no overlap, and $\vec{y}$ does not occur free in $D(\vec{x})$. Thus, n-applications of $\exists$-intro yield the sequent (5.6).

Consider now the case of $\exists$-intro. Suppose that $\phi \vdash \exists y\psi$ follows from $\phi \vdash \psi$ by an application of $\exists$-intro. Again, we will rewrite the former sequent as

$$\phi(x, y) \vdash \exists y\psi(x, y).$$

We wish to show that

$$D(\vec{x}), D(\vec{y}), F(\phi(x, y)) \vdash \exists\vec{y}F(\psi(x, y)).$$

By the inductive hypothesis, we have

$$D(\vec{x}), D(\vec{y}), F(\phi(x, y)) \vdash F(\psi(x, y)).$$

Thus, the result follows by repeated application of $\exists$-intro. □

The previous version of the substitution theorem applies only to the case of signatures without function symbols. Intuitively, however, the formal validity of proofs should also be maintained through uniform substitution of terms.

Example 5.4.8 Suppose that $\phi(x) \vdash \psi(x)$, which is equivalent to $\vdash \forall x(\phi(x) \vdash \psi(x))$. Now let $t(\vec{y})$ be a term with free variables $\vec{y} \equiv y_1, \ldots, y_n$, and suppose that each of these variables is free for x in ϕ and ψ, but none of them are themselves free in either one of these formulas. (In the simplest case, these variables simply do not occur in either one of the formulas.) Then $\forall$-elim and intro yield

$$\vdash \forall \vec{y}(\phi(t(\vec{y})) \to \psi(t(\vec{y}))),$$

which is equivalent to $\phi(t(\vec{y}) \vdash \psi(t(\vec{y}))$. In other words, a valid proof remains valid if a variable x is uniformly replaced by a term $t(\vec{y})$, so long as the relevant restrictions are respected. ⌟

At this stage, we have a definition of a generalized translation, and we've shown that it yields a generalized substitution theorem. What we would like to do now is to look at specific sorts of translations – and most particularly, at which translations should count as giving an equivalence of theories. It turns out, however, that giving a good definition of equivalence is a bit complicated. As many examples will show, it won't suffice to say that a translation $F : T \to T'$ is an equivalence just in case it has an inverse $G : T' \to T$, and not even a quasi-inverse in the sense of 4.5.15. For a good definition of equivalence, we need a notion of a "homotopy" between translations, and we need a notion of the composition of translations. We turn first to the second of these.

DEFINITION 5.4.9 (Composition of reconstruals) Suppose that $F : \Sigma \to \Sigma_1$ and $G : \Sigma_1 \to \Sigma_2$ are reconstruals. Define a reconstrual $H : \Sigma \to \Sigma_2$ as follows:

- Since $G : S_1 \to S_2^*$, there is a unique morphism $G^* : S_1^* \to S_2^*$ such that $G = G^* \circ \eta_{S_1}$. In other words, G^* acts on a sequence of S_1 sorts by applying G to each element and then concatenating. We then define $H = G^* \circ F : S \to S_2^*$.
- We use the same idea to associate each variable x of Σ with a (double) sequence $X_1, \ldots, X_n$ of variables of Σ_2. In short,

$$\begin{aligned} H(x) &= G^*(F(x)) \\ &= X_1, \ldots, X_n \\ &= (x_{11}, \ldots, x_{1m_1}), \ldots, (x_{n1}, \ldots, x_{nm_n}), \end{aligned}$$

where $F(x) = x_1, \ldots, x_n$, and $G(x_i) = X_i = (x_{i1}, \ldots, x_{im_i})$.

- Let $D_F(\vec{x})$ denote the domain formula of F corresponding to the Σ-variable x. Let $D_G(X_i)$ denote the domain formula of G corresponding to the Σ_1-variable x_i. Then we define

$$D_H(X_1, \dots, X_n) := G(D_F(\vec{x})).$$

Recall that any free variable in $G(D_F(\vec{x}))$ occurs in the double sequence $X_1, \dots, X_n$, and that $G(\phi) \vdash D_G(Y)$ if y is free in ϕ. Thus, $D_H(X_1, \dots, X_n) \vdash D_G(X_i)$ for each $i = 1, \dots, n$.

- For a relation symbol p of Σ, we define

$$(Hp)(X_1, \dots, X_n) = G((Fp)(\vec{x}_1, \dots, \vec{x}_n)).$$

PROPOSITION 5.4.10 *If F and G are translations, then $G \circ F$ is a translation.*

Proof This result follows trivially once we recognize that $G \circ F$ is a legitimate reconstrual. □

For some philosophers, it may seem that we have already greatly overcomplicated matters by using category theory to frame our discussion of theories. I'm sorry to say that matters are worse than that. The collection of theories really has more interesting structure than a category has; in fact, it's most naturally thought of as a **2-category**, where there are 0-cells (objects), 1-cells (arrows), and 2-cells (arrows between arrows). In particular, our 2-category of theories, **Th**, has first-order theories as the 0-cells, and translations as the 1-cells. We now define the 2-cells, which we call **t-maps**.

Let F and G be translations from T to T'. Since the definition of a t-map is heavily syntactic, we begin with an intuitive gloss in the special case where Σ has a single sort σ. In this case $F(\sigma)$ is a sequence $\sigma_1, \dots, \sigma_m$ of Σ'-sorts, and $G(\sigma)$ is a sequence $\sigma'_1, \dots, \sigma'_n$ of Σ'-sorts. Then a t-map $\chi : F \Rightarrow G$ consists of a formula $\chi(\vec{x}, \vec{y})$ that links m-tuples to n-tuples. This formula $\chi(\vec{x}, \vec{y})$ should have the following features:

1. The theory T' implies that $\chi(\vec{x}, \vec{y})$ is a functional relation from D_F to D_G, relative to the notion of equality given by the equivalence relations E_F and E_G.
2. For each formula ϕ of Σ, χ maps the extension of $F(\phi)$ into the extension of $G(\phi)$.

We now turn to the details of the definition.

DEFINITION 5.4.11 A **t-map** $\chi : F \Rightarrow G$ is a family of Σ'-formulas $\{\chi_\sigma\}$, where σ runs over the sorts of Σ, where each χ_σ has $d_K(\sigma) + d_L(\sigma)$ free variables, and such that T' entails the following (which we label with suggestive acronyms):

$$\chi_\sigma(\vec{x}, \vec{y}) \to (D_F(\vec{x}) \wedge D_G(\vec{y})) \qquad \text{(dom-ran)}$$

$$(E_F(\vec{x}, \vec{w}) \wedge E_G(\vec{y}, \vec{z}) \wedge \chi_\sigma(\vec{w}, \vec{z})) \to \chi_\sigma(\vec{x}, \vec{y}) \qquad \text{(well-def)}$$

$$D_F(\vec{x}) \to \exists \vec{y}(D_G(\vec{y}) \wedge \chi_\sigma(\vec{x}, \vec{y})) \qquad \text{(exist)}$$

$$(\chi_\sigma(\vec{x}, \vec{y}) \wedge \chi_\sigma(\vec{x}, \vec{z})) \to E_G(\vec{y}, \vec{z}) \qquad \text{(unique)}$$

Furthermore, for any Σ-formula $\phi(x_1,\ldots,x_n)$ with $x_1 : \sigma_1,\ldots,x_n : \sigma_n$, the theory must T' entail that

$$\chi_{\vec{\sigma}}(X,Y) \rightarrow (F(\phi)(X) \rightarrow G(\phi)(Y)),$$

where we abbreviate $X = \vec{x}_1,\ldots,\vec{x}_n$, $Y = \vec{y}_1,\ldots,\vec{y}_n$, and $\chi_{\vec{\sigma}}(X,Y) = \chi_{\sigma_1}(\vec{x}_1,\vec{y}_1) \wedge \ldots \wedge \chi_{\sigma_n}(\vec{x}_n,\vec{y}_n)$.

We are especially interested in what it might mean to say that two translations $F : T \rightarrow T'$ and $G : T \rightarrow T'$ are isomorphic – i.e., the conditions under which a t-map $\chi : F \Rightarrow G$ is an isomorphism.

DEFINITION 5.4.12 We say that a t-map $\chi : F \Rightarrow G$ is a **homotopy** (or an **isomorphism of translations**) if each of the functions χ establishes a bijective correspondence, relative to the equivalence relations E_F and E_G. More precisely, the theory T' entails

$$D_G(\vec{y}) \rightarrow \exists\vec{x}(D_F(\vec{x}) \wedge \chi(\vec{x},\vec{y})) \qquad \text{(onto)}$$

$$(\chi(\vec{x},\vec{y}) \wedge \chi(\vec{w},\vec{y})) \rightarrow E_F(\vec{x},\vec{w}) \qquad \text{(one-to-one)}$$

Furthermore, for each formula ϕ of Σ, the theory T' entails that

$$\chi(X,Y) \rightarrow (G(\phi)(Y) \rightarrow F(\phi)(X)).$$

Here we have omitted the sort symbol σ from χ_σ merely in the interest of notational simplicity.

DISCUSSION 5.4.13 Note that F and G can be isomorphic translations even if they have different dimension functions – i.e., if they encode Σ-variables into different-length strings of Σ'-variables. We will see an example below of a single sorted theory T, and a two-dimensional translation $F : T \rightarrow T$ that is isomorphic to the identity translation $1_T : T \rightarrow T$. In this case, the theory T might be glossed as saying: "pairs of individuals correspond uniquely to individuals."

DEFINITION 5.4.14 We say that two theories T and T' are **weakly intertranslatable** (also **homotopy equivalent**) if there are translations $F : T \rightarrow T'$ and $G : T' \rightarrow T$, and homotopies $\chi : GF \Rightarrow 1_T$ and $\chi' : 1_{T'} \Rightarrow FG$.

NOTE 5.4.15 Here the word "weakly" in "weakly equivalent" shouldn't be taken to hold any deep philosophical meaning – as if it indicates that the theories aren't fully equivalent. Instead, the use of that word traces back to category theory and topology, where it has proven to be interesting to "weaken" notions of strict equality, isomorphism, or homeomorphism. In many such cases, the weakened notion is a more interesting and useful notion than its strict counterpart. One thing we like about this proposed notion of theoretical equivalence is precisely its connection with the sorts of notions that prove to be fruitful in contemporary mathematical practice. If we were to wax metaphysical, we might say that such notions carve mathematical reality at the joints.

Example 5.4.16 (Quantifier variance) We can now complete the discussion of Example 5.4.4 by showing that Mette the Mereologist and Niels the Nihilist have equivalent theories – at least by the standard of "weak intertranslatability." Recall that the translation $F : T \to T'$ includes Niels' theory as a subtheory of Mette's, restricted to the atoms. The translation $G : T' \to T$ maps Mette's variables to pairs of Niels' variables (up to permutation), and it translates the parthood relation as the relation that holds between a diagonal pair and non-diagonal pair that matches in one place.

We give an informal description of the homotopy maps $\varepsilon : GF \Rightarrow 1_T$ and $\eta : FG \Rightarrow 1_{T'}$. First, $GF\sigma = \sigma, \sigma$. That is, GF translates Niels' variables into pairs of Niels' variables, and the domain formula is the diagonal $x = y$. It's easy enough then to define a functional relation

$$\varepsilon(x, y; z) \leftrightarrow (x = y) \wedge (x = z),$$

from the diagonal of σ, σ to σ. For the homotopy map η, note that FG translates Mette's variables into pairs of Mette's variables, and the domain formula is $a(x) \wedge a(y)$ – i.e., both x and y are atoms. We then define $\eta(x, y; z)$ to be the functional relation such that if $x = y$ then $z = x$, and if $x \neq y$, then z is the composite of x and y. A tedious verification shows that ε and η satisfy the definition of homotopy maps, and therefore F, G form a homotopy equivalence.

Thus, there is a precise notion of theoretical equivalence that validates the claim of quantifier variance. However, this fact just pushes the debate back one level – to a debate over what we should take to be the "correct" notion of theoretical equivalence. Perhaps weak intertranslatability seems more mathematically natural than its strong counterpart. Or perhaps weak intertranslatability is closer to the notion that mathematicians use in practice. But these kinds of considerations could hardly be expected to move someone who antecedently rejects the claim of quantifier variance. ⌟

Example 5.4.17 Let's look now at an example that is relevant to the debate between Carnap and Quine.

Suppose that $\Sigma = \{\sigma_1, \sigma_2, p, q\}$, with p a unary predicate symbol of sort σ_1, and q a unary predicate symbol of sort σ_2. Let T be the empty theory in Σ. For simplicity, we will suppose that T implies that there are at least two things of sort σ_1, and at least two things of sort σ_2. In order to get a more intuitive grasp on this example, let's suppose that the T-theorist is intending to use σ_1 to model the domain of mathematical objects, and σ_2 to model the domain of physical objects. As Carnap might say, "mathematical object" and "physical object" are *Allwörter* to mark out domains of inquiry. Let's suppose also that $p(x)$ stands for "x is prime," and $q(x)$ stands for "x is massive" (i.e., has nonzero mass).

Now, Quine thinks that there's no reason to use sorts. Instead, he says, we should suppose that there is a single domain that can be divided by the predicates, "being a mathematical object" and "being a physical object." He says,

> since the philosophers [viz. Carnap] who would build such categorial fences are not generally resolved to banish from language all falsehoods of mathematics and like absurdities, I fail to see much benefit in the partial exclusions that they do undertake; for the forms concerned would remain still quite under control if admitted rather, like self-contradictions, as false. (Quine, 1960, p. 229)

Quine's proposal seems to be the following:

1. Unify the sorts σ_1 and σ_2 into a single sort σ; and
2. For each formula ϕ with a type-mismatch, such as "There is a massive number," declare that ϕ is false.

For example, in the signature Σ, the predicate symbols p and q are of different sorts, hence they cannot be applied to the same variable, and $\phi \equiv \exists x(p(x) \wedge \neg q(x))$ is ill-formed. Quine suggests then that ϕ should be taken to be false. But what then are we to do about the fact that $\neg\phi \vdash \forall x(p(x) \to q(x))$? If ϕ is false, then it follows that all prime numbers are massive. Something has gone wrong here.

Of course, Quine is right to think that the many-sorted theory T is equivalent to a single-sorted theory T_1. Nonetheless, there are a couple of problems for Quine's suggestion that we simply throw away T in favor of T_1. First, there is another single-sorted theory T_2 that is equivalent to T, but T_1 and T_2 disagree on how to extend the ranges of predicates in T. Quine provides no guidance about whether to choose T_1 or T_2, and it seems that the choice would have to be *conventional*. The second problem is that T leaves open possibilities for specification that would be prematurely settled by passing to T_1 (or to T_2).

To be more specific, we will construct these theories T_1 and T_2. First let $\Sigma_i = \{\sigma, u, p', q'\}$, where σ is a sort symbol, and u, p', q' are unary predicate symbols. Let T_1 be the theory in Σ_1 with axioms:

$$\begin{aligned} T_1 &\vdash \exists x u(x) \wedge \exists x \neg u(x) \\ T_1 &\vdash \forall x(\neg u(x) \to \neg p'(x)) \\ T_1 &\vdash \forall x(u(x) \to \neg q'(x)). \end{aligned}$$

The first axiom ensures that the domains u and $\neg u$ are nonempty. The second axiom implements Quine's requirement that physical objects are not prime, and the third axiom implements Quine's requirement that mathematical objects are not massive. It then follows that

$$\begin{aligned} T_1 &\vdash \neg\exists x(p'(x) \wedge q'(x)) \\ T_1 &\vdash \forall x(p'(x) \to \neg q'(x)). \end{aligned}$$

It's not difficult to see that T can be translated into T_1. Indeed, we can set $F(\sigma_1) = \sigma = F(\sigma_2)$, taking the domain formula for σ_1 variables to be u, and the domain formulas for σ_2 variables to be $\neg u$. We can then set $F(p) = p'$ and $F(q) = q'$. It is not difficult to see that F is a translation. In fact, there is also a translation G from T_1 to T, but it is more difficult to define. The problem here is determining how to translate a variable x of the signature Σ_1 into variables of the signature Σ. In particular, x ranges over things that satisfy $u(x)$ as well as things that satisfy $\neg u(x)$, but each variable of Σ is held fixed to one of the sorts, either σ_1 or σ_2.

Consider now the theory T_2 that is just like T_1 except that it replaces the axiom $\forall x(u(x) \to \neg q'(x))$ with the axiom $\forall x(u(x) \to q'(x))$. The theory T_2 differs from T_1 precisely in that it adopts a different convention for how to extend the predicates $q'(x)$ and $\neg q'(x)$ to the domain $u(x)$. T_1 says that q' should be restricted to $\neg u(x)$, and T_2 says that $\neg q'$ should be restricted to $\neg u(x)$. Quine's original proposal seems to say that we should restrict *all* predicates of sort σ_2 to $\neg u(x)$, but that proposal is simply incoherent.

Thus, the many-sorted theory T could be replaced with the single-sorted theory T_1, or it could be replaced with the single-sorted theory T_2. In one sense, it shouldn't make any difference which of these two single-sorted theories we choose. (In fact, T_1 and T_2 are intertranslatable in the strict, single-sorted sense.) But in another sense, either choice could block us from adding new truths to the theory T.

Suppose, for example, that we decided to hold on to T, instead of replacing it with T_1 or T_2. Suppose further that we come to discover that

$$\psi \equiv \exists x p(x) \wedge \neg \exists y \neg q(y).$$

But if we take the translation manual $p \mapsto p'$ and $q \mapsto q'$, then T_1 rules out ψ since $T_1 \vdash \forall x(p'(x) \to \neg q'(x))$. In this case, then, it would have been disastrous to follow Quine's recommendation to replace T by T_1, because we would have thereby stipulated as false something that T allows to be true. One of the important lessons of this example is that equivalent theories aren't equally good in all ways. ⌟

5.5 Symmetry

Philosophers of science, and especially philosophers of physics, are fascinated by the topic of symmetry. And why so? For one, because contemporary physics is chock full of symmetries and symmetry groups. Moreover, philosophers of physics have taken it upon themselves to *interpret* the theories of physics – by which they mean, among other things, to say what those theories *really mean*, and to lay bare their *ontological commitments*. In the famous words of Bas van Fraassen, the goal of interpreting a theory is to say how the world might be such that the theory is true.

Symmetry is now thought to play a special role in interpretation, in particular as a tool to winnow the ontological wheat from the formal chaff (sometimes affectionately called "descriptive fluff" or "surplus structure"). Here's how the process is supposed to work: we are given a theory T that says a bunch of things. Some of the things that T says, we should take seriously. But some of the other things that T says – or seems, prima facie, to say – should not be taken seriously. So what rule should we use to factorize T into the pure descriptive part T_0, and the superfluous part T_1? At this point, we're supposed to look to the symmetries of T. In rough-and-ready formulation, T_0 is the part of T that is invariant under symmetries, and T_1 is the part of T that is not invariant under symmetries.

Philosophers didn't make this idea out of thin air; instead, they abstracted it from well-known examples of theories in physics.

- If you describe space by a three-dimensional vector space V, then you must associate the origin $0 \in V$ with a particular point in space. But all points in space were created equal, so the representation via V says something misleading. We can then wash out this superfluous structure by demanding that translation $x \mapsto x + a$ be a symmetry, which amounts to replacing V with the affine space A over V.
- In classical electrodynamics, we can describe the electromagnetic field via potentials. However, the values of these potentials don't matter; only the gradients (rates of change) of the potentials matter. There is, in fact, a group G of symmetries that changes the values of the potentials but leaves their gradients (and, hence, the Maxwell tensor F_{ab}) invariant.
- In quantum field theory, there is an algebra $\mathscr{F}$ of field operators and a group G of symmetries. Not all field operators are invariant under the group G. Those field operators that are invariant under G are called *observables*, and it is a common opinion that only the observables are "real."

Based on these examples, and others like them, it's tempting for philosophers to propose methodological rules, such as: "if two things are related by a symmetry, then they are the same," or "a thing is real only if it is invariant under symmetries." Such principles are tendentious, but my goal here isn't to attack them directly. Even before we can discuss the merits of these principles, we need to be clearer about what symmetries are.

What is a symmetry of a theory? Sometimes we hear talk of permutations of models. Other times we hear talk of permutations of spacetime points. And yet other times we hear talk about transformations of coordinates. The goal of this section, stated bluntly, is to clear away some of the major sources of confusion. These confusions come from conflating things that ought to be kept distinct. The first thing to distinguish are theories and individual models. Even if one is a firm believer in the semantic view of theories, still a collection of models is a very different thing from an individual model; and a symmetry of an individual model is a very different thing from a symmetry on the class of models. The second thing to distinguish is, yet again, syntax and semantics. One can look at symmetries from either point of view, but confusion can arise when we aren't clear about which point of view we're taking.

In physics itself, one occasionally hears talk of symmetries of equations. Such talk is especially prominent in discussions of spacetime theories, where one says things like, "X transforms as a tensor." Nonetheless, in recent years, philosophers of science have tended to look at symmetries as transformations of models. Certainly, it is possible to develop a rigorous mathematical theory of symmetries of models – as we shall discuss in the following two chapters. However, transformations of models aren't the only kind of symmetries that can be defined in a mathematically rigorous fashion. In this section, we discuss **syntactic symmetries** – i.e., symmetries of a theory considered as a syntactic object.

Some examples of syntactic symmetries are quite obvious and trivial.

Example 5.5.1 Let $\Sigma = \{p, q\}$ be a propositional logic signature, and let T be the empty theory in Σ. It seems intuitively correct to say that T cannot distinguish between the propositions p and q. And, indeed, we can cash this intuition out in terms of a "self-translation" $F : T \to T$. In particular, let F be the translation given by $Fp = q$ and $Fq = p$. It's easy to see then that F is its own inverse. Thus, F is a "self-equivalence" of the theory T. ⌟

In the previous example, F is its own inverse, and it is an exact inverse – i.e., $FF\phi$ is literally the formula ϕ. To formulate a general definition of a syntactic symmetry, both of these conditions can be loosened. First, the inverse of F may be a different translation $G : T \to T$. Second, G need not be an inverse in the strict sense, but only an inverse up to provable equivalence. Thus, we require only that there is a $G : T \to T$ such that $GF \simeq 1_T$ and $FG \simeq 1_T$ – i.e., $F : T \to T$ is an equivalence of theories.

DEFINITION 5.5.2 Let $F : T \to T$ be a translation of a theory T to itself. We say that F is a **syntactic symmetry** just in case F is an equivalence of theories.

DISCUSSION 5.5.3 The previous definition can make one's head spin. Isn't T trivially equivalent to itself? What does it mean to say that $F : T \to T$ is an equivalence? Just remember that whenever we say that two theories are equivalent, that is shorthand for saying that there is at least one equivalence between them. There may be, and typically will be, many different equivalences between them.

Example 5.5.4 Let's slightly change the previous example. Suppose now that T' is the theory in Σ with the single axiom $\vdash p$. Then intuitively, there should not be a symmetry of T' that takes p to q and vice versa. And that intuition can indeed be validated, although we leave the details to the reader. ⌟

Example 5.5.5 Now for a predicate logic example. Let Σ consist of a single binary relation symbol r. As shorthand, let's write $\phi(x, y) \equiv r(y, x)$, which is the "opposite" relation r^{op} of r. Let T be the empty theory in Σ. Now we define a translation $F : T \to T$ by setting $Fr = \phi$. To be more precise,

$$(Fr)(x, y) = \phi(x, y) = r(y, x).$$

In effect, F flips the order of the variables in r. It is easy to see then that $F : T \to T$ is a syntactic symmetry. ⌟

Example 5.5.6 Let's slightly change the previous example. Suppose now that T' is the theory in Σ with the single axiom

$$\vdash \forall x \exists y\, r(x, y).$$

Then there is no syntactic symmetry $F : T' \to T'$ such that $Fr = r^{op}$. Indeed, if there were such a symmetry F, then we would have

$$\forall x \exists y\, r(x, y) \vdash \forall x \exists y\, r(y, x),$$

which is intuitively not the case (and which can indeed be shown not to be the case).

Incidentally, this example shows yet again why it's not always good to identify things that are related by a symmetry. In the previous example, the relations $r(x, y)$ and $r^{op}(x, y)$ are related by a symmetry. A person with Ockhamist leanings may be sorely tempted to say that there is redundancy in the description provided by T, and that a better theory would treat $r(x, y)$ and $r^{op}(x, y)$ as a single relation. However, treating $r(x, y)$ and $r^{op}(x, y)$ as the same relation would foreclose certain possibilities – e.g., the possibility that $\forall x \exists y\, r(x, y)$ holds but $\forall x \exists y\, r^{op}(x, y)$ does not. In summary, redundancy in ideology isn't directly analogous to redundancy in ontology, and we should think twice before applying Ockham's razor at the ideological level. (For discussion of an analogous concrete case, see Belot [1998].) ⌟

EXERCISE 5.5.7 Suppose now that T is the theory in Σ with the single axiom

$$r(x, y) \vdash \neg r(y, x),$$

which says that r is asymmetric. This axiom can be rewritten as

$$r(x, y) \vdash \neg r^{op}(x, y).$$

Show that $Fr = r^{op}$ defines a symmetry of T.

EXERCISE 5.5.8 Show that the theory of a partial order (Example 4.1.1) has a symmetry that maps $\leq$ to the converse relation $\geq$.

Example 5.5.9 In the nineteenth century, mathematicians discovered a neat feature of projective geometry: points and lines play a dual role in the theory. Thus, they realized, every theorem in projective geometry automatically has a dual theorem, where the roles of points and lines have been reversed. In terms of first-order logic, projective geometry is most conveniently formulated within a many-sorted framework. We shall describe it as such in Section 7.4. One can also present projective geometry as a single-sorted theory T, with predicates for "is a point" and "is a line." In this case, the duality of projective geometry is a syntactic symmetry F of T that exchanges these two predicates. The duality of theorems amounts to the fact that $T \vdash \phi$ iff $T \vdash F\phi$.

A similar duality holds for the first-order theory of categories (see 5.1.8). In that case, the symmetry permutes the domain and codomain functions on arrows. One speaks intuitively of "flipping the arrows." However, that way of speaking can be misleading, since it suggests an action on a model (i.e., on a category), and not an action on syntactic objects. As we will soon see (Section 7.2), every syntactic symmetry of a theory does induce a functor on the category of models of that theory. In the case of the theory of categories, this dual functor takes each category $\mathbf{C}$ to its opposite category $\mathbf{C}^{op}$. ⌟

We now consider a special type of syntactic symmetry – a type that we might want to call **inner symmetry** or **continuous symmetry**. (The analogy here is an element of a Lie group that is connected by a continuous path to the identity element.) Suppose that $F : T \to T$ is a self-translation with the feature that $F \simeq 1_T$. That last symbol means, intuitively and loosely, that there is a formula $\chi(x, y)$ of Σ that establishes a bijective correspondence between the original domain of the quantifiers and the restricted domain $D_F(y)$. This bijective correspondence also matches up the extension of ϕ with the extension of $F\phi$, for each formula ϕ of Σ. (All these statements are relative to the theory T.)

The reason we might want to call F an "inner symmetry" is because the theory T itself can "see" that the formulas ϕ and $F\phi$ are coextensive: $T \vdash \phi \leftrightarrow F\phi$. In the general case of a syntactic symmetry, ϕ and $F\phi$ need not be coextensive. (In the first example, we have $Fp = q$, but $T \nvdash p \leftrightarrow q$.)

We claim that whenever this condition holds, i.e., when $F \simeq 1_T$, then F is a syntactic symmetry. Indeed, it's easy to check that $FF \simeq 1_T$, and hence F is an equivalence.

Example 5.5.10 Let Σ be a signature with a single propositional constant p. Let T be the empty theory in Σ. Define a reconstrual F of Σ by setting $Fp = \neg p$. Since Σ is empty, F is a translation. Moreover, since $FFp = \neg\neg p$ and $T \vdash p \leftrightarrow \neg\neg p$, it follows that F is its own quasi-inverse. Therefore, F is a syntactic symmetry. This result is not at all surprising: from the point of view of the empty theory T, p and $\neg p$ play the same sort of role.

Indeed, recall from Chapter 3 that translations between propositional theories correspond to homomorphisms between the corresponding Lindenbaum algebras. In this case, $F : T \to T$ corresponds to an automorphism $f : B \to B$. Moreover, B is the four-element Boolean algebra, and f is the automorphism that flips the two middle elements.

Although F is a syntactic symmetry, it is not the case that $T \vdash p \leftrightarrow Fp$. Therefore, F is not inner. Using the correspondence with Lindenbaum algebras, it's easy to see that T has no nontrivial inner symmetries. Or, to be more precise, if G is an inner symmetry of T, then $G \simeq 1_T$. For example, for $G = FF$, we have $Gp = \neg\neg p$. Here G is not strictly equal to the identity translation 1_T. Rather, for each formula ϕ, we have $T \vdash \phi \leftrightarrow G\phi$. ⌟

Example 5.5.11 Let T be Mette the Mereologist's theory, and let T' be Niels the Nihilist's theory. Recall from 5.4.16 that there is a pair of translations $F : T \to T'$ and $G : T' \to T$ that forms an equivalence. Thus, $GF \simeq 1_T$ and GF is an inner symmetry of Mette's theory. Here GF is the mapping that (intuitively speaking) relativizes Mette's quantifier to the domain of atoms. ⌟

Example 5.5.12 Let $\Sigma = \{\sigma_1, \sigma_2\}$, and let T be the empty theory in Σ. Define a reconstrual $F : \Sigma \to \Sigma$ by setting $F(\sigma_1) = \sigma_2$ and $F(\sigma_2) = \sigma_1$. Then F is a symmetry of T. This symmetry F is the only nontrivial symmetry of T, and it is not deformable to the identity 1_T. (If F were deformable to 1_T, then T would define an isomorphism between σ_1 and σ_2.) In contrast, the empty theory T' in signature $\Sigma' = \{\sigma\}$ has no nontrivial symmetries. It follows that T and T' are not equivalent in the category **Th**. Finally, let T'' be the theory in $\Sigma \cup \{f\}$, where f is a function symbol, and where T'' says that $f : \sigma_1 \to \sigma_2$ is an isomorphism. Then F is still a symmetry of T'', and it *is* is contractible to $1_{T''}$. In fact, it is not difficult to see that T' and T'' are equivalent. This equivalence will send the isomorphism f of T'' to the equality relation for T'. ⌟

The examples we have given were all drawn from first-order logic, and not even from the more complicated parts thereof (e.g., it would be interesting to investigate the syntactic symmetries of first-order axiomatizations of special relativity). The goal has been merely to illustrate the fact that it would be a mistake to consider syntactic symmetries as trivial symmetries; in fact, the syntactic symmetries of a theory tell us a lot about the structure of that theory, and even about the relations between theories. For example, if two theories are equivalent, then they have the same group of syntactic symmetries.

We have also been keen to emphasize that having "redundant syntactic structure" – in particular, having nontrivial syntactic symmetry – is by no means a defect of a theory. Indeed, one of the reasons to allow syntactic redundancy in a theory is to leave open the possibility of future developments of that theory.

5.6 Notes

- For more details on many-sorted logic, see Feferman (1974), Manzano (1993), and Manzano (1996). The last of these also discusses a sense in which second-order logic (with Henkin semantics) is reducible to many-sorted first-order logic. For an application of many-sorted logic in recent metaphysics, see Turner (2010, 2012).
- The concept of Morita equivalence – if not the name – is already familiar in certain circles of logicians. See Andréka et al. (2008) and Mere and Veloso (1992). The name "Morita equivalence" descends from Kiiti Morita's work on rings with equivalent categories of modules. Two rings R and S are said to be Morita equivalent just in case there is an equivalence $\text{Mod}(R) \cong \text{Mod}(S)$ between their categories of modules. The notion was generalized from rings to algebraic theories by Dukarm (1988). See also Adámek et al. (2006). There is also a notion of Morita equivalence for C^*-algebras, see Rieffel (1974). More recently, topos theorists have defined theories to be Morita equivalent just in case their classifying toposes are equivalent (Johnstone, 2003). See Tsementzis (2017b) for a comparison of the topos-theoretic notion of Morita equivalence with ours.
- Price (2009) discusses Quine's criticism of Carnap's *Allwörter*, coming to a similar conclusion as ours – but approaching it from a less technical angle. We agree with Price that in citing the technical result, Quine didn't settle the philosophical debate.

- The notion of a generalized translation between first-order theories seems to have been first described in van Benthem and Pearce (1984), who mention antecedent work by Szczerba (1977) and Gaifman. Our treatment is essentially a generalization of what can be found in Visser (2006); Friedman and Visser (2014); Rooduijn (2015). Our notion of homotopy is inspired by similar notions in Ahlbrandt and Ziegler (1986).
- The implementation of Morita equivalence to first-order logic comes from Barrett and Halvorson (2016b). We claim no originality for the notion of defining new sorts. For example, Burgess (1984) uses "extension by abstractions," which is the same thing as our quotient sorts. See also Mere and Veloso (1992); Andréka et al. (2008).
- Quine's argument for the dispensability of many-sorted logic is discussed by Barrett and Halvorson (2017b).
- For recent considerations on quantifier variance, see Warren (2014); Dorr (2014); Hirsch and Warren (2017).
- For more on symmetry, see Weatherall (2016b); Dewar (2017b); Barrett (2018b).

6 Semantic Metalogic

6.1 The Semantic Turn

Already in the nineteenth century, geometers were proving the relative consistency of theories by interpreting them into well-understood mathematical frameworks – e.g., other geometrical theories or the theory of real numbers. At roughly the same time, the theory of sets was under active development, and mathematicians were coming to realize that the things they were talking about (numbers, functions, etc.) could be seen to be constituted by sets. However, it was only in the middle of the twentieth century that Alfred Tarski gave a precise definition of an *interpretation* of a theory in the universe of sets.

Philosophers of science were not terribly quick to latch onto the new discipline of logical semantics. Early adopters included the Dutch philosopher Evert Beth and, to a lesser extent, Carnap himself. It required a generational change for the semantic approach to take root in philosophy of science. Here we are using "semantic approach" in the broadest sense – essentially for any approach to philosophy of science that is reactionary against Carnap's syntax program, but that wishes to use precise mathematical tools (set theory, model theory, etc.) in order to explicate the structure of scientific theories.

What's most interesting for us is how the shift to the semantic approach influenced shifts in philosophical perspective. Some of the cases are fairly clear. For example, with the rejection of the syntactic approach, many philosophers stopped worrying about the "problem of theoretical terms" – i.e., how scientific theories (with their abstract theoretical terms) connect to empirical reality. According to Putnam, among others, if you step outside the confines of Carnap's *Wissenschaftslogik* program, there is no problem of theoretical terms. (Interestingly, debates about the conventionality of geometry all but stopped around the 1970s, just when the move to the semantic view was in full swing.) Other philosophers diagnosed the situation differently. For example, van Fraassen saw the semantic approach as providing the salvation of empiricism – which, he thought, was incapable of an adequate articulation from a syntactic point of view.

In reading twentieth-century analytic philosophy, it can seem that logical semantics by itself is supposed to obviate many of the problems that exercised the previous generation of philosophers. For example, van Fraassen (1989, p. 222) says that "the semantic view of theories makes language largely irrelevant to the subject [philosophy of science]." Indeed, the picture typically presented to us is that logical semantics deals with mind-independent things (viz. set-theoretic structures), which can stand in mind-independent relations to concrete reality, and to which we have unmediated epistemic access. Such a picture suggests that logical semantics provides a bridge over which we can safely cross the notorious mind–world gap.

But something is fishy with this picture. How could logical semantics get any closer to "the world" than any other bit of mathematics? And why think that set-theoretic structures play this privileged role as intermediaries in our relation to empirical reality? For that matter, why should our philosophical views on science be tied down to some rather controversial view of the nature of mathematical objects? Why the set-theoretic middleman?

In what follows, we will attempt to put logical semantics back in its place. The reconceptualization we're suggesting begins with noting that logical semantics is a particular version of a general mathematical strategy called "representation theory." There is a representation theory for groups, for rings, for C^*-algebras, etc., and the basic idea of all these representation theories is to study one category **C** of mathematical objects by studying the functors from **C** to some other mathematical category, say **S**. It might seem strange that such an indirect approach could be helpful for understanding **C**, and yet, it has proven to be very frutiful. For example, in the representation theory of groups, one studies the representations of a group on Hilbert spaces. Similarly, in the representation theory of rings, one studies the modules over a ring. In all such cases, there is no suggestion that a represented mathematical object is less linguistic than the original mathematical object. If anything, the represented mathematical object has superfluous structure that is not intrinsic to the original mathematical object.

To fully understand that logical semantics is representation theory, one needs to see theories as objects in a category, and to show that "interpretations" are functors from that category into some other one. We carried out that procedure for propositional theories in Chapter 3, where we represented each propositional theory as a Boolean algebra. We could carry out a similar construction for predicate logic theories, but the resulting mathematical objects would be something more complicated than Boolean algebras. (Tarski himself suggested representing predicate logic theories as cylindrical algebras, but a more elegant approach involves syntactic categories in the sense of Makkai and Reyes [1977].) Thus, we will proceed in a different manner and directly define the arrows (in this case, translations) between predicate logic theories. We begin, however, with a little crash course in traditional model theory.

Example 6.1.1 Let T be the theory, in empty signature, that says, "there are exactly two things." A **model** of T is simply a set with two elements. However, every model of T has "redundant information" that is not specified by T itself. To the question "how many models does T have?" there are two correct answers: (1) more than any cardinal number and (2) exactly one (up to isomorphism). ⌟

Example 6.1.2 Let T_1 be the theory of groups, as axiomatized in Example 4.5.3. Then a model M of T_1 is a set S with a binary function $\cdot^M : S \times S \to S$ and a preferred element $e^M \in S$ that satisfy the conditions laid out in the axioms. Once again, every such model

M carries all the structure that T_1 requires of it and then some more structure that T_1 doesn't care about. ┘

In order to precisely define the concept of a model of a theory, we must first begin with the concept of a Σ-structure.

DEFINITION 6.1.3 A **Σ-structure** M is a mapping from Σ to appropriate structures in the category **Sets**. In particular, M fixes a particular set S, and then

- M maps each n-ary relation symbol $p \in \Sigma$ to a subset $M(p) \subseteq S^n = S \times \cdots \times S$.
- M maps each n-ary function symbol $f \in \Sigma$ to a function $M(f) : S^n \to S$.

A Σ-structure M extends naturally to all syntactic structures built out of Σ. In particular, for each Σ-term t, we define $M(t)$ to be a function, and for each Σ-formula ϕ, we define $M(\phi)$ to be a subset of S^n (where n is the number of free variables in ϕ). In order to do so, we need to introduce several auxiliary constructions.

DEFINITION 6.1.4 Let Γ be a finite set of Σ-formulas. We say that $\vec{x} = x_1, \ldots, x_n$ is a **context** for Γ just in case $\vec{x}$ is a duplicate-free sequence that contains all free variables that appear in any of the formulas in Γ. We say that $\vec{x}$ is a **minimal context** for Γ just in case every variable x_i in $\vec{x}$ occurs free in some formula in Γ. Note: we also include, as a context for sentences, the zero-length string of variables.

DEFINITION 6.1.5 Let $\vec{x}$ and $\vec{y}$ be duplicate-free sequences of variables. Then $\vec{x}.\vec{y}$ denotes the result of concatenating the sequences, then deleting repeated variables in order from left to right. Equivalently, $\vec{x}.\vec{y}$ results from deleting from $\vec{y}$ all variables that occur in $\vec{x}$, and then appending the resulting sequence to $\vec{x}$.

DEFINITION 6.1.6 For each term t, we define the **canonical context** $\vec{x}$ of t as follows. First, for a variable x, the canonical context is x. Second, suppose that for each term t_i, the canonical context $\vec{x}_i$ has been defined. Then the canonical context for $f(t_1, \ldots, t_n)$ is $(\cdots((\vec{x}_1.\vec{x}_2)\cdots).\vec{x}_n$.

EXERCISE 6.1.7 Suppose that $\vec{x} = x_1, \ldots, x_n$ is the canonical context for t. Show that $FV(t) = \{x_1, \ldots, x_n\}$.

DEFINITION 6.1.8 For each formula ϕ, we define the **canonical context** $\vec{x}$ of ϕ as follows. First, if $\vec{x}_i$ is the canonical context for t_i, then the canonical context for $t_1 = t_2$ is $\vec{x}_1.\vec{x}_2$, and the canonical context for $p(t_1, \ldots, t_n)$ is $(\cdots((\vec{x}_1.\vec{x}_2)\cdots).\vec{x}_n$. For the Boolean connnectives, we also use the operation $\vec{x}_1.\vec{x}_2$ to combine contexts. Finally, if $\vec{x}$ is the canonical context for ϕ, then the canonical context for $\forall x\phi$ is the result of deleting x from $\vec{x}$, if it occurs.

EXERCISE 6.1.9 Show that the canonical context for ϕ does, in fact, contain all and only those variables that are free in ϕ.

If a Σ-structure M has a domain set S, then it assigns relation symbols to subsets of the Cartesian products,

$$S, S \times S, S^3, \ldots$$

Of course, these sets are all connected to each other by projection maps, such as the projection $S \times S \to S$ onto the first coordinate. We will now develop some apparatus to handle these projection maps. To this end, let $[n]$ stand for the finite set $\{1, \dots, n\}$.

LEMMA 6.1.10 *For each injective function $p : [m] \to [n]$, there is a unique projection $\pi_p : S^n \to S^m$ defined by*

$$\pi_p \langle x_1, \dots, x_n \rangle = \langle x_{p(1)}, \dots, x_{p(m)} \rangle.$$

Furthermore, if $q : [\ell] \to [m]$ is also injective, then $\pi_{p \circ q} = \pi_q \circ \pi_p$.

Proof The first claim is obvious. For the second claim, it's easier if we ignore the variables $x_1, \dots, x_n$ and note that π_p is defined by the coordinate projections:

$$\pi_i \circ \pi_p = \pi_{p(i)},$$

for $i = 1, \dots, m$. Thus, in particular,

$$\pi_i \circ \pi_q \circ \pi_p = \pi_{q(i)} \circ \pi_p = \pi_{p(q(i))} = \pi_i \circ \pi_{p \circ q},$$

which proves the second claim. □

DEFINITION 6.1.11 Let $\vec{x} = x_1, \dots, x_m$ and $\vec{y} = y_1, \dots, y_n$ be duplicate-free sequences of variables. We say that $\vec{x}$ is a **subcontext** of $\vec{y}$ just in case each element in $\vec{x}$ occurs in $\vec{y}$. In other words, for each $i \in [m]$, there is a unique $p(i) \in [n]$ such that $x_i = y_{p(i)}$. Since $i \mapsto y_i$ is injective, $p : [m] \to [n]$ is also injective. Thus, p determines a unique projection $\pi_p : S^n \to S^m$. We say that π_p is the **linking projection** for contexts $\vec{y}$ and $\vec{x}$. If $\vec{x}$ and $\vec{y}$ are canonical contexts of formulas or terms, then we say that π_p is the **linking projection** for these formulas or terms.

We are now ready to complete the extension of the Σ-structure M to all Σ-terms.

DEFINITION 6.1.12 For each term t with n-free variables, we define $M(t) : S^n \to S$.

1. Recall that a constant symbol $c \in \Sigma$ is really a special case of a function symbol, viz. a 0-ary function symbol. Thus, $M(c)$ should be a function from S^0 to S. Also recall that the 0-ary Cartesian product of any set is a one-point set $\{*\}$. Thus, $M(c) : \{*\} \to S$, which corresponds to a unique element $c^M \in S$.
2. For each variable x, we let $M(x) : S \to S$ be the identity function. This might seem like a strange choice, but its utility will soon be clear.
3. Let $t \equiv f(t_1, \dots, t_n)$, where $M(t_i)$ has already been defined. Let n_i be the number of free variables in t_i. The context for t_i is a subcontext of the context for t. Thus, there is a linking projection $\pi_i : S^n \to S^{n_i}$. Whereas the $M(t_i)$ may have different domains (if $n_i \neq n_j$), precomposition with the linking projections makes them functions of a common domain S^n. Thus, we define

$$M[f(t_1, \dots, t_n)] = M(f) \circ \langle M(t_1) \circ \pi_1, \dots, M(t_n) \circ \pi_n \rangle,$$

which is a function from S^n to S.

We illustrate the definition of $M(t)$ with a couple of examples.

Example 6.1.13 Suppose that f is a binary function symbol, and consider the two terms $f(x,y)$ and $f(y,x)$. The canonical context for $f(x,y)$ is x,y, and the canonical context for $f(y,x)$ is y,x. Thus, the linking projection for $f(x,y)$ and x is the projection $\pi_0 : S \times S \to S$ onto the first coordinate; and the linking projection for $f(y,x)$ and x is $\pi_1 : S \times S \to S$ onto the second coordinate. Thus,

$$M(f(x,y)) = M(f) \circ \langle \pi_0, \pi_1 \rangle = M(f).$$

A similar calculation shows that $M(f(y,x)) = M(f)$, which is as it should be: $f(x,y)$ and $f(y,x)$ should correspond to the same function $M(f)$.

However, it does *not* follow that the formula $f(x,y) = f(y,x)$ should be regarded as a semantic tautology. Whenever we place both $f(x,y)$ and $f(y,x)$ into the *same* context, this context serves as a reference point by which the order of inputs can be distinguished. ⌟

DEFINITION 6.1.14 For each formula ϕ of Σ with n distinct free variables, we define $M(\phi)$ to be a subset of $S^n = S \times \cdots \times S$.

1. $M(\bot)$ is the empty set $\emptyset$, considered as a subset of the one-element set 1.
2. Suppose that $\phi \equiv (t_1 = t_2)$, where t_1 and t_2 are terms. Let n_i be the number of free variables in t_i. Since the context for t_i is a subcontext of that for $t_1 = t_2$, there is a linking projection $\pi_i : S^n \to S^{n_i}$. We define $M(t_1 = t_2)$ to be the equalizer of the functions $M(t_1) \circ \pi_1$ and $M(t_2) \circ \pi_2$.
3. Suppose that $\phi \equiv p(t_1, \ldots, t_m)$, where p is a relation symbol and $t_1, \ldots, t_m$ are terms. Let n be the number of distinct free variables in ϕ. Since the context of t_i is a subcontext of that of ϕ, there is a linking projection $\pi_i : S^n \to S^{n_i}$. Then $\langle \pi_1, \ldots, \pi_m \rangle$ is a function from S^n to $S^{n_1} \times \cdots \times S^{n_m}$. We define $M[p(t_1, \ldots, t_m)]$ to be the pullback of $M(p) \subseteq S^m$ along the function
$$\langle M(t_1) \circ \pi_1, \ldots, M(t_m) \circ \pi_m \rangle.$$
4. Suppose that M has already been defined for ϕ. Then we define $M(\neg\phi) = S^n \backslash M(\phi)$.
5. Suppose that ϕ is a Boolean combination of ϕ_1, ϕ_2, and that $M(\phi_1)$ and $M(\phi_2)$ have already been defined. Let π_i be the linking projection for ϕ_i and ϕ, and let π_i^* be the corresponding pullback (preimage) map that takes subsets to subsets. Then we define
$$M(\phi_1 \wedge \phi_2) = \pi_1^*(M(\phi_1)) \cap \pi_2^*(M(\phi_2)),$$
$$M(\phi_1 \vee \phi_2) = \pi_1^*(M(\phi_1)) \cup \pi_2^*(M(\phi_2)),$$
$$M(\phi_1 \to \phi_2) = (S^n \backslash \pi_1^*(M(\phi_1))) \cup \pi_2^*(M(\phi_2)).$$
6. Suppose that $M(\phi)$ is already defined as a subset of S^n. Suppose first that x is free in ϕ, and let $\pi : S^{n+1} \to S$ be the linking projection for ϕ and $\exists x\phi$. Then we define $M(\exists x\phi)$ to be the image of $M(\phi)$ under π, i.e.,
$$M(\exists x\phi) = \{\vec{a} \in S^n \mid \pi^{-1}(\vec{a}) \cap M(\phi) \neq \emptyset\}.$$

If x is not free in ϕ, then we define $M(\exists x\phi) = M(\phi)$.
Similarly, if x is free in ϕ, then we define

$$M(\forall x\phi) = \{\vec{a} \in S^n \mid \pi^{-1}(\vec{a}) \subseteq M(\phi)\}.$$

If x is not free in ϕ, then we define $M(\forall x\phi) = M(\phi)$.

Example 6.1.15 Let's unpack the definitions of $M(x = y)$ and $M(x = x)$. For the former, the canonical context for $x = y$ is x, y. Thus, the linking projection for $x = y$ and x is $\pi_0 : S \times S \to S$ onto the first coordinate, and the linking projection for $x = y$ and y is $\pi_1 : S \times S \to S$ onto the second coordinate. By definition, $M(x) \equiv 1_S \equiv M(y)$, and $M(x = y)$ is the equalizer of $1_s \circ \pi_0$ and $1_S \circ \pi_1$. This equalizer is clearly the diagonal subset of $S \times S$:

$$M(x = y) \equiv \{\langle a, b\rangle \in S \times S \mid a = b\} \equiv \{\langle a, a\rangle \mid a \in S\}.$$

In contrast, the canonical context for $x = x$ is x, and the linking projection for $x = x$ and x is simply the identity. Thus, $M(x = x)$ is defined to be the equalizer of $M(x)$ and $M(x)$, which is the entire set S. That is, $M(x = x) \equiv S$. ⌟

EXERCISE 6.1.16 Describe $M(f(x, y) = f(y, x))$, and explain why it won't neccesarily be the entire set $S \times S$.

We are now going to define a relation $\phi \vDash_M \psi$ of semantic entailment in a structure M; and we will use that notion to define the absolute relation $\phi \vDash \psi$ of semantic entailment. (In short: $\phi \vDash \psi$ means that $\phi \vDash_M \psi$ in every structure M.) Here ϕ and ψ are formulas (not necessarily sentences), so we need to take a bit of care with their free variables. One thing we could do is to consider the sentence $\forall\vec{x}(\phi \to \psi)$, where $\vec{x}$ is any sequence that includes all variables free in ϕ or ψ. However, even in that case, we would have to raise a question about whether the definition depends on the choice of the sequence $\vec{x}$. Since we have to deal with that question in any case, we will instead look more directly at the relation between the formulas ϕ and ψ, which might share some free variables in common.

As a first proposal, we might try saying that $\phi \vDash_M \psi$ just in case $M(\phi) \subseteq M(\psi)$. But the problem with this proposal is that $M(\phi)$ and $M(\psi)$ are typically defined to be subsets of different sets. For example: the definition of $\vDash_M$ should imply that $p(x) \vDash_M (p(x) \vee q(y))$. However, for any Σ-structure M, $M(p(x))$ is a subset of S whereas $M(p(x) \vee q(y))$ is a subset of $S \times S$. The way to fix this problem is to realize that $M(p(x))$ can also be considered to be a subset of $S \times S$. In particular, $p(x)$ is equivalent to $p(x) \wedge (y = y)$, and intuitively $M(p(x) \wedge (y = y))$ should be the subset of $S \times S$ of things satisfying $p(x)$ and $y = y$. In other words, $M(p(x) \wedge (y = y))$ should be $M(p(x)) \times S$.

Here's what we will do next. First we will extend the definition of M so that it assigns a formula ϕ an extension $M_{\vec{x}}(\phi)$ relative to a context $\vec{x}$. Then we will define $\phi \vDash_M \psi$

to mean that $M_{\vec{x}}(\phi) \subseteq M_{\vec{x}}(\psi)$, where $\vec{x}$ is an arbitrarily chosen context for ϕ, ψ. Then we will show that this definition does not depend on which context we chose.

In order to define $M_{\vec{y}}(\phi)$ where $\vec{y}$ is an arbitrary context for ϕ, we will first fix the canonical context $\vec{x}$ for ϕ, and we will set $M_{\vec{x}}(\phi) = M(\phi)$. Then for any other context $\vec{y}$ of which $\vec{x}$ is a subcontext, we will use the linking projection π_p to define $M_{\vec{y}}(\phi)$ as a pullback of $M_{\vec{x}}(\phi)$.

DEFINITION 6.1.17 Let $\vec{y} = y_1, \dots, y_n$ be a context for ϕ, let $\vec{x} = x_1, \dots, x_m$ be the canonical context for ϕ, and let $p : [m] \to [n]$ be the corresponding injection. We define $M_{\vec{y}}(\phi)$ to be the pullback of $M(\phi)$ along π_p. In particular, when $\vec{y} = \vec{x}$, then $p : [n] \to [n]$ is the identity, and $M_{\vec{x}}(\phi) = M(\phi)$.

Now we are ready to define the relation $\phi \vDash_M \psi$.

DEFINITION 6.1.18 For each pair of formulas ϕ, ψ, let $\vec{x}$ be the canonical context for $\phi \to \psi$. We say that $\phi \vDash_M \psi$ just in case $M_{\vec{x}}(\phi) \subseteq M_{\vec{x}}(\psi)$.

We will now show that the definition of $\phi \vDash_M \psi$ is independent of the chosen context $\vec{x}$ for ϕ, ψ. In particular, we show that for any two contexts $\vec{x}$ and $\vec{y}$ for ϕ, ψ, we have $M_{\vec{x}}(\phi) \subseteq M_{\vec{x}}(\psi)$ if and only if $M_{\vec{y}}(\phi) \subseteq M_{\vec{y}}(\psi)$. As the details of this argument are a bit tedious, the impatient reader may wish to skip to Definition 6.1.23.

We'll first check the compatibility of the definitions $M_{\vec{y}}(\phi)$ and $M_{\vec{z}}(\phi)$, where $\vec{y}$ and $\vec{z}$ are contexts for ϕ.

LEMMA 6.1.19 *Suppose that $\vec{x} = x_1, \dots, x_\ell$ is a subcontext of $\vec{y} = y_1, \dots, y_m$, and that $\vec{y}$ is a subcontext of $\vec{z} = z_1, \dots, z_n$. Suppose that $p : [\ell] \to [m]$, $q : [m] \to [n]$, and $r : [\ell] \to [n]$ are the corresponding injections. Then $r = q \circ p$.*

Proof By definition of p, $y_{p(i)} = x_i$ for $i \in [\ell]$. By definition of r, $z_{r(i)} = x_i$ for $i \in [\ell]$. Thus, $y_{p(i)} = z_{r(i)}$. Furthermore, by definition of q, $z_{q(p(i))} = y_{p(i)}$. Therefore, $z_{q(p(i))} = z_{r(i)}$, and $q(p(i)) = r(i)$. □

LEMMA 6.1.20 *Suppose that $\vec{x}$ is a context for ϕ, and that $\vec{x}$ is a subcontext of $\vec{y}$. Let $\pi^r : S^n \to S^m$ be the projection connecting the contexts $\vec{y}$ and $\vec{x}$. Then $M_{\vec{y}}(\phi)$ is the pullback of $M_{\vec{x}}(\phi)$ along π_r.*

Proof Let π_p be the projection connecting $\vec{x}$ to the canonical context for ϕ, and let π_q be the projection connecting $\vec{y}$ to the canonical context for ϕ. Thus, $M_{\vec{x}}(\phi) = \pi_p^*[M(\phi)]$, where π_p^* denotes the operation of pulling back along π_p. Similarly, $M_{\vec{y}}(\phi) = \pi_q^*[M(\phi)]$. Furthermore, $\pi_q = \pi_p \circ \pi_r$, and since pullbacks commute, we have

$$M_{\vec{y}}(\phi) = \pi_q^*[M(\phi)] = \pi_r^*[\pi_p^*[M(\phi)]] = \pi_r^*[M_{\vec{x}}(\phi)],$$

as was to be shown. □

PROPOSITION 6.1.21 *Suppose that $\vec{x}$ is a context for ϕ, ψ, and that $\vec{x}$ is a subcontext of $\vec{y}$. If $M_{\vec{x}}(\phi) \subseteq M_{\vec{x}}(\psi)$ then $M_{\vec{y}}(\phi) \subseteq M_{\vec{y}}(\psi)$.*

Proof Suppose that $M_{\vec{x}}(\phi) \subseteq M_{\vec{x}}(\psi)$. Let $\pi_r : S^n \to S^m$ be the projection connecting the contexts $\vec{y}$ and $\vec{x}$. By the previous lemma, $M_{\vec{y}}(\phi) = \pi_r^*[M_{\vec{x}}(\phi)]$ and $M_{\vec{y}}(\psi) = \pi_r^*[M_{\vec{x}}(\psi)]$. Since pullbacks preserve set inclusion, $M_{\vec{y}}(\phi) \subseteq M_{\vec{y}}(\psi)$. □

Since we defined $\phi \vDash_M \psi$ using a minimal context $\vec{x}$ for ϕ, ψ, we now have the first half of our result: if $\phi \vDash_M \psi$, then $M_{\vec{y}}(\phi) \subseteq M_{\vec{y}}(\psi)$ for any context $\vec{y}$ for ϕ, ψ. To complete the result, we now show that redundant variables can be deleted from contexts.

LEMMA 6.1.22 *Let $\vec{x}$ be a context for ϕ, and suppose that y does not occur in $\vec{x}$. Then $M_{\vec{x}.y}(\phi) = M_{\vec{x}}(\phi) \times S$.*

Proof Let $\vec{x} = x_1, \ldots, x_n$, and let $p : [n] \to [n+1]$ be the injection corresponding to the inclusion of $\vec{x}$ in $\vec{x}.y$. In this case, $p(i) = i$ for $i = 1, \ldots, n$, and $\pi_p : S^{n+1} \to S^n$ projects out the last coordinate. By Lemma 6.1.20, $M_{\vec{x}.y}(\phi)$ is the pullback of $M_{\vec{x}}(\phi)$ along π_p. However, the pullback of any set A along π_p is simply $A \times S$. □

Now suppose that $M_{\vec{x}.y}(\phi) \subseteq M_{\vec{x}.y}(\psi)$, where $\vec{x}$ is a context for ϕ, ψ, and y does not occur in $\vec{x}$. Then the previous lemma shows that $M_{\vec{x}.y}(\phi) = M_{\vec{x}}(\phi) \times S$ and $M_{\vec{x}.y}(\psi) = M_{\vec{x}}(\psi) \times S$. Thus, $M_{\vec{x}.y}(\phi) \subseteq M_{\vec{x}.y}(\psi)$ if and only if $M_{\vec{x}}(\phi) \subseteq M_{\vec{x}}(\psi)$. A quick inductive argument then shows that any number of appended empty variables makes no difference.

We can now conclude the argument that $M_{\vec{x}}(\phi) \subseteq M_{\vec{x}}(\psi)$ if and only if $M_{\vec{y}}(\phi) \subseteq M_{\vec{y}}(\psi)$, where $\vec{x}$ is a subcontext of $\vec{y}$. The "if" direction was already shown in Prop. 6.1.21. For the "only if" direction, suppose that $M_{\vec{y}}(\phi) \subseteq M_{\vec{y}}(\psi)$. First use Prop. 6.1.21 again to move any variables not in $\vec{x}$ to the end of the sequence $\vec{y}$. (Recall that $\vec{y}$ is a subcontext of any permutation of $\vec{y}$.) Then use the previous lemma to eliminate these variables. The resulting sequence is a permutation of $\vec{x}$, hence a subcontext of $\vec{x}$. Finally, use Prop. 6.1.21 one more time to show that $M_{\vec{x}}(\phi) \subseteq M_{\vec{x}}(\psi)$. Thus, we have shown that the definition of $\phi \vDash_M \psi$ is independent of the context chosen for ϕ, ψ.

DEFINITION 6.1.23 We say that ϕ **semantically entails** ψ, written $\phi \vDash \psi$, just in case $\phi \vDash_M \psi$ for every Σ-structure M. We write $\vDash \psi$ as shorthand for $\top \vDash \psi$.

NOTE 6.1.24 The canonical context $\vec{x}$ for the pair $\{\top, \phi\}$ is simply the context for ϕ. By definition, $M_{\vec{x}}(\top)$ is the pullback of 1 along the unique map $\pi : S^n \to 1$. Thus, $M_{\vec{x}}(\top) = S^n$, and $\top \vDash_M \phi$ if and only if $M(\phi) = S^n$.

We're now ready for two of the most famous definitions in mathematical philosophy.

Truth in a Structure

A sentence ϕ has zero free variables. In this case, $M(\phi)$ is defined to be a subset of $S^0 = 1$, a one-element set. We say that ϕ is **true** in M if $M(\phi) = 1$, and we say that ϕ is **false** in M if $M(\phi) = \emptyset$.

Model

Let T be a theory in signature Σ, and let M be a Σ-structure. We say that M is a **model** of T just in case: for any sentence ϕ of Σ, if $T \vdash \phi$, then $M(\phi) = 1$.

6.2 The Semantic View of Theories

In Chapter 4, we talked about how Rudolf Carnap used syntactic metalogic to explicate the notion of a scientific theory. By the 1960s, people were calling Carnap's picture the "syntactic view of theories," and they were saying that something was fundamentally wrong with it. According to Suppe (2000), the syntactic view of theories died in the late 1960s (March 26, 1969, to be precise) after having met with an overwhelming number of objections in the previous two decades. Upon the death of the syntactic view, it was unclear where philosophy of science would go. Several notable philosophers – such as Feyerabend and Hanson – wanted to push philosophy of science away from formal analyses of theories. However, others such as Patrick Suppes, Bas van Fraassen, and Fred Suppe saw formal resources for philosophy of science in other branches of mathematics, most particularly set theory and model theory. Roughly speaking, the "semantic view of theories" designates proposals to explicate theory-hood by means of semantic metalogic.

We now have the technical resources in place to state a preliminary version of the semantic view of theories:

(SV) A scientific theory is a class of Σ-structures for some signature Σ.

Now, proponents of the semantic view will balk at SV for a couple of reasons. First, semanticists stress that a scientific theory has two components:

1. A theoretical definition and
2. A theoretical hypothesis.

The theoretical definition, roughly speaking, is intended to replace the first component of Carnap's view of theories. That is, the theoretical definition is intended to specify some abstract mathematical object – the thing that will be used to do the representing. Then the theoretical hypothesis is some claim to the effect that some part of the world can be represented by the mathematical object given by the theoretical definition. So, to be clear, SV here is only intended to give one-half of a theory, viz. the theoretical definition. I am not speaking yet about the theoretical hypothesis.

But proponents of the semantic view will balk for a second reason: SV makes reference to a signature Σ. And one of the supposed benefits of the semantic view was to free us from the language dependence implied by the syntactic view. So, how are we to modify SV in order to maintain the insight that a scientific theory is independent of the language in which it is formulated?

I will give two suggestions, the first of which I think cannot possibly succeed. The second suggestion works, but it shows that the semantic view actually has no advantage over the syntactic view in being "free from language dependence."

How then to modify SV? The first suggestion is to formulate a notion of mathematical structure that makes no reference to language. At first glance, it seems simple enough to do so. The paradigm case of a mathematical structure is supposed to be an ordered n-tuple $\langle X, R_1, \ldots R_n \rangle$, where X is a set, and $R_1, \ldots, R_n$ are relations on X. (This notion of mathematical structure follows in the footsteps of Bourbaki [1970], which, incidentally, has been rendered obsolete by category theory.) Consider, for example, the proposal made by Lisa Lloyd:

> In our discussion, a *model* is not such an interpretation [i.e., not an Σ-structure], matching statements to a set of objects which bear certain relations among themselves, but the set of objects itself. That is, models should be understood as structures; in the cases we shall be discussing, they are mathematical structures, i.e., a set of mathematical objects standing in certain mathematically representable relations. (Lloyd, 1984, p. 30)

However, it's difficult to make sense of this proposal. Consider the following example.

Example 6.2.1 Let a be an arbitrary set, and consider the following purported example of a mathematical structure:

$$M = \big\langle \{a, b, \langle a, a \rangle\}, \{\langle a, a \rangle\} \big\rangle.$$

That is, the domain X consists of three elements $a, b, \langle a, a \rangle$, and the indicated structure is the singleton set containing $\langle a, a \rangle$. But how are we supposed to understand this structure? Are we supposed to consider $\{\langle a, a \rangle\}$ to be a subset of X or as a subset of $X \times X$? The former is a structure for a signature Σ with a single unary predicate symbol; the latter is a structure for a signature Σ' with a single binary relation symbol. In writing down M as an ordered n-tuple, we haven't yet fully specified an intended mathematical structure.

We conclude then that a mathematical structure cannot simply be, "a set of mathematical objects standing in certain mathematically representable relations." To press the point further, consider another purported example of a mathematical structure:

$$N = \big\langle \{a, b, \langle a, b \rangle\}, \{\langle a, b \rangle\} \big\rangle.$$

Are M and N isomorphic structures? Once again, the answer is underdetermined. If M and N are supposed to be structures for a signature Σ with a single unary predicate symbol, then the answer is yes. If M and N are supposed to be structures for a signature Σ' with a single binary relation symbol, then the answer is no. ⌟

Thus, it's doubtful that there is any "language-free" account of mathematical structures, and hence no plausible language-free semantic view of theories. I propose then that we embrace the fact that we are "suspended in language," to borrow a phrase from Niels Bohr. To deal with our language dependence, we need to consider notions of equivalence of theory-formulations – so that the same theory can be formulated in different languages. And note that this stratagem is available for both semantic and syntactic views of theories. Thus, "language independence" is not a genuine advantage of the semantic view of theories as against the syntactic view of theories.

Philosophical Moral

It is of crucial importance that we do not think of a Σ-structure M as representing the world. To say that the world is isomorphic to, or even partially isomorphic to, or even similar to, M, would be to fall into a profound confusion.

A Σ-structure M is *not* a "set-theoretic structure" in any direct sense of that phrase. Rather, M is a function whose domain is Σ and whose range consists of some sets, subsets, and functions between them. If one said that "M represents the world," then one would be saying that the world is represented by a mathematical object of type $\Sigma \to$ **Sets**. Notice, in particular, that M has "language" built into its very definition.

6.3 Soundness, Completeness, Compactness

We now prove versions of four central metalogical results: soundness, completeness, compactness, and Löwnheim–Skölem theorems. For these results, we will make a couple of simplifying assumptions, merely for the sake of mathematical elegance. We will assume that Σ is fixed signature that is countable and that has no function symbols. This assumption will permit us to use the topological techniques introduced by Rasiowa and Sikorski (1950).

Soundness

In its simplest form, the soundness theorem shows that for any sentence ϕ, if ϕ is provable ($\top \vdash \phi$), then ϕ is true in all Σ-structures ($\top \vDash \phi$). Inspired by categorical logic, we derive this version of soundness as a special case of a more general result for Σ-formulas. We show that: for any Σ-formulas ϕ and ψ, and for any context $\vec{x}$ for $\{\phi, \psi\}$, if $\phi \vdash_{\vec{x}} \psi$, then $M_{\vec{x}}(\phi) \subseteq M_{\vec{x}}(\psi)$.

The proof proceeds by induction on the construction of proofs – i.e., over the definition of the relation $\vdash$. Most cases are trivial verifications, and we leave them to the reader. We will just consider the case of the existential elimination rule, which we consider in the simple form:

$$\frac{\phi \vdash_{x,y} \psi}{\exists y \phi \vdash_x \psi}$$

assuming that y is not free in ψ. We assume then that the result holds for the top line – i.e., $M_{x,y}(\phi) \subseteq M_{x,y}(\psi)$. By definition, $M_x(\exists y \phi)$ is the image of $M_{x,y}(\phi)$ under the projection $X \times Y \to X$. And since y is not free in ψ, $M_{x,y}(\psi) = M_x(\psi) \times Y$.

To complete the argument, it will suffice to make the following general observation about sets: if $A \subseteq X \times Y$ and $B \subseteq X$, then the following inference is valid:

$$\frac{A \subseteq \pi^{-1}(B)}{\pi(A) \subseteq B}.$$

Indeed, suppose that $z \in \pi(A)$, which means that there is a $y \in Y$ such that $\langle z, y \rangle \in A$. By the top line, $\langle z, y \rangle \in \pi^{-1}(B)$, which means that $z = \pi\langle z, y \rangle \in B$. Now set $A = M_{x,y}(\phi)$ and $B = M_x(\psi)$, and it follows that existential elimination is sound.

We leave the remaining steps of this proof to the reader, and briefly comment on the philosophical significance (or lack thereof) of the soundness theorem. (The discussion here borrows from the ideas of Michaela McSweeney. See McSweeney [2016b].) Philosophers often gloss this theorem as showing that the derivation rules are "safe" – i.e., that they don't permit derivations which are not valid, or even more strongly, that the rules won't permit us to derive a false conclusion from true premises. But now we have a bit of a philosophical conundrum. What is this standard of validity against which we are supposed to measure $\vdash$? Moreover, why think that this other standard of validity is epistemologically prior to the standard of validity we have specified with the relation $\vdash$?

Philosophers often gloss the relation $\vDash$ in terms of "truth preservation." They say that $\phi \vDash \psi$ *means that* whenever ϕ is true, then ψ is true. Such statements can be highly misleading, if they cause the reader to think that $\vDash$ is the intuitive notion of truth preservation. No, the relation $\vDash$ is yet another attempt to capture, in a mathemtically precise fashion, our intuitive notion of logical consequence. We have two distinct ways of representing this intuitive notion: the relation $\vdash$ and the relation $\vDash$. The soundness and completeness theorems happily show that we've captured the same notion with two different definitions.

The important point here is that *logical syntax and logical semantics are enterprises of the same kind.* The soundness and completeness theorems are not theorems about how mathematics relates to the world, nor are they theorems about how a mathematical notion relates to an intuitive notion. No, these theorems demonstrate a relationship between mathematical things.

The soundness theorem has sometimes been presented as an "absolute consistency" result – i.e., that the predicate calculus is consistent *tout court.* But such presentations are misleading: The soundness theorem shows only that the predicate calculus is consistent relative to the relation $\vDash$, i.e., that the relation $\vdash$ doesn't exceed the relation $\vDash$. It doesn't prove that there is no sentence ϕ such that $\vDash \phi$ and $\vDash \neg\phi$. We agree, then, with David Hilbert: the only kind of formal consistency is relative consistency.

Completeness

In Chapter 3, we saw that the completeness theorem for propositional logic is equivalent to the Boolean ultrafilter axiom (i.e., every nonzero element in a Boolean algebra is contained in an ultrafilter). In many textbooks of logical metatheory, the completeness theorem for predicate logic uses Zorn's lemma, which is a variant of the axiom of choice (AC). It is known, however, that the completeness theorem does not require the full strength of AC. The proof we give here uses the Baire category theorem, which is derivable in ZF with the addition of the axiom of dependent choices, a slightly weaker choice principle. (Exercise: can you see where in the proof we make use of a choice principle?)

THEOREM 6.3.1 (Baire category theorem) *Let X be a compact Hausdorff space, and let $U_1, U_2, \dots$ be a countable family of sets, all of which are open and dense in X. Then $\bigcap_{i=1}^{\infty} U_i$ is dense in X.*

Proof Let $U = \bigcap_{i=1}^{\infty} U_i$, and let O be a nonempty open subset of X. We need only show that $O \cap U$ is nonempty. To this end, we inductively define a family O_i of open subsets of X as follows:

- $O_1 = O \cap U_1$, which is open, and nonempty since U_1 is dense;
- Assuming that O_n is open and nonempty, it has nonempty intersection with U_{n+1}, since the latter is dense. But any point $x \in O_n \cap U_{n+1}$ is contained in a neighborhood O_{n+1} such that $O_{n+1} \subseteq U_{n+1}$, and $\overline{O}_{n+1} \subseteq O_n$, using the regularity of X.

It follows then that the collection $\{\overline{O}_i : i \in \mathbb{N}\}$ satisfies the finite intersection property. Since X is compact, there is a p in $\bigcap_{i=1}^{\infty} \overline{O}_i$. Since $\overline{O}_{i+1} \subseteq O_i$, it also follows that $p \in O_i \subseteq U_i$, for all i. Therefore, $O \cap U$ is nonempty. □

Our proof of the completess theorem for predicate logic is similar in conception to the proof for propositional logic. First we construct a Boolean algebra B of provably-equivalent formulas. Using the definition of $\vdash$, it is not difficult to see that the equivalence relation is compatible with the Boolean operations. Thus, we may define Boolean operations as follows:

$$[\phi] \cap [\psi] = [\phi \wedge \psi], \qquad [\phi] \cup [\psi] = [\phi \vee \psi], \qquad -[\phi] = [\neg\phi].$$

If we let $0 = [\bot]$ and $1 = [\top]$, then it's easy to see that $\langle B, 0, 1, \cap, \cup, - \rangle$ is a Boolean algebra.

Now we want to show that if ϕ is not provably equivalent to a contradition, then there is a Σ-structure M such that $M(\phi)$ is not empty. In the case of propositional logic, it was enough to show that there is a homomorphism $f : B \to 2$ such that $f(\phi) = 1$. But that won't suffice for predicate logic, because once we have this homomorphism $f : B \to 2$, we need to use it to build a Σ-structure M, and to show that $M(\phi)$ is not empty. As we will now see, to ensure that $M(\phi)$ is not empty, we must choose a homomorphism $f : B \to 2$ that is "smooth on existentials."

DEFINITION 6.3.2 Let $f : B \to 2$ be a homomorphism. We say that f is **smooth on existentials** just in case for each formula ψ, if $f(\exists x \psi) = 1$, then $f(\psi[x_i/x]) = 1$ for some $i \in \mathbb{N}$.

We will see now that these "smooth on existentials" homomorphisms are dense in the Stone space X of B. In fact, the argument here is quite general. We first show that for any particular convergent family $a_i \to a$ in a Boolean algebra, the set of non-smooth homomorphisms is closed and has empty interior. By saying that $a_i \to a$ is convergent, we mean that $a_i \leq a$ for all i, and for any $b \in B$, if $a_i \leq b$ for all i, then $a \leq b$. That is, a is the least upper bound of the a_i.

Let's say that a homomorphism $f : B \to 2$ is **smooth** relative to the convergent family $a_i \to a$ just in case $f(a_i) \to f(a)$ in the Boolean algebra 2. Now let D be the set of homomorphisms $f : B \to 2$ such that f is *not* smooth on $a_i \to a$. Any homomorphism $f : B \to 2$ preserves order, and hence $f(a_i) \leq f(a)$ for all i. Thus, if $f(a_i) = 1$ for any i, then f is smooth on $a_i \to a$. It follows that

$$D = E_a \cap \left[\bigcap_{i \in I} E_{\neg a_i} \right].$$

As an intersecton of closed sets, D is closed. To see that D has empty interior, suppose that $f \in E_b \subseteq D$, where E_b is a basic open subset of X. Then we have $E_b \subseteq E_{\neg a_i}$, which implies that $a_i \leq \neg b$; and since $a_i \leq a$, we have $a_i \leq a \wedge \neg b$. Thus, $a \wedge \neg b$ is an upper bound for the family $\{a_i\}$. Moreover, if $a = a \wedge \neg b$, then $a \wedge b = 0$ in contradiction with the fact that $f(a \wedge b) = 1$. Therefore, a is not the upper bound of $\{a_i\}$, a contradiction. We conclude that D contains no basic open subsets, and hence it has empty interior.

Now, this general result about smooth homomorphisms is of special importance for the Boolean algebra of equivalence classes of formulas. For in this case, existential formulas are the least upper bound of their instances.

LEMMA 6.3.3 *Let ϕ be a Σ-formula, and let I be the set of indices such that x_i does not occur free in ϕ. Then in the Lindenbaum algebra, $E_{(\exists x \phi)}$ is the least upper bound of $\{E_{(\phi[x_i/x])} \mid i \in I\}$.*

Proof For simplicity, set $E = E_{(\exists x \phi)}$ and $E_i = E_{(\phi[x_i/x])}$. The $\exists$-intro rule shows that $E_i \leq E$. Now suppose that $E_\psi \in B$ such that $E_i \leq E_\psi$ for all $i \in \mathbb{N}$. That is, $\phi[x_i/x] \vdash \psi$ for all $i \in I$. Since ϕ and ψ have a finite number of free variables, there is some $i \in I$ such that x_i does not occur free in ψ. By the $\exists$-elim rule, $\exists x_i \phi[x_i/x] \vdash \psi$. Since x_i does not occur free in ϕ, $\exists x_i \phi[x_i/x]$ is equivalent to $\exists x \phi$. Thus, $\exists x \phi \vdash \psi$, and $E \leq E_\psi$. Therefore, E is the least upper bound of $\{E_i \mid i \in I\}$. □

Thus, for each existential Σ-formula ϕ, the clopen set E_ϕ is the union of the clopen subsets corresponding to the instances of ϕ, plus the meager set D_ϕ of homomorphisms that are not smooth relative to ϕ. Since the signature Σ is countable, there are countably many such existential formulas, and countably many of these sets D_ϕ of non-smooth homomorphisms. Since each D_ϕ is meager, the Baire category theorem entails that their union also is meager. Thus, the set U of homomorphisms that are smooth on *all* existentials is open and dense in the Stone space X.

We are now ready to continue with the completeness theorem. Let ϕ be our arbitrary formula that is not provably equivalent to a contradiction. We know that the set E_ϕ of homomorphisms $f : B \to 2$ such that $f([\phi]) = 1$ is open and nonempty. Hence, E_ϕ has nonempty intersection with U. Let $f \in E_\phi \cap U$. That is, $f([\phi]) = 1$, and f is smooth on all existentials. We now use f to define a Σ-structure M.

- Let the domain S of M be the set of natural numbers.
- For an n-ary relation symbol $R \in \Sigma$, let $\vec{a} \in M(R)$ if and only if $f(R(x_{a_1}, \ldots, x_{a_n})) = 1$.

LEMMA 6.3.4 *For any Σ-formula ϕ with canonical context $x_{c_1}, \ldots, x_{c_n}$, if $f(\phi) = 1$, then $\vec{c} \in M(\phi)$.*

Proof We prove this result by induction on the construction of ϕ. Note that an n-tuple $\vec{c}$ of natural numbers corresponds to a unique function $c : [n] \to \mathbb{N}$. Supposing that we are given a fixed enumeration $x_1, x_2, \ldots$ of the variables of Σ, each such function c also corresponds to an n-tuple $x_{c_1}, \ldots, x_{c_n}$, possibly with duplicate variables. Since each formula ϕ determines a canonical context (without duplicates), ϕ also determines an injection $a : [n] \to \mathbb{N}$. For any other function $c : [n] \to \mathbb{N}$, we let ϕ_c denote the result of replacing all free occurences of x_{a_i} in ϕ with x_{c_i}.

1. Suppose that $\phi \equiv R(x_{a_1}, \ldots, x_{a_m})$, and let $x_{c_1}, \ldots, x_{c_n}$ be the canonical context of ϕ. Thus, for each $i = 1, \ldots, m$, there is a $p(i)$ such that $x_{a_i} = x_{c_{p(i)}}$. Now, $M(\phi)$ is defined to be the pullback of $M(R)$ along π_p. Since $\pi_i \pi_p(\vec{c}) = c_{p(i)} = a_i$ and $\vec{a} \in M(R)$, it follows that $\vec{c} \in M(\phi)$.
2. Suppose that the result is true for ϕ and ψ, and suppose that $f(\phi \wedge \psi) = 1$. Let $\vec{x} = x_{c_1}, \ldots, x_{c_n}$ be the canonical context of $\phi \wedge \psi$. The context of ϕ is a subsequence of $\vec{x}$, i.e., it is of the form $x_{c_{p(1)}}, \ldots, x_{c_{p(m)}}$ where $p : [m] \to [n]$ is an injection. If $\pi_p : S^n \to S^m$ is the corresponding projection, then

$$\pi_p(\vec{c}) = \langle c_{p(1)}, \ldots, c_{p(m)} \rangle .$$

Similarly, if $x_{c_{q(1)}}, \ldots, x_{c_{q(\ell)}}$ is the context of ψ, then

$$\pi_q(\vec{c}) = \langle c_{q(1)}, \ldots, c_{q(\ell)} \rangle .$$

Since $f(\phi) = 1 = f(\psi)$, the inductive hypothesis entails that $\pi_p(\vec{c}) \in M(\phi)$ and $\pi_q(\vec{c}) \in M(\psi)$. By definition, $M(\phi \wedge \psi) = \pi_p^*(M(\phi)) \cap \pi_q^*(M(\psi))$, hence $\vec{c} \in M(\phi \wedge \psi)$ iff $\pi_p(\vec{c}) \in M(\phi)$ and $\pi_q(\vec{c}) \in M(\psi)$.
3. Suppose that $\phi \equiv \exists x_k \psi$, and that the result is true for ψ, as well as for any ψ' that results from uniform replacement of free variables in ψ. Suppose first that x_k is free in ψ. For notational simplicity, we will assume that x_k is the last variable in the canonical context for ψ. Thus, if the context for ϕ is $x_{c_1}, \ldots, x_{c_n}$, then the context for ψ is $x_{c_1}, \ldots, x_{c_n}, x_k$. (In the case where ϕ is a sentence, i.e., $n = 0$, the string $\vec{c}$ is empty.)

   Now suppose that $f(\exists x_k \psi) = 1$. Since f is smooth on existentials, there is a $j \in \mathbb{N}$ such that x_j is not free in ψ, and $f(\psi[x_j/x_k]) = 1$. The context of $\psi[x_j/x_k]$ is $x_{c_1}, \ldots, x_{c_n}, x_j$, and the inductive hypothesis entails that $\vec{c}, j \in M(\psi[x_j/x_k])$. By the definition of $M(\exists x_k \psi)$, if $\vec{c}, j \in M(\psi[x_j/x_k])$, then $\vec{c} = \pi(\vec{c}, j) \in M(\exists x_k \psi)$.

The remaining inductive steps are similar to the preceding steps, and are left to the reader. □

This lemma concludes the proof of the completeness theorem, and immediately yields two other important model-theoretic results.

THEOREM 6.3.5 (Downward Löwenheim–Skølem) *Let Σ be an countable signature, and let ϕ be a Σ-sentence. If ϕ has a model, then ϕ also has a countable model.*

Proof If ϕ has a model, then, by the soundness theorem, ϕ is not provably equivalent to a contradiction. Thus, by the completeness theorem, ϕ has a model whose domain is the natural numbers. □

DISCUSSION 6.3.6 The downward Löwenheim–Skølem theorem does not hold for arbitrary sets of sentences in uncountable signatures. Indeed, let $\Sigma = \{c_r \mid r \in \mathbb{R}\}$, and let T be the theory with axioms $c_r \neq c_s$ when $r \neq s$. Then T has a model (for example, the real numbers $\mathbb{R}$) but no countable model.

The Löwenheim–Skølem theorem has sometimes been thought to be paradoxical, particularly in application to the case where T is the theory of sets. The theory of sets implies a sentence ϕ whose intended interpretation is, "there is an uncountable set." The LS theorem implies that if T has any model, then it has a countable model M, and hence that $\vDash_M \phi$. In other words, there is a countable model M that makes true the sentence, "there is an uncountable set."

THEOREM 6.3.7 (Compactness) *Suppose that T is a set of Σ-sentences. If each finite subset of T has a model, then T has a model.*

It would be nice to be able to understand the compactness theorem for predicate logic directly in terms of the compactness of the Stone space of the Lindenbaum algebra. However, this Stone space isn't exactly the space of Σ-structures, and so its compactness isn't the same thing as compactness in the logical sense. We could indeed use each point $f \in X$ to define a Σ-structure M; but, in general, $f(\phi) = 1$ wouldn't entail that $M(\phi) = 1$. What's more, there are additional Σ-structures that are not represented by points in X, in particular, Σ-structures with uncountably infinite domains. Thus, we are forced to turn to a less direct proof of the compactness theorem.

Proof We first modify the proof of the completeness theorem by constructing the Boolean algebra B_T of equivalence classes of formulas modulo T-provable equivalence. This strengthened completeness theorem shows that if $T \vDash \phi$, then $T \vdash \phi$. However, if $T \vdash \phi$, then $T_0 \vdash \phi$ for some finite subset T_0 of T. □

DISCUSSION 6.3.8 The compactness theorem yields all sorts of surprises. For example, it shows that there is a model that satisfies all of the axioms of the natural numbers, but which has a number greater than all natural numbers. Let Σ consist of a signature for arithmetic and one additional constant symbol c. We assume that Σ has a name n for each natural number. Now let

$$T = Th(\mathbb{N}) \cup \{n < c \mid n \in \mathbb{N}\},$$

where $Th(\mathbb{N})$ consists of all Σ-sentences true in $\mathbb{N}$. It's easy to see that each finite subset of T of consistent. Therefore, by compactness, T has a model M. In the model M, $n^M < c^M$ for all $n \in \mathbb{N}$.

6.4 Categories of Models

There are many interesting categories of mathematical objects such as sets, groups, topological spaces, smooth manifolds, rings, etc. Some of these categories are of special interest for the empirical sciences, as the objects in thos categories are the "models" of a scientific theory. For example, a model of Einstein's general theory of relativity (GTR) is a smooth manifold with Lorentzian metric. Hence, the mathematical part of GTR can be considered to be some particular category of manifolds. (The choice of arrows for this category of models raises interesting theoretical questions. See, e.g., Fewster [2015].) Similarly, since a model of quantum theory is a complex vector space equipped with some particular dynamical evolution, the mathematical part of quantum theory can be considered to be some category of vector spaces.

Philosophers of science want to talk about real-life scientific theories – not imaginary theories that can be axiomatized in first-order logic. Nonetheless, we can benefit tremendously from considering tractable formal analogies, what scientists themselves would call "toy models." In this section, we pursue an analogy between models of a scientific theory and models of a first-order theory T. In particular, we show that any first-order theory T has a category $\mathrm{Mod}(T)$ of models, and intertranslatable theories have equivalent categories of models. Thus, we can think of the 2-category of all categories of models of first-order theories as a formal analogy to the universe of all scientific theories.

There are two natural definitions of arrows in the category $\mathrm{Mod}(T)$, one more liberal (homomorphism) and another more conservative (elementary embedding).

DEFINITION 6.4.1 Let Σ be a fixed signature, and let M and N be Σ-structures. We will use X and Y to denote their respective domain sets. A **Σ-homomorphism** $h : M \to N$ consists of a function $h : X \to Y$ that satisfies the following:

1. For each relation symbol $R \in \Sigma$, there is a commutative diagram:

$$\begin{array}{ccc} MR & \longrightarrow & NR \\ \downarrow & & \downarrow \\ X^n & \xrightarrow{h^n} & Y^n \end{array}$$

   Here the arrows $MR \rightarrowtail X^n$ and $NR \rightarrowtail Y^n$ are the subset inclusions, and $h^n : X^n \to Y^n$ is the map defined by $h^n\langle a_1, \ldots, a_n\rangle = \langle h(a_1), \ldots, h(a_n)\rangle$. The fact that the diagram commutes says that for any $\langle a_1, \ldots, a_n\rangle \in MR$, we have $\langle h(a_1), \ldots, h(a_n)\rangle \in NR$.
2. For each function symbol $f \in \Sigma$, the following diagram commutes:

$$\begin{array}{ccc} X^n & \xrightarrow{h^n} & Y^n \\ \downarrow{\scriptstyle Mf} & & \downarrow{\scriptstyle Nf} \\ X & \xrightarrow{h} & Y \end{array}$$

In other words, for each $\langle a_1, \ldots, a_n \rangle \in X^n$, we have $h(M(f)\langle a_1, \ldots, a_n \rangle) = N(f)\langle h(a_1), \ldots, h(a_n) \rangle$. When c is a constant symbol, this condition implies that $h(c^M) = c^N$.

DEFINITION 6.4.2 Let M and N be Σ-structures, and let $h : M \to N$ be a homomorphism. We say that h is a Σ-**elementary embedding** just in case for each Σ-formula ϕ, the following is a pullback diagram:

$$\begin{array}{ccc} M(\phi) & \longrightarrow & N(\phi) \\ \downarrow & & \downarrow \\ X^n & \xrightarrow{h^n} & Y^n \end{array}$$

In other words, for all $\vec{a} \in X^n$, $\vec{a} \in M(\phi)$ iff $h(\vec{a}) \in N(\phi)$. In particular, for the case where ϕ is a sentence, the following is a pullback:

$$\begin{array}{ccc} M(\phi) & \longrightarrow & N(\phi) \\ \downarrow & & \downarrow \\ 1 & \longrightarrow & 1 \end{array}$$

which means that $M \vDash \phi$ iff $N \vDash \phi$.

EXERCISE 6.4.3 Show that the composite of elementary embeddings is an elementary embedding.

Note that the conditions for being an elementary embedding are quite strict. For example, let ϕ be the sentence that says there are exactly n things. If $h : M \to N$ is an elementary embedding, then $M \vDash \phi$ iff $N \vDash \phi$. Thus, if the domain X of M has cardinality $n < \infty$, then Y also has cardinality $n < \infty$. Suppose, for example, that T is the theory of groups. Then for any two finite groups G, H, there is an elementary embedding $h : G \to H$ only if $|G| = |H|$. Therefore, the notion of elementary embedding is stricter than the notion of a group homomorphism.

Similarly, let Σ be the empty signature. Let M be a Σ-structure with one element, and let N be a Σ-structure with two elements. Then any mapping $h : M \to N$ is a homomorphism, since Σ is empty. However, $\vDash_M x = y$ but $\nvDash_N x = y$. Therefore, there is no elementary embedding $h : M \to N$.

The strictness of elementary embeddings leads to a little dilemma in choosing arrows in our definition of the category $\text{Mod}(T)$ of models of a theory T. Do we choose homomorphisms between models, of which there are relatively many, or do we choose elementary embeddings, of which there are relatively few? We have opted to play it safe.

DEFINITION 6.4.4 We henceforth use $\text{Mod}(T)$ to denote the category whose objects are models of T, and whose arrows are elementary embeddings between models. As with any category, we say that an arrow $f : M \to N$ in $\text{Mod}(T)$ is an isomorphism just in case there is an arrow $g : N \to M$ such that $g \circ f = 1_M$ and $f \circ g = 1_N$. In this particular case, we say that f is a Σ-**isomorphism**.

There is a clear sense in which elementary embeddings between models of T are structure that is definable in terms of T. In short, elementary embeddings between models should be considered to be part of the semantic content of the theory T. Accordingly, formally equivalent theories ought at least to have equivalent categories of models. We elevate this idea to a definition.

DEFINITION 6.4.5 Let T and T' be theories, not necessarily in the same signature. We say that T and T' are **categorically equivalent** just in case the categories $\text{Mod}(T)$ and $\text{Mod}(T')$ are equivalent.

Notice that if we had chosen all homomorphisms as arrows, then $\text{Mod}(T)$ would have more structure, and it would be more difficult for the categories $\text{Mod}(T)$ and $\text{Mod}(T')$ to be equivalent. In fact, there are theories T and T' that most mathematicians would consider to be equivalent, but which this criterion would judge to be inequivalent.

DEFINITION 6.4.6 If M is a Σ-structure, we let $Th(M)$ denote the theory consisting of all Σ-sentences ϕ such that $M \vDash \phi$.

DEFINITION 6.4.7 Let M and N be Σ-structures. We say that M and N are **elementarily equivalent**, written $M \equiv N$, just in case $Th(M) = Th(N)$.

EXERCISE 6.4.8 Show that if $h : M \to N$ is an isomorphism, then M and N are elementarily equivalent.

The converse to this exercise is not true. For example, let T be the empty theory in the signature $\{=\}$. Then for each cardinal number κ, T has a model M with cardinality κ; and if M and N are infinite models of T, then M and N are elementarily equivalent. (The signature $\{=\}$ has no formulas that can discriminate between two different infinite models.) Thus, T has models that are elementarily equivalent but not isomorphic.

6.5 Ultraproducts

The so-called ultraproduct construction is often considered to be a technical device for proving theorems. Here we will emphasize the structural features of ultraproducts, rather than the details of the construction. Note, however, that ultraproducts are not themselves limits or colimits in the sense of category theory. Thus, we cannot give a simple formula relating an ultraproduct to the models from which it is constructed. In one sense, ultraproducts are more like limits in the topological sense than they are in the category-theoretic sense. Indeed, in the case of propositional theories, the ultraproduct of models of a theory *is* the topological limit in the Stone space of the theory.

To see this, it helps to redescribe limits in a topological space X in terms of infinitary operations $X^\infty \to X$. Recall that a point $p \in X$ is said to be a limit point of a subset $A \subseteq X$ just in case every open neighborhood of p intersects A. When X is nice enough (e.g., second countable), these limit points can be detected by sequences. That is, in such cases, p is a limit point of A just in case there is a sequence $a_1, a_2, \ldots$ of elements in A

such that $\lim_i a_i = p$. This last equation is simply shorthand for the statement: for each neighborhood U of p, there is a $n \in \mathbb{N}$ such that $a_i \in U$ for all $i \geq n$.

Suppose now, more specifically, that X is a compact Hausdorff space. Consider the product $\prod_{i \in \mathbb{N}} X$, which consists of infinite sequences of elements of X. We can alternately think of elements of $\prod_{i \in \mathbb{N}} X$ as functions from $\mathbb{N}$ to X. Since $\mathbb{N}$ is discrete, every such function $f : \mathbb{N} \to X$ is continuous, i.e., f^{-1} maps open subsets of X to (open) subsets of $\mathbb{N}$. Of course, f^{-1} also preserves inclusions of subsets. Hence, for each filter $\mathscr{V}$ of open subsets of X, $f^{-1}(\mathscr{V})$ is a filter on $\mathbb{N}$. For each point $p \in X$, let $\mathscr{V}_p$ be the filter of open neighborhoods of p. Now, for each ultrafilter $\mathscr{U}$ on $\mathbb{N}$, we define an operation $\lim_{\mathscr{U}} : \prod_{i \in \mathbb{N}} X \to X$ by the following condition:

$$\lim_{\mathscr{U}} f = p \quad \Longleftrightarrow \quad f^{-1}(\mathscr{V}_p) \subseteq \mathscr{U}.$$

To show that this definition makes sense, we need to check that there is a unique p satisfying the condition on the right. For uniqueness, suppose that $f^{-1}(\mathscr{V}_p)$ and $f^{-1}(\mathscr{V}_q)$ are both contained in $\mathscr{U}$. If $p \neq q$, then there are $U \in \mathscr{V}_p$ and $V \in \mathscr{V}_q$ such that $U \cap V$ is empty. Then $f^{-1}(U) \cap f^{-1}(V)$ is empty, in contradiction with the fact that $\mathscr{U}$ is an ultrafilter. For existence, suppose first that $\mathscr{U}$ is a principal ultrafilter – i.e., contains all sets containing some $n \in \mathbb{N}$. Let $p = f(n)$. Then for each neighborhood V of p, $f^{-1}(V)$ contains $f(n)$, and hence is contained in $\mathscr{U}$. Suppose now that $\mathscr{U}$ is non-principal, hence contains the cofinite filter. Since X is Hausdorff, the sequence $f(1), f(2), \ldots$ has a limit point p. Thus, for each $V \in \mathscr{V}_p$, $f^{-1}(V)$ is a cofinite subset of $\mathbb{N}$, and hence is contained in $\mathscr{U}$. In either case, there is a $p \in X$ such that $f^{-1}(\mathscr{V}_p) \subseteq \mathscr{U}$.

Thus, the topological structure on a compact Hausdorff space X can be described in terms of a family of operations $\lim_{\mathscr{U}} : \prod_i X_i \to X$, where $\mathscr{U}$ runs through all the ultrafilters on $\mathbb{N}$. This result holds in particular when $X = \text{Mod}(T)$ is the Stone space of models of a propositional theory. A limit model $\lim_{\mathscr{U}} M_i$ is called an **ultraproduct** of the models M_i. Thus, in the propositional case, an ultraproduct of models is simply the limit relative to the Stone space topology.

We will now try to carry over this intuition to the case of general first-order theories, modifying details when necessary. To begin with, if T is a first-order theory $\text{Mod}(T)$ is too large to have a topology – it is a class and not a set. What's more, even if we pretend that $\text{Mod}(T)$ is a set, the ultraproduct construction couldn't be expected to yield a topology, but something like a "pseudo-topology" or "weak topology," where limits are defined only up to isomorphism.

The details of the ultraproduct construction run as follows. Let I be an index set, and suppose that for each $i \in I$, M_i is a Σ-structure. If $\mathscr{U}$ is an ultrafilter on I, then we define a Σ-structure $N := \lim_{\mathscr{U}} M_i$ as follows:

- First consider the set $\prod M_i$ of "sequences," where each $a_i \in M_i$. We say that two such sequences are equivalent if they eventually agree in the sense of the ultrafilter $\mathscr{U}$. That is, $(a_i) \sim (b_i)$ just in case $\{i \mid a_i = b_i\}$ is contained in $\mathscr{U}$. We let the domain of N be the quotient of $\prod M_i$ under this equivalence relation.
- For each relation symbol R of Σ, we let $N(R)$ consist of sequences on n-tuples that eventually lie in $M_i(R)$ in the sense of the ultrafilter $\mathscr{U}$. That is, $(a_i) \in N(R)$

just in case $\{i \mid a_i \in M_i(R)\}$ is contained in $\mathscr{U}$. (Here one uses the fact that $\mathscr{U}$ is a ultrafilter to prove that $N(R)$ is well-defined as a subset of N.)

The resulting model $\lim_{\mathscr{U}} M_i$ is said to be an **ultraproduct** of the models M_i. In the special case where each M_i is the same M, we call $\lim_{\mathscr{U}} M_i$ an **ultrapower** of M. In this case, there is a natural elementary embedding $h : M \to \lim_{\mathscr{U}} M_i$ that maps each $a \in M$ to the constant sequence $a, a, \ldots$

We now cite without proof a fundamental theorem for ultraproducts.

THEOREM 6.5.1 (Łos) *Let $\{M_i \mid i \in I\}$ be a family of Σ-structures, let $\mathscr{U}$ be an ultrafilter on I, and let $N = \lim_{\mathscr{U}} M_i$. Then for each Σ-sentence ϕ, $N \vDash \phi$ iff $\{i \mid M_i \vDash \phi\} \in \mathscr{U}$.*

Intuitively speaking, $\lim_{\mathscr{U}} M_i$ satisfies exactly those sentences that are eventually validated by M_i as i runs through the ultrafilter $\mathscr{U}$.

We saw before that elementarily equivalent models need not be isomorphic. Indeed, for M and N to be elementarily equivalent, it's sufficient that there is a third model L and elementary embeddings $h : M \to L$ and $j : N \to L$. The following result shows that this condition is necessary as well.

PROPOSITION 6.5.2 *Let M and N be Σ-structures. Then the following are equivalent.*

1. *$M \equiv N$, i.e., M and N are elementarily equivalent.*
2. *There is a Σ-structure L and elementary embeddings $h: M \to L$ and $j: N \to L$.*
3. *M and N have isomorphic ultrapowers.*

Sketch of proof (3 $\Rightarrow$ 2) Suppose that $j : \lim_{\mathscr{U}_1} M \to \lim_{\mathscr{U}_2} N$ is an isomorphism, and let $L = \lim_{\mathscr{U}_2} N$. Let $h : M \to \lim_{\mathscr{U}_1} M$ be the natural embedding, and similarly for $k : N \to \lim_{\mathscr{U}_2} N$. Then $j \circ h : M \to L$ and $k : N \to L$ are elementary embeddings.

(2 $\Rightarrow$ 1) Since elementary embeddings preserve truth-values of sentences, this result follows immediately.

(1 $\Rightarrow$ 3) This is a difficult result, known as the Keisler–Shelah isomorphism theorem. We omit the proof and refer the reader to Keisler (2010) for further discussion. □

6.6 Relations between Theories

In the previous two chapters, we analyzed theories through a syntactic lens. Thus, to explicate relations between theories – such as equivalence and reduction – we used a syntactic notion, viz. translation. In this chapter, we've taken up the semantic analysis of theories – i.e., thinking about theories in terms of their models. Accordingly, we would like to investigate precise technical relations between categories of models that correspond with our intuitive notions of the relations that can hold between theories. In the best-case scenario, the technical notions we investigate will be useful in honing our intuitions about specific, real-life cases.

This investigation takes on special philosophical significance when we remember that at a few crucial junctures, philosophers claimed a decisive advantage for semantic analyses of relations between theories. Let's recall just a couple of the most prominent such maneuvers.

- van Fraassen (1980) claims that while the empirical content of a theory cannot be isolated syntactically, it can be isolated semantically. Since the notion of empirical content is essential for empiricism, van Fraassen thinks that empiricism requires the semantic view of theories.
- Defenders of various dressed-up versions of physicalism claim that the mental–physical relationship cannot be explicated syntactically, but can be explicated semantically. For example, the non-reductive physicalists of the 1970s claimed that the mental isn't reducible (syntactically) to the physical, but it does supervene (semanatically) on the physical. Similarly, Bickle (1998) claims that the failure of mind–brain reduction can be blamed on the syntactic explication of reduction, and that the problems can be solved by using a semantic explication of reduction.

These claims give philosophers a good reason to investigate the resources of logical semantics.

Let's begin by setting aside some rather flat-footed attempts to use semantics to explicate relations between theories. In particular, there seems to be a common misconception that the models of a theory are language-free, and can provide the standard by which to decide questions of theoretical equivalence. The (mistaken) picture here is that two theories, T and T', in different languages, are equivalent just in case $\mathrm{Mod}(T) = \mathrm{Mod}(T')$. We can illustrate this idea with a picture:

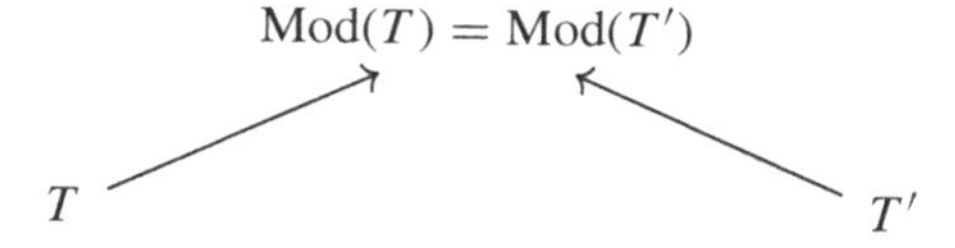

The picture here is that the theory formulations T and T' are language-bound, but the class $\mathrm{Mod}(T) = \mathrm{Mod}(T')$ of models is a sort of thing-in-itself that these different formulations intend to pick out.

If you remember that models are mappings from signatures, then you realize that there is something wrong with this picture. Yes, there are categories $\mathrm{Mod}(T)$ and $\mathrm{Mod}(T')$, but these categories are no more language-independent than the syntactic objects T and T'. In particular, if Σ and Σ' are different signatures, then there is no standard by which one can compare $\mathrm{Mod}(T)$ with $\mathrm{Mod}(T')$. A model of T is a function from Σ to **Sets**, and a model of T' is a function from Σ' to **Sets**. Functions with different domains cannot be equal; but it would also be misleading to say that they are *unequal*. In the world of sets, judgments of equality and inequality only make sense for things that live in the same set.

In a similar fashion, we can't make any progress in analyzing the relations between T and T' by setting their models side by side. One occasionally hears philosophers of science say things like

(I) There is a model of T that is not isomorphic to any model of T'; hence, T and T' are not equivalent.

(E) If T is a subtheory of T', then each model of T can be embedded in a model of T'.

However, if T and T' are theories in different signatures, then neither I nor E makes sense. The notions of isomorphism and elementary embedding are signature-relative: a function $h : M \to N$ is an elementary embedding just in case $h(M(\phi)) = N(\phi)$ for each Σ-formula ϕ. If T and T' are written in different signatures, then there is simply no way to compare a model M of T directly with a model N of T'. (And this lesson goes not only for theories in first-order logic, but for any mathematically formalized scientific theory – such as quantum mechanics, general relativity, Hamiltonian mechanics, etc.)

With these flat-footed analyses set aside, we can now raise some serious questions about the relations between Mod(T) and Mod(T'). For example, what mathematical relation between Mod(T) and Mod(T') would be a good explication of the idea that T is equivalent to T'? Is it enough that Mod(T) and Mod(T') are equivalent categories, or should we require something more? Similarly, what mathematical relation between Mod(T) and Mod(T') would be a good explication of the idea that T is reducible to T'? Finally, to return to the issue of empiricism, can the empirical content of a theory T be identified with some structure inside the category Mod(T)?

We will approach these questions from two directions. Our first approach will involve attempting to transfer notions from the syntactic side to the semantic side, as in the following picture:

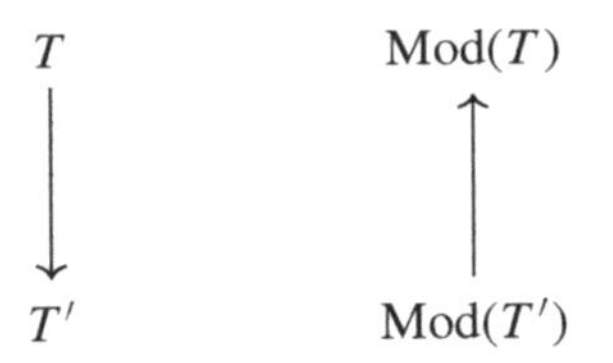

You will have noticed that we already followed this approach in Chapter 3, with respect to propositional theories. The goal is to take a syntactic relation between theories (such as "being reducible to") and to translate it over to a semantic relation between the models of those theories.

Of course, this first approach won't be at all satisfying to those who would be free from the "shackles of language." Thus, our second angle of attack is to ask directly about relations between Mod(T) and Mod(T'). Where do Mod(T) and Mod(T') live in the mathematical universe, and what are the mathematical relationships between them? Again, it will be no surprise to you that we think Mod(T) should, at the very least, be considered to be a *category*, whose mathematical structure includes not only models, but also arrows between them. Moreover, once we equip Mod(T) with sufficient structure, we will see that these two approaches converge – i.e., that the most interesting relations between Mod(T) and Mod(T') are those that correspond to some syntactic relation between T and T'. It is in this sense that logical semantics is *dual* to logical syntax.

We begin then with the first approach and, in particular, with showing that each translation $F : T \to T'$ gives rise to a functor $F^* : \mathrm{Mod}(T') \to \mathrm{Mod}(T)$. We will also provide a partial translation manual between properties of the translation F and properties of the functor F^*. To the extent that such a translation manual exists, each syntactic relation between T and T' corresponds to a unique semantic relation between $\mathrm{Mod}(T)$ and $\mathrm{Mod}(T')$, and vice versa.

DEFINITION 6.6.1 Suppose that $F : T \to T'$ is a translation, and let M be a model of T'. We define a Σ-structure F^*M as follows:

- Let F^*M have the same domain as M.
- For each relation symbol r of Σ, let

$$(F^*M)(r) = M(Fr).$$

- For each function symbol f of Σ, let $(F^*M)(f)$ be the function with graph $M(Ff)$.

We will now show that $(F^*M)(\phi) = M(F\phi)$ for each Σ-formula ϕ. However, we first need an auxiliary lemma. For this, recall that if $f : X \to Y$ is a function, then its graph is the subset $\{\langle x, f(x)\rangle \mid x \in X\}$ of $X \times Y$.

LEMMA 6.6.2 *For each Σ-term t, $M(Ft)$ is the graph of the function $(F^*M)(t)$.*

Proof We prove this by induction on the construction of t. Recall that if t is a term with n free variables, then Ft is a formula with $n + 1$ free variables, and $M(Ft)$ is a subset of S^{n+1}.

- Suppose that $t \equiv x$. Then $Ft \equiv (x = y)$ for some variable $y \not\equiv x$. In this case, $M(Ft)$ is the diagonal of $S \times S$, which is the graph of $1_S = (F^*M)(x)$.
- Now suppose that the result is true for $t_1, \ldots, t_m$, and let $t \equiv f(t_1, \ldots, t_m)$. Recall that Ft is defined as the composite of the relation Ff with the relations $Ft_1, \ldots, Ft_m$. Since M preserves the relevant logical structures, $M(Ft)$ is the composite of the relation $M(Ff)$ with the relations $M(Ft_1), \ldots, M(Ft_m)$. Moreover, $(F^*M)(t)$ is defined to be the composite of the function $(F^*M)(f)$ with the functions $(F^*M)(t_1), \ldots, (F^*M)(t_m)$. In general, the graph of a composite function is the composite of the graphs. Therefore, $M(Ft)$ is the graph of $(F^*M)(t)$.

□

PROPOSITION 6.6.3 *For each Σ-formula ϕ, $(F^*M)(\phi) = M(F\phi)$.*

Proof We prove this by induction on the construction of ϕ.

- Suppose that $\phi \equiv (t_1 = t_2)$. Then $F\phi$ is the formula $\exists y(Ft_1(\vec{x}, y) \wedge Ft_2(\vec{x}, y))$. Here, for simplicity, we write $\vec{x}$ for the canonical context of $F\phi$. Thus, $M(F\phi)$ consists of elements $\vec{a} \in S^n$ such that $\langle \vec{a}, b\rangle \in M(Ft_1)$ and $\langle \vec{a}, b\rangle \in M(Ft_2)$ for some $b \in S$. By the previous lemma, $M(Ft_i)$ is the graph of $(M^*F)(t_i)$. Thus, $M(F\phi)$ is the equalizer of $(M^*F)(t_1)$ and $(M^*F)(t_2)$. That is, $M(F\phi) = (M^*F)(\phi)$.

- Suppose that $\phi \equiv p(t_1, \dots, t_m)$. Then $F\phi$ is the formula

$$\exists z_1 \cdots \exists z_m (Fp(y_1, \dots, y_m) \wedge Ft_1(\vec{x}, y_1) \wedge \cdots \wedge Ft_m(\vec{x}, y_m)).$$

  Hence $M(F\phi)$ consists of those $\vec{a} \in S^n$ such that there are $b_1, \dots, b_m \in S$ with $\langle \vec{a}, b_i \rangle \in M(Ft_i)$ and $\vec{b} \in M(Fp)$. By the previous lemma, $M(Ft_i)$ is the graph of $(F^*M)(t_i)$. Hence, $M(F\phi)$ consists of those $\vec{a} \in S^n$ such that

$$\langle (F^*M)(t_1), \dots, (F^*M)(t_m) \rangle (\vec{a}) \in M(Fp) = (F^*M)(p).$$

  In other words, $M(F\phi) = (F^*M)(\phi)$. (Here we have ignored the fact that the terms $t_1, \dots, t_m$ might have different free variables. In that case, we need simply to prefix the $(F^*M)(t_i)$ with the appropriate projections to represent them on the same domain S^n.)
- Suppose that $\phi \equiv (\phi_1 \wedge \phi_2)$, and the result is true for ϕ_1 and ϕ_2. Now, $F(\phi_1 \wedge \phi_2) = F\phi_1 \wedge F\phi_2$. Hence $M(F(\phi_1 \wedge \phi_2))$ is the pullback of $M(F\phi_1)$ and $M(F\phi_2)$ along the relevant projections (determined by the contexts of ϕ_1 and ϕ_2). Since F preserves contexts of formulas, and $M(F\phi_i) = (F^*M)(\phi_i)$, it follows that $M(F(\phi_1 \wedge \phi_2)) = (F^*M)(\phi_1 \wedge \phi_2)$.
- We now deal with the existential quantifier. For simplicity, suppose that ϕ has free variables x and y. We suppose that the result is true for ϕ; that is,

$$(F^*M)(\phi) = M(F(\phi)),$$

  and we show that

$$(F^*M)(\exists x\phi) = M(F(\exists x\phi)).$$

  By definition, $(F^*M)(\exists x\phi)$ is the image of $(F^*M)(\phi)$ under the projection $\pi : X \times Y \to Y$. Moreover, $F(\exists x\phi) = \exists x F(\phi)$, which means that $M(F(\exists x\phi))$ is the image of $M(F(\phi))$ under the projection π.

□

PROPOSITION 6.6.4 *Suppose that $F : T \to T'$ is a translation. If M is a model of T' then F^*M is a model of T.*

Proof Suppose that $T \vdash \phi$. Since F is a translation, $T' \vdash F\phi$. Since M is a model of T', $M(F\phi) = S^n$. Therefore, $(F^*M)(\phi) = S^n$. Since ϕ was an arbitrary Σ-formula, we conclude that F^*M is a model of T. □

DEFINITION 6.6.5 Let $F : T \to T'$ be a translation. We now extend the action of F^* from models of T' to elementary embeddings between these models. Let M and N be models of T' with corresponding domains X and Y. Let $h : M \to N$ be an elementary embedding. Since F^*M has the same domain as M, and similarly for F^*N and N, this h is a candidate for being an elementary embedding of F^*M into F^*N. We need only check that the condition of Defn. 6.4.2 holds – i.e., that for each Σ-formula ϕ, the following diagram is a pullback:

$$\begin{array}{ccc} (F^*M)(\phi) & \longrightarrow & (F^*N)(\phi) \\ \downarrow & & \downarrow \\ X^n & \xrightarrow{h^n} & Y^n \end{array}$$

But $(F^*M)(\phi) = M(F\phi)$ and $(F^*N)(\phi) = N(F\phi)$. Since $h : M \to N$ is elementary, the corresponding diagram is a pullback. Therefore, $h : F^*M \to F^*N$ is elementary.

Now, the underlying function of $F^*h : F^*M \to F^*N$ is the same as the underlying function of $h : M \to N$. Thus, F^* preserves composition of functions, as well as identity functions; and $F^* : \text{Mod}(T') \to \text{Mod}(T)$ is a functor.

We have shown that each translation $F : T \to T'$ corresponds to a functor $F^* : \text{Mod}(T') \to \text{Mod}(T)$. Now we would like to compare properties of F with properties of F^*. The fundamental result here is that if F is a homotopy equivalence, then F^* is an equivalence of categories.

PROPOSITION 6.6.6 *If T and T' are intertranslatable, then* $\text{Mod}(T)$ *and* $\text{Mod}(T')$ *are equivalent categories. In particular, if $F : T \to T'$ and $G : T' \to T$ form a homotopy equivalence, then F^* and G^* are inverse functors.*

Sketch of proof In the following chapter, we prove a stronger result: if T and T' are Morita equivalent, then $\text{Mod}(T)$ and $\text{Mod}(T')$ are equivalent categories. In order to avoid duplicating work, we will just sketch the proof here. One shows that $(FG)^* = G^*F^*$, and that for any two translations F and G, if $F \simeq G$, then $F^* = G^*$. Since $GF \simeq 1_T$, it follows that

$$F^*G^* = (GF)^* = 1_T^* = 1_{\text{Mod}(T)}.$$

Similarly, $G^*F^* = 1_{\text{Mod}(T')}$, and therefore $\text{Mod}(T)$ and $\text{Mod}(T')$ are equivalent categories. □

COROLLARY 6.6.7 *If T and T' are intertranslatable, then T and T' are categorically equivalent.*

One upshot of this result is that categorical properties of $\text{Mod}(T)$ are **invariants** of intertranslatability. For example, if $\text{Mod}(T)$ has all finite products and $\text{Mod}(T')$ does not, then T and T' are not intertranslatable. We should think a bit, then, about which features of a category are invariant under categorical equivalence.

Recall that the identity of a category **C** has nothing to do with the identity of its objects. All that matters is the relations that these objects have to each other. Thus, if we look at $\text{Mod}(T)$ qua category, then we are forgetting that its objects are models. Instead, we are focusing exclusively on the arrows (elementary embeddings) that relate these models, including the symmetries (automorphisms) of models. Here are some of the properties that can be expressed in the language of category theory:

1. **C** has products.
2. **C** has coproducts.
3. **C** has all small limits.

The list could go on, but the real challenge is to say which of the properties of the category Mod(T) corresponds to an interesting feature of the theory T. For example, might it be relevant that Mod(T) has products – i.e., that for any two models M, N of T, there is a model $M \times N$, with the relevant projections, etc.? Keep in mind that these mathematical statements don't have an obvious interpretation in terms of what the theory T might be saying about the world. For example, to say that Mod(T) has products doesn't tell us that there is an operation that takes two possible worlds and returns another possible world.

Recall, in addition, that category theorists ignore properties that are not invariant under categorical equivalence. For example, the property "**C** has exactly two objects," is not invariant under all categorical equivalences. Although the notion of a "categorical property" is somewhat vague, the practicing category theorist knows it when he sees it – and fortunately, work is in progress in explicating this notion more precisely (see Makkai, 1995; Tsementzis, 2017a).

To be clear, we don't mean to say that Mod(T) should be seen *merely* as a category. If we did that, then we would lose sight of some of the most interesting information about a theory. Consider, in particular, the following fact:

PROPOSITION 6.6.8 *If T is a propositional theory, then* Mod(T) *is a discrete category – i.e., the only arrows in* Mod(T) *are identity arrows.*

This result implies that for any two propositional theories T and T', if they have the same number of models, then they are categorically equivalent. But don't let this make you think that the space Mod(T) of models of a propositional theory T has no interesting structure. We saw in Chapter 3 that it has interesting topological structure, which represents a notion of "closeness" of models.

At the time of writing, there is no canonical account of the structure that is possessed by Mod(T) for a general first-order theory T. However, there has been much interesting mathematical research in this direction. The first main proposal, due to Makkai (1985), defines the "ultraproduct structure" on Mod(T) – i.e., which models are ultraproducts of which others. Interestingly, as we saw in the previous section, the ultraproduct construction looks like a topological limiting construction – and the coincidence is exact for the case of propositional theories. The second proposal for identifying the structure of Mod(T) is due originally to Butz and Moerdijk (1998), and has been recently developed by Awodey and Forssell (2013). According to this second proposal, Mod(T) is a topological groupoid, i.e., a groupoid in the category of topological spaces. Thus, according to both proposals, Mod(T) is like a category with a topology on it, where neither bit of structure – categorical or topological – is dispensable.

In the case of predicate logic theories, the categorical structure of Mod(T) does occasionally tell us something about T. We first show that the completeness or incompleteness of a theory can be detected by its category of models. Recall that a theory T in signature Σ is said to be **complete** just in case for each Σ-sentence ϕ, either $T \vdash \phi$ or $T \vdash \neg\phi$. Obviously every inconsistent theory is incomplete. So when we talk about a complete theory T, we usually mean a complete, consistent theory. In this case, the following conditions are equivalent:

1. T is complete.
2. $\mathrm{Cn}(T) = \mathrm{Th}(M)$ for some Σ-structure M.
3. T has a unique model, up to elementary equivalence – i.e., if M, N are models of T, then $M \equiv N$.
4. $\mathrm{Mod}(T)$ is directed in the sense that for any two models M_1, M_2 of T, there is a model N of T and elementary embeddings $h_i : M_i \to N$.

EXERCISE 6.6.9 Prove that the four conditions are equivalent. Hint: use Prop. 6.5.2.

The last property is a categorical property: if **C** and **D** are categorically equivalent, then **C** is directed iff **D** is directed. Therefore, completeness of theories is an invariant of categorical equivalence.

Now, it is well known that complete theories can nonetheless have many non-isomorphic models. It has occasionally been thought that an ideal theory T would be **categorical** in the sense that every two models of T are isomorphic. (The word "categorical" here has nothing to do with category theory.) However, the Löwenheim–Skølem theorem destroys any hope of finding a nontrivial categorical theory: if T has an infinite model, then it has models of other infinite cardinalities, and these models cannot be isomorphic. For the purposes, then, of classifying more of less "nice" theories, logicians found it useful to weaken categoricity in the following way:

DEFINITION 6.6.10 Let κ be a cardinal number, and let T be a theory in signature Σ. We say that T is κ-**categorical** just in case any two models M and N of T, if $|M| = |N| = \kappa$, then there is an isomorphism $h : M \to N$.

Example 6.6.11 Let T be the empty theory in signature $\{=\}$. A model of T is simply a set, and two models of T are isomorphic if they have the same cardinality. Therefore, T is κ-categorical for each cardinal number κ. ┘

Example 6.6.12 Let $\Sigma = \{<\}$, where $<$ is a binary relation symbol. Let T be the theory in Σ that says that $<$ is a discrete linear order without endpoints. Then T is not $\aleph_0$-categorical. For example, the set $\mathbb{N}$ of natural numbers (with its standard ordering) is a model of T, but so is the disjoint union $\mathbb{N} \amalg \mathbb{N}$, where every element of the second copy is greater than every element of the first. ┘

Thus, if T is categorical for all cardinal numbers, then $\mathrm{Mod}(T)$ has a relatively simple structure as a category: it is like a tower, with a unique (up to isomorphism) model M_κ for each cardinal number κ. (A generalization of the Löwenheim–Skølem theorem shows that for each infinite model M of T, there is a model N of T of higher cardinality and an elementary embedding $h : M \to N$.) Nonetheless, it is well known that there are many inequivalent categorical theories, and these theories are differentiated by the topological groups of symmetries of their models.

We now set aside the discussion of equivalence to look at other types of relations between theories. Recall that a translation $F : T \to T'$ is said to be **essentially surjective** just in case for each Σ-sentence ψ there is a Σ-sentence ϕ such that $T' \vdash \psi \leftrightarrow F\phi$.

A paradigmatic case of an essentially surjective translation is the translation from a theory T to a theory T' with some new axioms in the same signature. Recall also that a functor $F^* : \mathbf{C} \to \mathbf{D}$ is said to be **full** just in case for any objects M, N of $\mathbf{C}$, and for any arrow $f : F^*M \to F^*N$, there is an arrow $g : M \to N$ such that $F^*g = f$. In the special case of groups (i.e., categories with only one object, and only isomorphisms), a functor is full iff it is a surjective homomorphism.

PROPOSITION 6.6.13 *If* $F : T \to T'$ *is essentially surjective then* $F^* : \mathrm{Mod}(T') \to \mathrm{Mod}(T)$ *is full.*

Proof Let $h : F^*M \to F^*N$ be a Σ-elementary embedding. We need to show that $h = F^*j$ where $j : M \to N$ is a Σ'-elementary embedding. Finding the function j is easy, since h is already a function from the domain of M to the domain of N. Thus, we need only show that h is Σ'-elementary – i.e., that for any Σ'-formula ψ, the following is a pullback:

$$\begin{array}{ccc} M(\psi) & \longrightarrow & N(\psi) \\ \downarrow & & \downarrow \\ X^n & \xrightarrow{h^n} & Y^n \end{array}$$

Since F is eso, there is a Σ-formula ϕ such that $T' \vdash \psi \leftrightarrow F\phi$. Since M and N are models of T', $M(\psi) = M(F\phi) = F^*M(\phi)$ and $N(\psi) = N(F\phi) = (F^*N)(\phi)$. Since h is Σ-elementary, the diagram is a pullback. Therefore, $j : M \to N$ is Σ'-elementary, and F^* is full. □

The preceding result can be quite useful in showing that there is no essentially surjective translation from T to T'.

Example 6.6.14 Let T be the theory in signature $\{=\}$ that says there are exactly two things. Let T' be the theory in signature $\{=, c\}$ that says there are exactly two things. These two theories consist of exactly the same sentences; and yet, we will now see that they are not intertranslatable.

The theory T is categorical: i.e., it has a unique model $M = \{*, \star\}$ up to isomorphism, and $\mathrm{Aut}(M) = \mathbb{Z}_2$ is the permutation group on two elements. Thus, $\mathrm{Mod}(T)$ is equivalent to the group $\mathbb{Z}_2$. The theory T' is also categorical; however, its models are rigid, i.e., have no nontrivial automorphisms. Hence, $\mathrm{Mod}(T')$ is equivalent to the group (e). Clearly there is no full functor $G : \mathrm{Mod}(T') \to \mathrm{Mod}(T)$, and, therefore, Prop. 6.6.13 entails that there is no essentially surjective translation $F : T \to T'$.

It would hardly to make sense to think of either T or T' as an actual scientific theory. However, in the spirit of constructing toy models, we could raise a fanciful question: if Jack accepted T and Jill accepted T', then what would be the locus of their disagreement? They both assert precisely the same sentence: there are exactly two things. We cannot say that they disagree about whether there are constant symbols, because symbols aren't things "in the world," but are devices used to speak about things in the world. So perhaps Jack and Jill disagree about whether the two things in the world are interchangeable? ⌟

The next pair of results derive properties of F from properties of F^*. We first recall the syntactic notion of a conservative extension.

DEFINITION 6.6.15 A translation $F : T \to T'$ is said to be **conservative** just in case $T' \vdash F\phi$ only if $T \vdash \phi$, for each Σ-formula ϕ.

Thus, a conservative translation $F : T \to T'$ is one that does not create new consequences for T. Paradigm examples of this kind of translation can be generated by the inclusion $I : \Sigma \to \Sigma'$ where $\Sigma \subseteq \Sigma'$. Adding this new vocabulary to Σ does not generate new consequences for a theory T in Σ.

Now let's consider how the notion of a conservative extension might be formulated semantically. Recall that a functor $F^* : \text{Mod}(T') \to \text{Mod}(T)$ is said to be **essentially surjective** just in case for each model M of T, there is a model N of T' and an isomorphism $h : M \to F^*N$. In the case of an inclusion $I : \Sigma \to \Sigma'$, the functor I^* is essentially surjective iff each model of T can be expanded to a model of T'.

It's fairly easy to see that if F^* is essentially surjective, then F is conservative. In fact, we can weaken the condition on F^* as follows.

DEFINITION 6.6.16 Let $F : T \to T'$ be a translation. We say that $F^* : \text{Mod}(T') \to \text{Mod}(T)$ is **covering** just in case for each $M \in \text{Mod}(T)$, there is an $N \in \text{Mod}(T')$ and an elementary embedding $h : M \to F^*(N)$.

PROPOSITION 6.6.17 *Let $F : T \to T'$ be a translation. If F^* is covering then F is conservative.*

Proof Suppose that $T' \vdash F\phi$. Let M be an arbitrary model of T, and let $h : M \to F^*(N)$ be the promised elementary embedding. Since $N \vDash F\phi$, we have $F^*(N) \vDash \phi$, and since h is elementary, $M \vDash \phi$. Since M was an arbitrary model of T, it follows that $T \vdash \phi$. □

COROLLARY 6.6.18 *Let $F : T \to T'$ be a translation. If F^* is essentially surjective, then F is conservative.*

The following example shows that the condition of F^* being essentially surjective is strictly stronger than F being conservative. Thus, a translation $F : T \to T'$ may be conservative even though not every model of T can be expanded to a model of T'.

Example 6.6.19 Let $\Sigma = \{c_q \mid q \in \mathbb{Q}\}$, and let $\Sigma' = \{c_r \mid r \in \mathbb{R}\}$. Let T' be the theory with axioms $c_r \neq c_s$ when $r \neq s$, and let T be the restriction of T' to Σ. Obviously, for each model M of T, there is a model $M' = M \amalg N$ of T' and an elementary embedding $h : M \to I^*(M')$. By Prop. 6.6.17, T' is a conservative extension of T. However, a countable model M of T cannot be isomorphic to $I^*(M')$, for any model M' of T'. Therefore, I^* is not essentially surjective. ⌟

DISCUSSION 6.6.20 We have given a relatively weak condition on $F^* : \text{Mod}(T') \to \text{Mod}(T)$, which implies that $F : T \to T'$ is conservative. Unfortunately, we do not know if these conditions are equivalent. It seems, in fact, that F being conservative is equivalent to a slightly weaker (and more complicated) condition on F^*, as described by Breiner (2014).

The dual functor $F^* : \mathrm{Mod}(T') \to \mathrm{Mod}(T)$ has many additional uses. For example, we can now complete the proof that two theories T_1 and T_2 have a common definitional extension iff they are intertranslatable (i.e., homotopy equivalent).

THEOREM 6.6.21 (Barrett) *Suppose that T_i is a theory in Σ_i, where Σ_1 and Σ_2 are disjoint signatures. If T_1 and T_2 are intertranslatable, then T_1 and T_2 have a common definitional extension.*

Proof Suppose that T_1 and T_2 are intertranslatable, with $F : T_1 \to T_2$ and $G : T_2 \to T_1$ the relevant translations. We begin by defining definitional extensions T_1^+ and T_2^+ of T_1 and T_2 to the signature $\Sigma_1 \cup \Sigma_2$.

We define $T_1^+ = T_1 \cup \{\delta_s : s \in \Sigma_2\}$, where for each symbol $s \in \Sigma_2$ the Σ_2-sentence δ_s is an explicit definition of s. If $q \in \Sigma_2$ is an n-ary predicate symbol, then we let the definition $\delta_q \equiv \forall \vec{x}(q \leftrightarrow Gq)$. If $g \in \Sigma_2$ is an n-ary function symbol, then we let the definition $\delta_g \equiv \forall \vec{x} \forall y(g(\vec{x}) = y \leftrightarrow Gg(\vec{x}, y))$. It is straightforward to verify that T_1 satisfies the admissibility condition for δ_g.

We define $T_2^+ = T_2 \cup \{\delta_t : t \in \Sigma_1\}$ in the same manner. If $p \in \Sigma_1$ is an n-ary predicate symbol, then we let $\delta_p \equiv \forall \vec{x}(p \leftrightarrow Fp)$. If $f \in \Sigma_1$ is an n-ary function symbol, then we let $\delta_f \equiv \forall \vec{x} \forall y(f(\vec{x}) = y \leftrightarrow Ff(\vec{x}, y))$. It is also straightforward to verify that T_2 satisfies the admissibility condition for δ_f.

We show now that T_1^+ and T_2^+ are logically equivalent. Without loss of generality, we show that every model of T_2^+ is a model of T_1^+. The converse follows via an analogous argument. Let M be a model of T_2^+. We show that M is a model of T_1^+. There are two cases that need checking.

First, we show that $M(\phi) = 1$ when $T_1 \vdash \phi$. Since F^*M is a model of T_1, we have $1 = (F^*M)(\phi) = M(F\phi)$. One can then verify by induction that for every Σ_1 formula ψ, and for every model M of T_2^+, $M(\psi) = M(F\psi)$. Therefore, $M(\phi) = 1$.

Second, we show that $M(\delta_s) = 1$ for every $s \in \Sigma_2$. Let $q \in \Sigma_2$ be an n-ary predicate symbol. Then

$$M(q(\vec{x})) \;=\; M(FGq(\vec{x})) \;=\; M(Gq(\vec{x})).$$

The first equality follows from the fact that F and G are quasi-inverse and the fact that M is a model of T_2^+. The second equivalence follows from the argument of the previous paragraph. Thus, $M(\delta_q) = 1$. In a similar manner one can verify that $M(\delta_g) = 1$ for every function symbol $g \in \Sigma_2$.

We have therefore shown that each model of T_1^+ is a model T_2^+. Thus, T_1^+ and T_2^+ are logically equivalent, and T_1 and T_2 are definitionally equivalent. □

Example 6.6.22 Let $\Sigma \equiv \{=\}$, let T_1 be the theory in Σ that says there is exactly one thing, and let T_2 be the theory in Σ that says there are exactly two things. In one important sense, T_1 and T_2 have the same number of models: one (up to isomorphism).

Since T_1 and T_2 should not be considered to be equivalent, having the same number of models is not an adequate criterion for equivalence.

Perhaps we can strengthen that criterion by saying that two theories are equivalent if the models of the one can be *constructed* from the models of the other? But that criterion seems also to say that T_1 and T_2 are equivalent. From each model $\{*\}$ of T_1, we can construct a corresponding model $\{*, \{*\}\}$ of T_2; and we can recover the original model $\{*\}$ from the model $\{*, \{*\}\}$.

This criterion is alluring, but it is still far too liberal. We will need to do something to capture its intuition, but without making the criterion of equivalence too liberal.

One natural suggestion here is to consider $\text{Mod}(T_1)$ and $\text{Mod}(T_2)$ as categories, and to consider functors between them. There are then two proposals to consider:

1. Each functor $F : \text{Mod}(T_1) \to \text{Mod}(T_2)$ represents a genuine theoretical relation between T_1 and T_2.
2. Every genuine theoretical relation between T_1 and T_2 is represented by a functor $F : \text{Mod}(T_1) \to \text{Mod}(T_2)$.

There is immediate reason to question the first proposal. For example, in the case of propositional theories T_1 and T_2, the categories $\text{Mod}(T_1)$ and $\text{Mod}(T_2)$ are discrete. Hence, functors $F : \text{Mod}(T_1) \to \text{Mod}(T_2)$ correspond one-to-one with functions on objects (in this case, models). But we have seen cases where intuitively inequivalent propositional theories have categories with the same number of models. Thus, it seems that not every functor (or function) between $\text{Mod}(T_1)$ and $\text{Mod}(T_2)$ represents a legitimate relation between the theories.

There's another, more concrete, worry about the first proposal. Consider the case where T_1 and T_2 are fairly expressive theories in first-order logic. For example, T_1 might be Peano arithmetic, and T_2 might be ZF set theory. Setting aside worries about the size of sets, a function from $\text{Mod}(T_1)$ to $\text{Mod}(T_2)$ is simply a pairing $\langle M, N\rangle$ of models of T_2 with models of T_1. But there need not be any "internal" relation between M and N. This goes against an intuition that for theories T_1 and T_2 to be equivalent, there needs to be relations between their individual models, and not just their categories of models qua categories. In the case at hand, we want to say that for any model M of T_1, there is a model N of T_2, and some relation $\Phi(M, N)$ between M and N. But what relations Φ are permitted? And does the *same* relation Φ need to hold for every model M and the corresponding N, or can the relation itself depend on the input model M? ⌟

6.7 Beth's Theorem and Implicit Definition

[T]here is an argument, based on an application of Beth's renowned definability theorem, which might appear to render simultaneous support for physicalism and anti-reductionism impossible. (Hellman and Thompson, 1975)

The logical positivists vacillated between being metaphysically neutral and being committed to metaphysical naturalism. One particular instance of the latter commitment was their view on the mind–body problem. With the new symbolic logic as their tool, they had a clear story to tell about how the mental is related to the physical: it is **reducible** to it. For example, suppose that $r(x)$ denotes some kind of mental property, say the property of being in pain. In this case, the reductionist says that there is a predicate $\phi(x)$ in the language of basic physics such that $\forall x(r(x) \leftrightarrow \phi(x))$ – i.e., something is in pain iff it instantiates the physical property ϕ.

Of course, we should be clearer when we say that $\forall x(r(x) \leftrightarrow \phi(x))$, for even a Cartesian dualist might say that this sentence is contingently true. That is, a Cartesian dualist might say that there is a purely physical description $\phi(x)$ that happens, as a matter of contingent fact, to pick out exactly those things that are in pain. The reductionist, in contrast, wants to say more – that there is some sort of lawlike connection between being in pain and being in a certain physical state. At the very least, a reductionist would say that

$$T \vdash \forall x(r(x) \leftrightarrow \phi(x)),$$

where T is our best scientific theory (perhaps the ideal future scientific theory). That is, according to the best theory, to be in pain is nothing more or less than to instantiate the physical property ϕ.

By the third quarter of the twentieth century, this sort of hard-core reductionism had fallen out of fashion. In fact, some of the leading lights in analytic philosophy – such as Hilary Putnam – had devised master arguments which were taken to demonstrate the utter implausibility of the reductionist point of view. Nonetheless, what had not fallen out of favor among analytic philosophers was the naturalist stance that had found its precise explication in the reductionist thesis. Thus, analytic philosophers found themselves on the hunt for a new, more plausible way to express their naturalistic sentiments.

In the 1970s, philosophers with naturalistic sentiments often turned to the concept of "supervenience" in order to describe the relationship between the mental and the physical. Now, there has been much debate in the ensuing years about how to cash out the notion of supervenience, and we don't have anything to add to that debate. Instead, we'll opt for the most obvious explication of supervenience in the context of first-order logic, in which case supervenience amounts to the model theorist's notion of implicit definability:

> Given a fixed background theory T, a predicate r is implicitly definable in terms of others $p_1, \ldots, p_n$ just in case for any two models M, N of T, if M and N agree on the extensions of $p_1, \ldots, p_n$, then M and N agree on the extension of r.

Now, there is a relevant theorem from model theory, viz. **Beth's theorem**, which shows that if T implicitly defines r in terms of $p_1, \ldots, p_n$, then T explicitly defines r in terms of $p_1, \ldots, p_n$; that is

$$T \vdash \forall x(r(x) \leftrightarrow \phi(x)),$$

where ϕ is a formula built from the predicates $p_1, \ldots, p_n$. In other words, if r supervenes on $p_1, \ldots, p_n$, then r is reducible to $p_1, \ldots, p_n$. According to Hellman and Thompson, this result "might appear to render simultaneous support for physicalism and anti-reductionism impossible."

We begin the technical exposition with a description of the background assumptions of Beth's theorem. To be clear, philosophers can take exception with these background assumptions. They might say that we have stacked the deck against non-reductive physicalism by means of these assumptions, and that a different account of supervenience will permit it to be distinguished from reducibility. Although such a response is completely reasonable, it suggests that physicalism isn't a sharp hypothesis but a stance that can be held "come what may."

Fixed Assumptions of Svenonius' and Beth's Theorems

- T is a theory in signature Σ.
- $\Sigma^+ = \Sigma \cup \{r\}$, where r is an n-ary relation symbol.
- T^+ is a theory in Σ^+.
- T^+ is a conservative extension of T.

Svenonius' and Beth's theorems are closely related. Svenonius' theorem begins with an assumption about symmetry and invariance:

In each model M of T^+, the subset $M(r)$ is invariant under Σ-automorphisms.

It then shows that for each model M of T^+, there is a Σ-formula ϕ such that $M(r) = M(\phi)$. The formula ϕ may differ from model to model. Beth's theorem begins with the assumption that T^+ implicitly defines r in terms of Σ.

DEFINITION 6.7.1 We say that T^+ **implicitly defines** r in terms of Σ just in case for any two models M, N of T^+, if $M|_\Sigma = N|_\Sigma$, then $M = N$.

(Here $M|_\Sigma$ is the Σ-structure that results from "forgetting" what M assigns to the relation symbol $r \in \Sigma^+ \backslash \Sigma$.) Beth's theorem then shows that T^+ explicitly defines r in terms of Σ – i.e., there is a single Σ-formula ϕ such that $T^+ \vdash \forall \vec{x}(r(\vec{x}) \leftrightarrow \phi(\vec{x}))$, hence, in every model M of T^+, the relation r is coextensive with ϕ.

There are a variety of ways that one can prove the theorems of Beth and Svenonius. The reader may like, for example, to study the fairly straightforward proof of Beth's theorem in Boolos et al. (2002, chapter 20). However, our goal here is not merely to convince you that these theorems are true. We want to give you a feel for why they are true and to help you see that they are instances of certain general mathematical patterns. To achieve these ends, it can help to expand the mathematical context – even if that requires a bit more work. Accordingly, we will present a proof of Beth's theorem with a more topological slant.

We begin with the notion of a **type** in a model M of a theory T. (The terminology here is not particularly intuitive, but it has become standard. A better phrase might be been "ideal element.") As a quick overview, each element $a \in M$ corresponds to a family Γ of formulas $\phi(x)$, viz. those formulas that it satisfies. That is,

$$\Gamma = \{\phi(x) \mid a \in M(\phi(x))\}.$$

In fact, this set Γ is a filter relative to implication in M. That is, if $\phi(x) \in \Gamma$ and $M \models \forall x(\phi(x) \to \psi(x))$, then $\psi(x) \in \Gamma$. Similarly, if $\phi(x), \psi(x) \in \Gamma$, then $\phi(x) \wedge \psi(x) \in \Gamma$. Finally, for any $\phi(x)$, either $\phi(x) \in \Gamma$ or $\neg\phi(x) \in \Gamma$.

However, it's also possible to have an ultrafilter Γ of formulas for which there is no corresponding element $a \in M$. The obvious cases here are where the formulas "run off to infinity." For example, consider the family of formulas

$$\Gamma = \{r < x \mid r \in \mathbb{R}\},$$

with $\mathbb{R}$ the real numbers. Then Γ is a filter, but no real number a satisfies all formulas in Γ. Intuitively speaking, the filter Γ is satisfied by an ideal point at infinity that is greater than any real number. (While Γ is a filter, it is not an ultrafilter. It is contained in infinitely many distinct ultrafilters, each of which corresponds to a point at infinity.)

It's also possible for a model M to have ideal points "in the interstices" between the real points. For example, in the case of the real numbers $\mathbb{R}$, let's say that a filter Γ of formulas is *centered on* 0 just in case Γ contains each formula $-\delta < x < \delta$. Then a simple counting argument (with infinite cardinalities) shows that there are infinitely many incompatible filters, all of which are centered on 0. Each such filter corresponds to an ideal element that is smaller than any finite real number.

DEFINITION 6.7.2 Let M be a Σ-structure, and let p be a set of Σ-formulas in context $\vec{x} = x_1, \ldots, x_n$. We call p an ***n*-type** if $p \cup \mathrm{Th}(M)$ is satisfiable. We say that p is a **complete *n*-type** if $\phi \in p$ or $\neg\phi \in p$ for all Σ-formulas ϕ in context $\vec{x}$. We let S_n^M be the set of all complete n-types.

Each element a in a model gives rise to a complete 1-type:

$$\mathrm{tp}^M(a) = \{\phi(x) \mid a \in M(\phi(x))\}.$$

Similarly, each n-tuple $\vec{a} = a_1, \ldots, a_n$ gives rise to a complete n-type $\mathrm{tp}^M(\vec{a}) \in S_n^M$. We say that $\vec{a}$ **realizes the type** $p \in S_1^M$ when $p = \mathrm{tp}^M(\vec{a})$.

DEFINITION 6.7.3 We now equip the set S_n^M of complete n-types with a topology, and we show that this topology makes S_n^M a Stone space. For each Σ-formula ϕ in context $\vec{x}$, let

$$E_\phi = \{p \in S_n^M \mid \phi \in p\}.$$

The definition here is similar to that which we used in defining the Stone space of a propositional theory. In that case, E_ϕ was the set of models of the sentence ϕ. In the present case, S_n^M are not quite models of a theory. But if M is a model of a theory T, then the types S_n^M are essentially all elements of M^n along with ideal elements.

In order to show that S_n^M is a compact topological space, we will need to adduce a central theorem of model theory – the so-called realizing types theorem. We cite the result without proof, referring the interested reader to Marker (2006, chapter 4).

THEOREM 6.7.4 (Realizing types) *Suppose that F is a finite subset of S_n^M. Then there is an elementary extension N of M such that each $p \in F$ is realized in N.*

PROPOSITION 6.7.5 *S_n^M is a compact topological space.*

Proof Recall that a topological space is compact just in case any family of closed sets with the finite intersection property (fip) has nonempty intersection. Suppose then that $\mathcal{F}$ is a collection of closed subsets of S_n^M that has the fip. It will suffice to consider the case where the elements of $\mathcal{F}$ are each of the form E_ϕ for some Σ-formula ϕ. Let $\mathcal{F}_0$ denote the corresponding family of formulas. Since $\mathcal{F}$ has the fip, for each $\phi_1, \ldots, \phi_n \in \mathcal{F}_0$, there is some $p \in S_n^M$ such that $\phi_1, \ldots, \phi_n \in p$, hence $\phi_1 \wedge \cdots \wedge \phi_n \in p$. By the realizing types theorem, there is an elementary extension N of M and $a \in N(\phi_1 \wedge \cdots \wedge \phi_n)$. Thus, $\mathrm{Th}(M) \cup \mathcal{F}_0$ is finitely satisfiable. By the compactness theorem, $\mathrm{Th}(M) \cup \mathcal{F}_0$ is satisfiable, and hence $\mathcal{F}_0$ is an n-type. Since each n-type is contained in a complete n-type, we are done. □

We now look at the relationship between types and symmetries of models.

DEFINITION 6.7.6 Let M be a Σ-structure, and let $a, b \in M$. We say that a and b are **indiscernible** in M just in case $\mathrm{tp}^M(a) = \mathrm{tp}^M(b)$. In other words, for every Σ-formula ϕ, $a \in M(\phi)$ iff $b \in M(\phi)$.

DEFINITION 6.7.7 Let $a, b \in M$. We say that a and b are **co-orbital** just in case there is an automorphism $h : M \to M$ such that $h(a) = b$.

Since automorphisms are invertible and closed under composition, being co-orbital is an equivalence relation on M, and it partitions M into a family of equivalence classes. We call these equivalence classes the **orbits** under the symmetry group $\mathrm{Aut}(M)$.

PROPOSITION 6.7.8 *Let $h : M \to N$ be an elementary embedding. Then $\mathrm{tp}^M(a) = \mathrm{tp}^N(h(a))$.*

Proof Since h is an elementary embedding $a \in M(\phi)$ iff $h(a) \in N(\phi)$, for all Σ-formulas ϕ. □

The preceeding result leads immediately to the following.

PROPOSITION 6.7.9 *If two elements a, b are co-orbital, then they are indistinguishable. That is, if there is an automorphism $h : M \to M$ such that $h(a) = b$, then $\mathrm{tp}^M(a) = \mathrm{tp}^M(b)$.*

Example 6.7.10 We now show that the converse to the previous proposition is not generally true – i.e., indistinguishable elements are not necessarily co-orbital. Let $\Sigma = \{<, c_1, c_2, \ldots, d_1, d_2, \ldots\}$, where $<$ is a binary relation, and the c_i and d_j are constant symbols. Define a Σ-structure M as follows: the domain of M is the rational numbers $\mathbb{Q}$; $<$ is given its standard interpretation on $\mathbb{Q}$; $M(c_i) = -\frac{1}{i}$ and $M(d_i) = 1 + \frac{1}{i}$ for $i = 1, 2, \ldots$

- We claim first that $[0,1]$ is invariant under all automorphisms of M. Indeed, for each $i \in \mathbb{N}$, let $(c_i, d_i) = M(c_i < x < d_i)$. Then
$$[0,1] = \bigcap_{i=1}^{\infty} (c_i, d_i).$$
If $h : M \to M$ is an automorphism, then (c_i, d_i) is invariant under h, hence $[0,1]$ is invariant under h.
- We claim that there is no Σ-formula ϕ such that $[0,1] = M(\phi)$. Indeed, it's easy to see that for any formula ϕ, if $1 \in M(\phi)$, then there is a $\delta > 0$ such that $1 + \delta \in M(\phi)$.
- We claim that $\text{tp}^M(a) = \text{tp}^M(b)$ for all $a, b \in [0,1]$. For this, we can argue in two steps. First, for any $a, b \in (0,1)$, there is an automorphism $h : M \to M$ such that $h(a) = b$. Second, choose $a \in (0,1)$, and show that $\text{tp}^M(a) = \text{tp}^M(1)$. Let $\phi \in \text{tp}^M(1)$. By an argument similar to the preceding one, there is some $\delta > 0$ and some $c \in (1 - \delta, 1)$ such that $\phi \in \text{tp}^M(c)$. Since there is an automorphism h such that $h(a) = c$, it follows that $\phi \in \text{tp}^M(a)$. Therefore, $\text{tp}^M(1) \subseteq \text{tp}^M(a)$.
- We claim that there is no automorphism $h : M \to M$ such that $h(0) = 1$. Suppose, to the contrary, that h is such an automorphism, and let $a \in (0,1)$. Since $0 < a$ and h is order preserving, $1 = h(0) < h(a)$. Thus, there is an $i \in \mathbb{N}$ such that $h(a) \in (d_i, \infty)$. But $\text{tp}^M(a) = \text{tp}^M(h(a))$, and, therefore, $a \in (d_i, \infty)$ – a contradiction.
- Notice, finally, that the element $1 \in M$ has the following feature: for every formula ϕ, if $M \vDash \phi(1)$, then $M \vdash \phi(a)$ for some $a > 1$.

┘

The previous considerations show that M has a partition $\mathbb{O}$ into orbits and a partition $\mathbb{I}$ into indiscernables, and that $\mathbb{I} \subseteq \mathbb{O}$.

We will also need the following result, which shows that indiscernibles in M always lie on the same orbit in some elementary extension N of M. We again refer the reader to Marker (2006, chapter 4) for a proof.

PROPOSITION 6.7.11 *Let M be a Σ-structure, and suppose that $\text{tp}^M(a) = \text{tp}^M(b)$. Then there is a Σ-structure N, an elementary embedding $h : M \to N$, and an automorphism $s : N \to N$ such that $s(h(a)) = h(b)$.*

In the case of finite structures, most of these subtle distinctions evaporate. For example, in finite structures, indistinguishable elements are automatically co-orbital.

PROPOSITION 6.7.12 *Let M be a finite Σ-structure, and suppose that $\text{tp}^M(a) = \text{tp}^M(b)$. Then there is an automorphism $k : M \to M$ such that $k(a) = b$.*

Proof Suppose that $\text{tp}^M(a) = \text{tp}^M(b)$. By Prop 6.7.11, there is an elementary embedding $h : M \to N$ and an automorphism $j : N \to N$ such that $j(h(a)) = h(b)$. Since M is finite, h is an isomorphism. Define $k = h^{-1} \circ j \circ h$. Then $k(a) = h^{-1}(j(h(a))) = h^{-1}(h(b)) = b$. □

Let's talk now about **invariant subsets** of a model M. A subset $A \subseteq M$ is said to be **invariant** just in case $h(A) = A$ for every automorphism $h : M \to M$. By definition, the automorphisms of a Σ-structure preserve the extensions of Σ formulas. That is, if ϕ is a Σ-formula (with a single free variable), then $M(\phi)$ is invariant under all automorphisms of M. The converse, however, is not true – i.e., not all invariant subsets are extensions of some formula. We already saw one example of this situation in 6.7.10. Other examples are easy to come by. Consider, for example, the natural numbers $\mathbb{N}$ as a model of Peano arithmetic. This model is **rigid** – i.e., there are no nontrivial automorphisms. Hence, every subset of $\mathbb{N}$ is invariant under automorphisms. Nonetheless, the language Σ of Peano arithmetic only has a countable number of formulas. Thus, there are many invariant subsets of $\mathbb{N}$ that are not of the form $\mathbb{N}(\phi)$ for some formula ϕ.

Once again, finite structures don't have as much subtlety. Indeed, in finite structures, all invariant subsets are definable.

THEOREM 6.7.13 (finite Svenonius) *If M is a finite Σ-structure, and A is an invariant subset of M, then there is a Σ-formula θ such that $A = M(\theta)$.*

Proof Let $\mathscr{B}$ be the Boolean algebra of representable subsets of M, i.e., sets of the form $M(\phi)$ for some formula ϕ. For each $a \in M$, the set

$$\{M(\phi) \mid \phi \in \mathrm{tp}^M(a)\} = \{M(\phi) \mid a \in M(\phi)\},$$

is an ultrafilter on $\mathscr{B}$. Thus, if X is the Stone space of $\mathscr{B}$, there is a map $\pi \equiv \mathrm{tp}^M : M \to X$ such that $\pi(a)[M(\phi)] = 1$ iff $a \in M(\phi)$. In this case, since $\mathscr{B}$ is finite, each ultrafilter is principal, i.e., is the up-set of some $M(\phi)$. Hence $\pi : M \twoheadrightarrow X$ is surjective.

Since A is invariant under $\mathrm{Aut}(M)$, Prop. 6.7.12 entails that $\mathrm{tp}^M(a) \neq \mathrm{tp}^M(b)$ whenever $a \in A$ and $b \notin A$. Thus, A descends along π, i.e., $\pi^{-1}[\pi(A)] = A$. Since X is finite, $\pi(A)$ is clopen – i.e., there is a formula θ such that $A = M(\theta)$. □

The following key result will lead very quickly to a proof of Svenonius' theorem.

PROPOSITION 6.7.14 *Let M be a Σ^+-structure, and suppose that for every elementary extension N of M, any automorphism of $N|_\Sigma$ preserves $N(r)$. Then there is a Σ-formula ϕ such that $M \vDash \forall x(r(x) \leftrightarrow \phi(x))$.*

Proof We first claim that in every elementary extension N of M, if $a,b \in N$ such that $\mathrm{tp}^N(a)|_\Sigma = \mathrm{tp}^N(b)|_\Sigma$, then $a \in N(r)$ iff $b \in N(r)$. Suppose not, i.e., that there is an elementary extension N of M with $a,b \in N$ satisfying the same Σ-formulas, but $a \in N(r)$ and $b \notin N(r)$. By using an argument similar to the realizing types theorem, we can show that there is an elementary extension $i : N \to N'$, and an automorphism s of $N'|_\Sigma$ such that $s(i(a)) = i(b)$. Thus, s does not leave $N'(r)$ invariant, contradicting the assumptions of the proposition.

Now since any finite subset of S_1^M is realized in some elementary extension of M (Prop 6.7.4), it follows that for all $p,q \in S_1^M$, if $p|_\Sigma = q|_\Sigma$, then $p \in E_r$ iff $q \in E_r$. Conversely, if $p \in E_r$ and $q \notin E_r$, then there is some Σ-formula ϕ such that $p \in E_\phi$ and $q \notin E_\phi$. From this, it follows that the intersection of all E_ϕ such that $p \in E_\phi$ lies in E_ϕ. By the compactness of S_1^M, there are finitely many Σ-formulas $\phi_1, \dots, \phi_n$ such that $p \in E_{\phi_i}$ and

$$E_{\phi_i} \cap \cdots \cap E_{\phi_n} \subseteq E_r.$$

If we let $\psi_p \equiv \phi_1 \wedge \cdots \wedge \phi_n$, then $p \in E_{\psi_p}$ and $E_{\psi_p} \subseteq E_r$. The family $\{E_{\psi_p} \mid p \in E_r\}$ covers E_r, hence by compactness again has a finite subcover. Taking the conjunction of the corresponding formulas gives an explicit definition of $r(x)$ in terms of Σ. □

THEOREM 6.7.15 (Svenonius) *Suppose that in every model M of T^+, the set $M(r)$ is invariant under all Σ-automorphisms. Then there are Σ-formulas $\phi_1, \dots, \phi_n$ such that*

$$T \vdash \forall x(r(x) \leftrightarrow \phi_1(x)) \vee \cdots \vee \forall x(r(x) \leftrightarrow \phi_n(x)).$$

Proof By the previous proposition, for each model M of T, there is a Σ-formula ϕ_M such that $M \vDash \forall x(r(x) \leftrightarrow \phi_M(x))$. Let $\psi_M \equiv \forall x(r(x) \leftrightarrow \phi_M(x))$, and let Δ be the set of $\neg\psi_M$, where M runs over all models of T. (To deal with size issues, we could consider isomorphism classes of models bounded by a certain size, depending on the signature Σ.) Then $T \cup \Delta$ is inconsistent. By compactness, there is a finite subset $\neg\psi_1, \dots, \neg\psi_n$ of Δ such that $T \vdash \psi_1 \vee \cdots \vee \psi_n$. □

If, in addition, the theory T is complete, then the assumptions of Svenonius' theorem entail that T explicitly defines r in terms of Σ. Beth's theorem derives the same conclusion, without the completeness assumption, but with a stronger notion of implicit definability. Consider the following explications of the notion of definability:

IE Invariance under elementary embeddings: For any models M and N of T, and for any Σ-elementary embedding $h : M \to N$, $h(M(r)) = N(r)$.

IA Invariance under automorphisms: For any model M of T, and for any Σ-automorphism $h : M \to M$, $h(M(r)) = M(r)$.

IS For any models M and N of T, if $M|_\Sigma = N|_\Sigma$ then $M = N$. (This version is very close to the metaphysician's notion of global supervenience.)

ID Let T' be the result of uniformly replacing r in T with r'. Then $T \cup T' \vdash \forall x(r(x) \leftrightarrow r'(x))$.

The implication IE $\Rightarrow$ IA is immediate. To see that IE $\Rightarrow$ IS, suppose that $M|_\Sigma = N|_\Sigma$, and let $h : M \to N$ be the identity function. Then h is a Σ-elementary embedding, hence IE implies that $h(M(r)) = N(r)$, that is, $M(r) = N(r)$. To see that IS $\Rightarrow$ ID, let M be a model of $T \cup T'$. Define a $\Sigma \cup \{r\}$ structure N to agree with M on Σ, and such that $N(r) = M(r')$. Because M is a model of T', it follows that N is a model of T. Hence $M(r) = N(r) = M(r')$, and $M \vDash \forall x(r(x) \leftrightarrow r'(x))$.

We now show that $\neg$IE $\Rightarrow$ $\neg$ID. If $\neg$IE, then there are models M and N of T, and an elementary embedding $h : M \to N$ such that $h(M(r)) \neq N(r)$. We use N to define a $\Sigma \cup \{r, r'\}$ structure N': let N' agree with N on $\Sigma \cup \{r\}$, and let $N'(r') = h(M(r))$. Obviously N' is a model of T. To see that N' is a model of T', first let M' be the $\Sigma \cup \{r'\}$ structure that looks just like M except that $M'(r') = M(r)$. Then $M' \vDash T'$, and $N'(r') = h(M(r)) = h(M'(r'))$. That is, N' is the push-forward of M', and hence

$N' \models T'$. Finally, $N'(r) \neq N'(r')$, and hence $T \cup T' \not\models \forall x(r(x) \leftrightarrow r'(x))$. This completes the proof that $\neg$IE $\Rightarrow$ $\neg$ID.

All told, we have the following chain of implications:

$$\begin{array}{ccccc} \text{IE} & \Longleftrightarrow & \text{ID} & \Longleftrightarrow & \text{IS} \\ & & \Downarrow & & \\ & & \text{IA} & & \end{array}$$

What's more, the implication ID $\Rightarrow$ IA cannot be reversed.

We now sketch the proof that the stronger notion of implicit definability (IE,ID,IS) implies explicit definability.

THEOREM 6.7.16 (Beth's theorem) *If T implicitly defines r in terms of Σ, then T explicitly defines r in terms of Σ.*

Proof We follow the outlines of the proof by Poizat (2012, 185). Assume that T implicitly defines r in terms of Σ. Since IS $\Rightarrow$ IA, Svenonius' theorem implies that there are Σ-formulas $\phi_1, \dots, \phi_n$ such that

$$T \vdash \forall x(r(x) \leftrightarrow \phi_1(x)) \vee \cdots \vee \forall x(r(x) \leftrightarrow \phi_n(x)).$$

If T were inconsistent, or consistent with only a single one of these disjuncts, then T would explicitly define r in terms of Σ. So suppose that $n > 1$, and T is consistent with all n disjuncts. For each disjunct $\forall x(r(x) \leftrightarrow \phi_i(x))$, let T_i be the theory that results from replacing r in T with ϕ_i. Implicit definability then yields $T_i \cup T_j \vdash \forall x(\phi_i(x) \leftrightarrow \phi_j(x))$. Notice that r does not occur in $T_i \cup T_j$. Using the compactness theorem, we can then use Σ-formulas $\theta_1, \dots, \theta_m$ to divide the space of models of T into cells with the feature that for each k, we have $T, \theta_k \vdash \forall x(r(x) \leftrightarrow \phi_{i(k)})$ for some $i(k)$. One can then use the formulas $\theta_1, \dots, \theta_m$ to construct an explicit definition of r in terms of Σ. □

Example 6.7.17 Petrie (1987) argues that global supervenience does not entail reducibility. We first state his definition verbatim:

> Let $\mathscr{A}$ and $\mathscr{B}$ be sets of properties. We say that $\mathscr{A}$ **globally supervenes** on $\mathscr{B}$ just in case worlds which are indiscernible wih regard to $\mathscr{B}$ are also indiscernible with regard to $\mathscr{A}$.

Switching to the formal mode – i.e., speaking of predicates rather than properties – and restricting to the context of first-order logic, it appears that global supervenience is just another name for implicit definability. We will use $\Sigma = \{p\}$ for the subvenient predicate symbol, and we let $\Sigma^+ = \Sigma \cup \{r\}$. Petrie describes his example as follows (with notation adapted):

> There are two structures M and N, both of which have domain $\{a, b\}$. In M, the extension of p is $\{a, b\}$ and the extension of r is $\{a\}$. In N, the extension of p is $\{a\}$ and the extension of r is empty.

Petrie points out that this example does not satisfy strong supervenience. However, since $M|_\Sigma \neq N|_\Sigma$, it trivially satisfies global supervenience – i.e., r is implicitly defined in terms of p.

Here we need to slow down: implicit definability is defined in terms of some background theory T. In this case, however, there can be no theory T such that M and N are models of T, and such that T implicitly defines r in terms of p. Indeed, for any theory T in Σ^+, if M is a model of T, then so is M' where $M'(p) = \{a,b\}$ and $M'(r) = \{b\}$. But then $M|_\Sigma = M'|_\Sigma$, whereas $M(r) \neq M'(r)$. Therefore, T does not implicitly define r in terms of Σ.

Thus, Petrie's space of possible worlds is not of the form Mod(T) for any theory T. One of the key assumptions of formal logic is that possibilities are specified only up to isomorphism – i.e., if M is possible, and M' results from permuting some of the (featureless) elements of M, then M' is also possible. (Why would it be possible that a is an r, but not possible that b is an r?) Thus, for one to grant the force of Petrie's counterexample, one has to abandon a key assumption of formal logic. Is it worth sacrificing formal logic in order to defend non-reductive physicalism? ┘

6.8 Notes

- Within mathematics, the study of logical semantics is called **model theory**, and there are several excellent textbooks. Some of our favorites: Hodges (1993); Marker (2006); Poizat (2012).
- The classic sources on the semantic view of theories are Suppe (1974, 1989).
- The completeness of the predicate calculus was first proven by Kurt Gödel in his PhD thesis (Gödel, 1929).

  The typical textbook proof of the theorem proceeds as follows: supposing that Γ is proof-theoretically consistent, show that Γ can be expanded to a maximally consistent set Γ^*. This expansion invokes Zorn's lemma, which is a variant of the axiom of choice. The resulting set Γ^* is then used to construct a model of Γ.

  The topological proof in this chapter has several advantages over the typical textbook proof. For example, the topological theorem makes it clear that completeness doesn't require the full axiom of choice. It is known that the Baire category theorem is strictly weaker than AC (see Herrlich and Keremedis, 2000; Herrlich, 2006). The topological completeness proof was first given by Rasiowa and Sikorski (1950). See also (Rasiowa and Sikorski, 1963).

  An even more elegant proof of completeness is provided by Deligne's embedding theorem for coherent categories (see Makkai and Reyes, 1977). If $T \nvdash \bot$, then T corresponds to a (Boolean) coherent category C_T. By Deligne's theorem, there is an embedding $F : C_T \to$ **Sets**, which yields a model of T.
- The category **Sets** has tons of structure (limits, colimits, exponentials, etc.), and so is adequate to represent all syntactic structures of a first-order theory. If we're only interested in a fragment of first-order logic, it can also be interesting to look at representations in less structured categories. For example, Cartesian categories have enough structure to represent algebraic theories (such as the theory of groups). For more details, see Johnstone (2003).

- In Section 6.5, we redescribed topological structure on X as a family of operations $X^{\infty} \to X$, i.e., functions from infinite sequences to points of X. This description is not merely heuristic: the category **CHaus** of compact Hausdorff spaces is equivalent to the category of algebras for the ultrafilter monad on **Sets**. Thus, **CHaus** is the category of models of an (infinitary) algebraic theory. For more details, see (Mac Lane, 1971, VI.9) and (Manes, 1976, 1.5.24).
- For an interesting analysis of supervenience and reduction in terms of ultraproducts, see Dewar (2018b).
- Beth's theorem first appeared in (1956), and Svenonius' in (1959). In recent work, Makkai (1991); Zawadowski (1995); Moerdijk and Vermeulen (1999) show that Beth's theorem is equivalent to a result about *effective descent morphisms*, a notion of central importance in mainstream mathematics. This kind of unifying result shows that there is no clear boundary between mathematics and metamathematics.
- Our discussion of Beth's theorem draws from Barrett (2018b). The relevance of Beth's theorem to the prospects of non-reductive physicalism was first pointed out by Hellman and Thompson (1975), and has been subsequently discussed by Bealer (1978); Hellman (1985); Tennant (1985, 2015). According to Hellman (personal communication), the issue was first brought up by Hilary Putnam in a graduate seminar at Harvard. For more on supervenience and its history in analytic philosophy, see McLaughlin and Bennett (2018).

7 Semantic Metalogic Redux

In the previous chapter, we covered some of the standard topics in model theory – focusing on those parts we think are of most interest to philosophers. However, the semantic methods of the previous chapter are restricted to the special case of single-sorted logic. In this chapter, we cover semantics for many-sorted logic. But our aim here is not many-sorted logic for its own sake. Indeed, we feel that one first really understands single-sorted logic when one sees it as a special case of many-sorted logic. What's more, even for single-sorted theories, some of the most interesting relations between theories can only be explicated by means of many-sorted methods.

7.1 Structures and Models

For the most part, generalizing semantics to many-sorted logic is straightforward: where a single-sorted structure M has a single domain set, a many-sorted structure has a separate domain set $M(\sigma)$ for each separate sort symbol $\sigma \in \Sigma$. Moreover, if a Σ-formula ϕ has free variables of different sorts, then its extension $M(\phi)$ will be a subset of a Cartesian product of different domains.

DEFINITION 7.1.1 Let Σ be a signature. A **Σ-structure** M is a mapping from Σ to appropriate structures in the category **Sets**. In particular:

- M maps each sort symbol $\sigma \in \Sigma$ to a set $M(\sigma)$.
- M maps each n-ary relation symbol p of sort $\sigma_1 \times \cdots \times \sigma_n$ to a subset $M(p) \subseteq M(\sigma_1) \times \cdots \times M(\sigma_n)$.
- M maps each function symbol f of sort $\sigma_1 \times \cdots \times \sigma_n \to \sigma_{n+1}$ to a function $M(f) : M(\sigma_1) \times \cdots \times M(\sigma_n) \to M(\sigma_{n+1})$.

As was the case with single-sorted logic, each Σ-structure M extends to a map, still called M, from Σ-terms to functions, and from Σ-formulas to subsets. In particular:

- For each term t of type $\sigma_1 \times \cdots \times \sigma_n \to \sigma_{n+1}$,

$$M(t) : M(\sigma_1) \times \cdots \times M(\sigma_n) \to M(\sigma_{n+1}).$$

- For each formula ϕ of type $\sigma_1 \times \cdots \times \sigma_n$,

$$M(\phi) \subseteq M(\sigma_1) \times \cdots \times M(\sigma_n).$$

DEFINITION 7.1.2 Let M and N be Σ-structures, where Σ has sorts $\sigma_1, \ldots, \sigma_n$. An **elementary embedding** $h : M \to N$ consists of a family $\{h_i \mid \sigma_i \in \Sigma\}$ of functions $h_i : M(\sigma_i) \to N(\sigma_i)$ that preserves the extension of each Σ-formula ϕ.

It's easy to see that the composition of elementary embeddings is an elementary embedding. Thus, for a given theory T, we let Mod(T) be the category whose objects are models of T and whose arrows are elementary embeddings. Notice that this definition directly generalizes the definition we gave for single-sorted theories; hence, for a single-sorted theory T, there is no ambiguity when we write Mod(T).

7.2 The Dual Functor to a Translation

Intuitively speaking, a translation $F : T \to T'$ should induce a functor $F^* : \text{Mod}(T') \to \text{Mod}(T)$ going in the opposite direction. To see this, recall that models of a theory T' aren't static structures but are more like functors from T' into the category **Sets**. If we think of a model of T' as a functor $M : T' \to \mathbf{Sets}$, then we can precompose this functor with a translation $F : T \to T'$, giving a functor

$$T \overset{F}{\longmapsto} T' \overset{M}{\longmapsto} \mathbf{Sets},$$

i.e., we get a model $F^*M = M \circ F$ of T. However, since M and F are not actually functors, we have to do some work to validate this intuition.

DEFINITION 7.2.1 Let $F : T \to T'$ be a fixed translation. Given an arbitrary model M of T', we define a Σ-structure F^*M as follows:

- For a sort symbol σ of Σ, first consider the set

$$M(F(\sigma)) = M(F(\sigma)_1) \times \cdots \times M(F(\sigma)_n),$$

  and its subset $M(D_\sigma)$, where D_σ is any one of the domain formulas that F associates with σ. (These domain formulas are all equivalent.) Since F is a translation and M is a model of T', $M(E_\sigma)$ is an equivalence relation on $M(D_\sigma)$. Let $q : M(D_\sigma) \to Y$ be the corresponding quotient map, and let $(F^*M)(\sigma) = Y$.
- For a relation symbol $p : \sigma_1, \ldots, \sigma_n$ of Σ, we define

$$(F^*M)(p) = (q_1 \times \cdots \times q_n)(M(Fp)),$$

  where $q_i : M(D_{\sigma_i}) \to Y_i$ is the corresponding projection.
- For a function symbol $f : \sigma_1, \ldots, \sigma_n \to \sigma_{n+1}$ of Σ, we define $(F^*M)(f)$ be the function with graph

$$(q_1 \times \cdots \times q_n \times q_{n+1})(M(Ff)).$$

NOTE 7.2.2 Suppose that $F : T \to T'$ is a translation, and let $\phi(x)$ be a Σ-formula. Then $F(\phi)(\vec{x})$ is compatible with the relation $E(\vec{x}, \vec{y})$ in the following sense: T' implies that if $F(\phi)(\vec{x})$ and $E(\vec{x}, \vec{y})$, then $F(\phi)(\vec{y})$. It follows from this that in any model M of T', the subset $A \equiv M(F(\phi)(\vec{x}))$ of $M(D)$ is compatible with the equivalence relation

$M(E(\vec{x}, \vec{y}))$. That is, if $q : M(D) \to Y$ is the quotient map induced by $M(E(\vec{x}, \vec{y}))$, then $q^{-1}(q(A)) = A$.

PROPOSITION 7.2.3 *Suppose that* $F : T \to T'$ *is a translation, and that* ϕ *is a* Σ*-formula. Then* $(F^*M)(\phi)$ *is the image of* $M(F(\phi))$ *under the corresponding quotient map* q.

Proof To be precise, we prove that the result holds for each Σ-formula ϕ, and context $x_1, \dots, x_n$ for ϕ. Once a context $x_1, \dots, x_n$ for ϕ is fixed, we also fix the corresponding context $\vec{x}_1, \dots, \vec{x}_n$ for $F(\phi)$. Moreover, for a Σ-structure N, we take $N(\phi)$ to mean the extension of ϕ relative to the context $x_1, \dots, x_n$.

The base case follows immediately from the definition of F^*M. As for inductive cases, we will treat $\wedge$ and $\exists y$ and leave the others to the reader. We simplify notation as follows: let $N = F^*M$, and for each Σ-formula ϕ, let $\phi^* = F(\phi)$.

- Suppose that the result is true for ϕ_1 and ϕ_2, in any of their contexts. Let $x_1, \dots, x_n$ be a context for $\phi_1 \wedge \phi_2$, hence also for ϕ_1 and ϕ_2. Let $D = D(\vec{x}_1) \wedge \cdots \wedge D(\vec{x}_n)$ be the conjunction of domain formulas for $x_1, \dots, x_n$; let $E = E(\vec{x}_1, \vec{y}_1) \wedge \cdots \wedge E(\vec{x}_n, \vec{y}_n)$ be the conjunction of the corresponding equivalence relations; and let $q : M(D) \to Y$ be the quotient map determined by $M(E)$. If we let $A_i = M(\phi_i^*) \subseteq M(D)$, then the inductive hypothesis asserts that $N(\phi_i) = q(A_i)$. Since $(\phi_1 \wedge \phi_2)^* = \phi_1^* \wedge \phi_2^*$, it follows that $M((\phi_1 \wedge \phi_2)^*) = A_1 \cap A_2$. Thus,

$$\begin{aligned} q(M(\phi^*)) &= q(A_1 \cap A_2) \\ &= q(A_1) \cap q(A_2) \\ &= N(\phi_1) \cap N(\phi_2) \\ &= N(\phi). \end{aligned}$$

  The second equation follows from the preceding note; the third equation follows from the induction hypothesis; and the final equation by the fact that N is a Σ-structure.
- Suppose that the result is true for ϕ. That is,

$$N(\phi) = (q_1 \times q_2)(M(\phi^*)),$$

  where $q_1 : M(D_1) \to Y_1$ and $q_2 : M(D_2) \to Y_2$ are the quotient maps. We show that the result is also true for $\exists y \phi$. Consider the commuting diagram:

$$\begin{array}{ccc} M(D_1) \times M(D_2) & \xrightarrow{\pi} & M(D_2) \\ \big\downarrow{\scriptstyle q_1 \times q_2} & & \big\downarrow{\scriptstyle q_2} \\ Y_1 \times Y_2 & \xrightarrow{\quad\pi\quad} & Y_2 \end{array}$$

  where π is the projection onto the second coordinate. By definition,

$$M((\exists y \phi)^*) = M(\exists \vec{y}(\phi^*)) = \pi^*(M(\phi^*)).$$

Hence,

$$N(\exists y\phi) = \pi(N(\phi)) = \pi((q_1 \times q_2)(M(\phi^*)))$$
$$= q_2(\pi^*(M(\phi^*))) = q_2(M((\exists y\phi)^*)).$$

Thus, the result also holds for $\exists y\phi$. □

PROPOSITION 7.2.4 *Let $F : T \to T'$ be a translation. If M is a model of T', then F^*M is a model of T.*

Proof Let ϕ be a Σ-formula in context $x_1, \ldots, x_n$ such that $T \vdash \phi$. Since $F : T \to T'$ is a translation, $T' \vdash F(\phi)$. If M is a model of T', then

$$M(F(\phi)) = M(\vec{x}_1, \ldots, \vec{x}_n).$$

By the previous proposition, $(F^*M)(\phi)$ is the image of $M(F(\phi))$ under the quotient map $q : M(D(\vec{x}_1, \ldots, \vec{x}_n)) \to Y$ induced by the equivalence relation $M(E)$, where

$$E = E(\vec{x}_1, \vec{y}_1) \wedge \cdots \wedge E(\vec{x}_n, \vec{y}_n).$$

Thus, $(F^*M)(\phi) = Y = (F^*M)(x_1, \ldots, x_n)$, and F^*M is a model of T. □

Thus, if $F : T \to T'$ is a translation, then F gives rise to a mapping F^* from the objects of Mod(T') to the objects of Mod(T). We now define F^* on the arrows of Mod(T'). Let M and N be models of T', and let $h : M \to N$ be an elementary embedding. Recall that h consists of family $\{h_\sigma \mid \sigma \in \Sigma'\}$ of functions $h_i : M(\sigma) \to N(\sigma)$ that preserves the extension of each Σ'-formula ψ. Now let σ be a sort of Σ, and let $\vec{x}$ be a sequence of Σ'-variables of sort $F(\sigma) = \sigma_1, \ldots, \sigma_n$. Consider the following diagram:

$$\begin{array}{ccc} M(D_\sigma) & \xrightarrow{h} & N(D_\sigma) \\ \downarrow q_M & & \downarrow q_N \\ (F^*M)(\sigma) & \overset{F^*h}{\dashrightarrow} & (F^*N)(\sigma) \end{array}$$

where q_M and q_N are the quotient maps induced by $M(E)$ and $N(E)$, respectively, and h is the restriction of $h_1 \times \cdots \times h_n$ to $M(D_\sigma)$, which is well defined since h preserves the extensions of Σ'-formulas. Moreover, if $\langle \vec{a}, \vec{b} \rangle \in M(E)$, then $\langle h(\vec{a}), h(\vec{b}) \rangle \in N(E)$. Thus, h determines a unique function $F^*h : (F^*M)_\sigma \to (F^*N)_\sigma$ such that the previous diagram commutes.

Now let ϕ be a Σ-formula. Then $a \in (F^*M)(\phi)$ iff $a = q_M(b)$, for some $b \in M(F(\phi))$. Moreover, $b \in M(F(\phi))$ iff $h(b) \in N(F(\phi))$ iff $q_N(h(b)) \in (F^*N)(\phi)$. This shows that $a \in (F^*M)(\phi)$ iff $(F^*h)(a) \in (F^*N)(\phi)$. Therefore, F^*h is an elementary embedding.

It is easy to see that F^* preserves composition of elementary embeddings, as well as identity morphisms of models. Therefore, F^* is a functor from Mod(T') to Mod(T).

DISCUSSION 7.2.5 It is tempting to think that any functor $G : \text{Mod}(T') \to \text{Mod}(T)$ corresponds to some potentially interesting relationship between the theories T and T'. However, functors of the form $F^* : \text{Mod}(T') \to \text{Mod}(T)$, where $F : T \to T'$ is a translation, seem to be singled out by the fact that they preserve important theoretical structures. First, the functor F^* is "definable" in the sense that the resulting model F^*M is defined in terms of the model M, and in a uniform fashion. That is, the "same definition" of the new model works, regardless of which model we plug into F^*. (For more on the notion of definable functors, see Hudetz [2018a].) Second, in the case of propositional theories, a functor $G : \text{Mod}(T') \to \text{Mod}(T)$ is simply a function from the Stone space X' of T' to the Stone space X of T, and functors of the form F^* are precisely the continuous functions.

We now have the tools we need to do some work with the notion of Morita equivalence. We'll first show how similar Morita equivalence is to its poorer cousin, definitional equivalence. In particular, we'll show that Morita equivalent theories have equivalent categories of models.

7.3 Morita Equivalence Implies Categorical Equivalence

As with a definitional extension, a Morita extension T^+ should "say nothing more" than the original theory T. We will make this idea precise by proving three results about the relationship between $\text{Mod}(T^+)$ and $\text{Mod}(T)$. First, the models of T^+ are "determined" by the models of T.

THEOREM 7.3.1 (Barrett) *Let $\Sigma \subseteq \Sigma^+$ be signatures and T a Σ-theory. If T^+ is a Morita extension of T to Σ^+, then every model M of T has a unique expansion (up to isomorphism) M^+ that is a model of T^+.*

Before proving this result, we introduce some notation and prove a lemma. Suppose that a Σ^+-theory T^+ is a Morita extension of a Σ-theory T. Let M and N be models of T^+ with $h : M|_\Sigma \to N|_\Sigma$ an elementary embedding between the Σ-structures $M|_\Sigma$ and $N|_\Sigma$. The elementary embedding h naturally induces a map $h^+ : M \to N$ between the Σ^+-structures M and N.

We know that h is a family of maps $h_\sigma : M_\sigma \to N_\sigma$ for each sort $\sigma \in \Sigma$. (Here we have used M_σ for the domain $M(\sigma)$ that M assigns to the sort symbol σ, and similarly for N_σ.) In order to describe h^+, we need to describe the map $h_\sigma^+ : M_\sigma \to N_\sigma$ for each sort $\sigma \in \Sigma^+$. If $\sigma \in \Sigma$, we simply let $h_\sigma^+ = h_\sigma$. On the other hand, when $\sigma \in \Sigma^+ \backslash \Sigma$, there are four cases to consider. We describe h_σ^+, in the cases where the theory T^+ defines σ as a product sort or a subsort. The coproduct and quotient sort cases are described analogously.

First, suppose that T^+ defines σ as a product sort. Let $\pi_1, \pi_2 \in \Sigma^+$ be the projections of arity $\sigma \to \sigma_1$ and $\sigma \to \sigma_2$ with $\sigma_1, \sigma_2 \in \Sigma$. The definition of the function h_σ^+ is suggested by the following diagram.

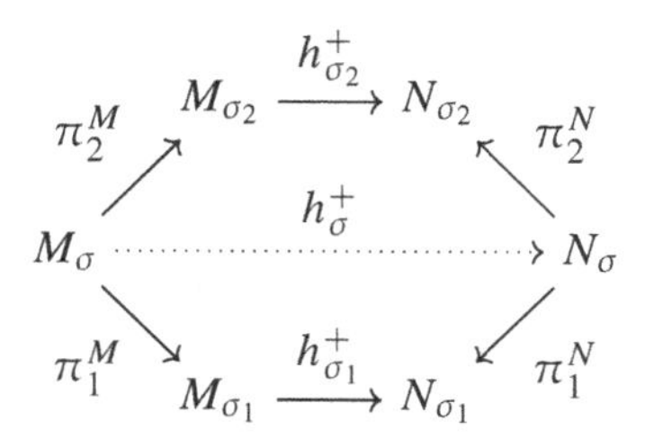

Let $m \in M_\sigma$. We define $h^+_\sigma(m)$ to be the unique $n \in N_\sigma$ that satisfies both $\pi^N_1(n) = h^+_{\sigma_1} \circ \pi^M_1(m)$ and $\pi^N_2(n) = h^+_{\sigma_2} \circ \pi^M_2(m)$. We know that such an n exists and is unique because N is a model of T^+ and T^+ defines the symbols σ, π_1, and π_2 to be a product sort. One can verify that this definition of h^+_σ makes the preceding diagram commute.

Suppose, on the other hand, that T^+ defines σ as the subsort of "elements of sort σ_1 that are ϕ." Let $i \in \Sigma^+$ be the inclusion map of arity $\sigma \to \sigma_1$ with $\sigma_1 \in \Sigma$. As before, the definition of h^+_σ is suggested by the following diagram.

$$\begin{array}{ccc} M_\sigma & \overset{h^+_\sigma}{\dashrightarrow} & N_\sigma \\ \downarrow i^M & & \downarrow i^N \\ M_{\sigma_1} & \xrightarrow{h^+_{\sigma_1}} & N_{\sigma_1} \end{array}$$

Let $m \in M_\sigma$. We see that following implications hold:

$$\begin{aligned} M \vDash \phi[i^M(m)] &\Rightarrow M|_\Sigma \vDash \phi[i^M(m)] \\ &\Rightarrow N|_\Sigma \vDash \phi[h^+_{\sigma_1}(i^M(m))] \Rightarrow N \vDash \phi[h^+_{\sigma_1}(i^M(m))] \end{aligned}$$

The first and third implications hold since $\phi(x)$ is a Σ-formula, and the second holds because $h_{\sigma_1} = h^+_{\sigma_1}$ and h is an elementary embedding. T^+ defines the symbols i and σ as a subsort and M is a model of T^+, so it must be that $M \vDash \phi[i^M(m)]$. By the preceding implications, we see that $N \vDash \phi[h^+_{\sigma_1}(i^M(m))]$. Since N is also a model of T^+, there is a unique $n \in N_\sigma$ that satisfies $i^N(n) = h^+_{\sigma_1}(i^M(m))$. We define $h^+_\sigma(m) = n$. This definition of h^+_σ again makes the diagram commute.

When T^+ defines σ as a coproduct sort or a quotient sort one describes the map h^+_σ analogously. We leave it to the reader to work out the details of these cases.

For the purposes of proving Theorem 7.3.1, we need the following simple lemma about this map h^+.

LEMMA 7.3.2 *If $h : M|_\Sigma \to N|_\Sigma$ is an isomorphism, then $h^+ : M \to N$ is an isomorphism.*

Proof We know that $h_\sigma : M_\sigma \to N_\sigma$ is a bijection for each $\sigma \in \Sigma$. Using this fact and the definition of h^+, one can verify that $h^+_\sigma : M_\sigma \to N_\sigma$ is a bijection for each sort $\sigma \in \Sigma^+$. So h^+ is a family of bijections. And furthermore, the commutativity of the preceding diagrams implies that h^+ preserves any function symbols that are used to define new sorts.

It only remains to check that h^+ preserves predicates, functions, and constants that have arities and sorts in Σ. Since $h : M|_\Sigma \to N|_\Sigma$ is a isomorphism, we know that h^+ preserves the symbols in Σ. So let $p \in \Sigma^+\backslash\Sigma$ be a predicate symbol of arity $\sigma_1 \times \ldots \times \sigma_n$ with $\sigma_1, \ldots, \sigma_n \in \Sigma$. There must be a Σ-formula $\phi(x_1, \ldots, x_n)$ such that $T^+ \vDash \forall_{\sigma_1} x_1 \ldots \forall_{\sigma_n} x_n (p(x_1, \ldots, x_n) \leftrightarrow \phi(x_1, \ldots, x_n))$. We know that $h : M|_\Sigma \to N|_\Sigma$ is an elementary embedding, so in particular it preserves the formula $\phi(x_1, \ldots, x_n)$. This implies that $(m_1, \ldots, m_n) \in p^M$ if and only if $(h_{\sigma_1}(m_1), \ldots, h_{\sigma_n}(m_n)) \in p^N$. Since $h^+_{\sigma_i} = h_{\sigma_i}$ for each $i = 1, \ldots, n$, it must be that h^+ also preserves the predicate p. An analogous argument demonstrates that h^+ preserves functions and constants. □

We now turn to the proof of Theorem 7.3.1.

Proof of Theorem 7.3.1 Let M be a model of T. First note that if M^+ exists, then it is unique up to isomorphism. For if N is a model of T^+ with $N|_\Sigma = M$, then by letting h be the identity map (which is an isomorphism) Lemma 7.3.2 implies that $M^+ \cong N$. We need only to define the Σ^+-structure M^+. To guarantee that M^+ is an expansion of M we interpret every symbol in Σ the same way that M does. We need to say how the symbols in $\Sigma^+\backslash\Sigma$ are interpreted. There are a number of cases to consider.

Suppose that $p \in \Sigma^+\backslash\Sigma$ is a predicate symbol of arity $\sigma_1 \times \ldots \times \sigma_n$ with $\sigma_1, \ldots, \sigma_n \in \Sigma$. There must be a Σ-formula $\phi(x_1, \ldots, x_n)$ such that $T^+ \vDash \forall_{\sigma_1} x_1 \ldots \forall_{\sigma_n} x_n (p(x_1, \ldots, x_n) \leftrightarrow \phi(x_1, \ldots, x_n))$. We define the interpretation of the symbol p in M^+ by letting $M^+(p) = M(\phi)$. Obviously this definition implies that $M^+ \vDash \delta_p$. The cases of function and constant symbols are handled similarly.

Let $\sigma \in \Sigma^+\backslash\Sigma$ be a sort symbol. We describe the cases where T^+ defines σ as a product sort or a subsort. The coproduct and quotient sort cases follow analogously. Suppose first that σ is defined as a product sort with π_1 and π_2 the projections of arity $\sigma \to \sigma_1$ and $\sigma \to \sigma_2$, respectively. We define $M^+_\sigma = M^+_{\sigma_1} \times M^+_{\sigma_2}$ with $\pi_1^{M^+} : M^+_\sigma \to M^+_{\sigma_1}$ and $\pi_2^{M^+} : M^+_\sigma \to M^+_{\sigma_2}$ the canonical projections. One can easily verify that $M^+ \vDash \delta_\sigma$. On the other hand, suppose that σ is defined as a subsort with defining Σ-formula $\phi(x)$ and inclusion i of arity $\sigma \to \sigma_1$. We define $M^+_\sigma = M(\phi) \subseteq M_{\sigma_1}$ with $i^{M^+} : M^+_\sigma \to M^+_{\sigma_1}$ the inclusion map. One can again verify that $M^+ \vDash \delta_\sigma$. □

The previous result immediately yields an important corollary:

THEOREM 7.3.3 (Barrett) *If T^+ is a Morita extension of T, then T^+ is a conservative extension of T.*

Proof Suppose that T^+ is not a conservative extension of T. One can easily see that $T \vdash \phi$ implies that $T^+ \vdash \phi$ for every Σ-sentence ϕ. So there must be some Σ-sentence ϕ such that $T^+ \vdash \phi$, but $T \nvdash \phi$. This implies that there is a model M of T such that $M \vDash \neg\phi$. This model M has no expansion that is a model of T^+ since $T^+ \vdash \phi$, contradicting Theorem 7.3.1. □

Theorems 7.3.1 and 7.3.3 are natural generalizations from definition extensions to Morita extensions. In the case that T^+ is a definitional extension of T, there are natural maps $I : T \to T^+$ and $R : T^+ \to T$ that form a homotopy equivalence. We now define a reduction map $R : T^+ \to T$ for the case where T^+ is a Morita extension of T.

Lemma 4.6.11 shows that if T^+ is a definitional extension of T to Σ^+, then for every Σ^+-formula ϕ there is a corresponding Σ-formula $R\phi$ such that $T^+ \vdash \phi \leftrightarrow R\phi$. The following example demonstrates that this result does not generalize to the case of Morita extensions in a perfectly straightforward manner.

Example 7.3.4 Recall the theories T and T^+ from Example 5.2.3, and consider the Σ^+-formula $\phi(x,z)$ defined by $i(z) = x$. One can easily see that there is no Σ-formula $\phi^*(x,z)$ that is equivalent to $\phi(x,z)$ according to the theory T^+. Indeed, the variable z cannot appear in any Σ-formula since it is of sort $\sigma^+ \in \Sigma^+ \backslash \Sigma$. A Σ-formula simply cannot say how variables with sorts in Σ relate to variables with sorts in Σ^+. ⌟

In order to define $R : T^+ \to T$, therefore, we need a way of specifying how variables with sorts in $\Sigma^+ \backslash \Sigma$ relate to variables with sorts in Σ. We do this by defining the concept of a "code" (see Szczerba, 1977).

DEFINITION 7.3.5 Let $\Sigma \subseteq \Sigma^+$ be signatures with T a Σ-theory and T^+ a Morita extension of T to Σ^+. We define a **code** formula $\xi(x, y_1, y_2)$ for each variable x of sort $\sigma \in \Sigma^+ \backslash \Sigma$ as follows:

- Suppose that T^+ defines σ as a product sort with π_1 and π_2 the corresponding projections. Then $\xi(x, y_1, y_2)$ is the Σ^+-formula

$$(y_1 = \pi_1(x)) \wedge (y_2 = \pi_2(x)).$$

- Suppose that T^+ defines σ as a coproduct sort with corresponding function symbols $\rho_1 : \sigma_1 \to \sigma$ and $\rho_2 : \sigma_2 \to \sigma$. Then $\xi(x, y_1, y_2)$ is either the Σ^+-formula $\rho_1(y_1) = x$ or the Σ^+-formula $\rho_2(y_2) = x$, where y_i is a variable of sort σ_i. (Note: $\xi(x, y_1, y_2)$ is *not* the disjunction of these two formulas.)
- Suppose that T^+ defines σ as a subsort with $i : \sigma \to \sigma'$ the corresponding function symbol. Then $\xi(x, y)$ is the formula $i(x) = y$, where y is a variable of sort $\sigma' \in \Sigma$.
- Suppose that T^+ defines σ as a quotient sort with $\epsilon : \sigma' \to \sigma$ the corresponding function symbol. Then $\xi(x, y)$ is the Σ^+-formula $\epsilon(y) = x$, where y is again a variable of sort $\sigma' \in \Sigma$.
- Given the empty sequence of variables, we let the **empty code** be the tautology $\exists x(x =_\sigma x)$, where $\sigma \in \Sigma$ is a sort symbol.

Given the conjuncts $\xi_1, \ldots, \xi_n$, we will use the notation $\xi(x_1, \ldots, y_{n2})$ to denote the code $\xi_1(x_1, y_{11}, y_{12}) \wedge \ldots \wedge \xi_n(x_n, y_{n1}, y_{n2})$ for the variables $x_1, \ldots, x_n$. Note that the variables y_{i1} and y_{i2} have sorts in Σ for each $i = 1, \ldots, n$. One should think of a code $\xi(x_1, \ldots, y_{n2})$ for $x_1, \ldots, x_n$ as encoding one way that the variables $x_1, \ldots, x_n$ with sorts in $\Sigma^+ \backslash \Sigma$ might be related to variables $y_{11}, \ldots, y_{n2}$ that have sorts in Σ. One additional piece of notation will be useful in what follows. Given a Σ^+-formula

ϕ, we will write $\phi(x_1, \ldots, x_n, \overline{x}_1, \ldots, \overline{x}_m)$ to indicate that the variables $x_1, \ldots, x_n$ have sorts $\sigma_1, \ldots, \sigma_n \in \Sigma^+ \backslash \Sigma$ and that the variables $\overline{x}_1, \ldots, \overline{x}_m$ have sorts $\overline{\sigma}_1, \ldots, \overline{\sigma}_m \in \Sigma$.

LEMMA 7.3.6 (Functionality of codes) *Let T be a Σ-theory and T^+ a Morita extension of T to the signature Σ^+. Let $\vec{x}, \vec{z}$ be n-tuples of variables of the same sorts in Σ^+ and let $\xi(\vec{x}, \vec{y})$ be a code for $\vec{x}$. Then we have*

$$T^+ \vdash (\xi(\vec{x}, \vec{y}) \wedge \xi(\vec{z}, \vec{y})) \rightarrow \vec{x} = \vec{z},$$

where $\vec{x} = \vec{z}$ is shorthand for $(x_1 =_{\sigma_1} z_1) \wedge \cdots \wedge (x_n =_{\sigma_n} z_n)$.

Proof This fact follows immediately from the definition of codes. □

We can now state our generalization of Lemma 4.6.11.

THEOREM 7.3.7 (Barrett) *Let $\Sigma \subseteq \Sigma^+$ be signatures and T a Σ-theory. Suppose that T^+ is a Morita extension of T to Σ^+ and that $\phi(x_1, \ldots, x_n, \overline{x}_1, \ldots, \overline{x}_m)$ is a Σ^+-formula. Then for every code $\xi(x_1, \ldots, y_{n2})$ for the variables $x_1, \ldots, x_n$ there is a Σ-formula $\phi^*(\overline{x}_1, \ldots, \overline{x}_m, y_{11}, \ldots, y_{n2})$ such that T^+ entails*

$$\xi(x_1, \ldots, y_{n2}) \rightarrow (\phi(x_1, \ldots x_n, \overline{x}_1, \ldots, \overline{x}_m) \leftrightarrow \phi^*(\overline{x}_1, \ldots, \overline{x}_m, y_{11}, \ldots, y_{n2})).$$

The idea behind Theorem 7.3.7 is simple. Although one might not initially be able to translate a Σ^+-formula ϕ into an equivalent Σ-formula ϕ^*, such a translation is possible after one specifies how the variables in ϕ with sorts in $\Sigma^+ \backslash \Sigma$ are related to variables with sorts in Σ.

We first prove the following lemma. Given a Σ^+-term t, we will again write $t(x_1, \ldots, x_n, \overline{x}_1, \ldots, \overline{x}_m)$ to indicate that the variables $x_1, \ldots, x_n$ have sorts $\sigma_1, \ldots, \sigma_n \in \Sigma^+ \backslash \Sigma$ and that the variables $\overline{x}_1, \ldots, \overline{x}_m$ have sorts $\overline{\sigma}_1, \ldots, \overline{\sigma}_m \in \Sigma$.

LEMMA 7.3.8 *Let $t(x_1, \ldots, x_n, \overline{x}_1, \ldots, \overline{x}_m)$ be a Σ^+-term of sort σ and x a variable of sort σ. Let $\xi(x, x_1, \ldots, x_n, y_1, y_2, y_{11}, \ldots, y_{n2})$ be a code for the variables $x, x_1, \ldots, x_n$, where the variables y_1 and y_2 are used for coding the variable x. Then there is a Σ-formula $\phi_t(x, \overline{x}_1, \ldots, \overline{x}_m, y_{01}, \ldots, y_{n2})$ such that T^+ implies*

$$\xi(x, \ldots, y_{n2}) \rightarrow \big(t(x_1, \ldots, \overline{x}_m) = x \leftrightarrow \phi_t(x, \overline{x}_1, \ldots, \overline{x}_m, y_1, \ldots, y_{n2})\big).$$

If $\sigma \in \Sigma$, then x will not appear in the code ξ. If $\sigma \in \Sigma^+ \backslash \Sigma$, then x will not appear in the Σ-formula ϕ_t.

Proof We induct on the complexity of t. First, suppose that t is a variable x_i of sort σ. If $\sigma \in \Sigma$, then there are no variables in t with sorts in $\Sigma^+ \backslash \Sigma$. So ξ must be the empty code. Let $\phi_t(x, x_i)$ be the Σ-formula $x = x_i$. This choice of ϕ_t trivially satisfies the desired property. If $\sigma \in \Sigma^+ \backslash \Sigma$, then there are four cases to consider. We consider the cases where σ is a product sort and a subsort. The coproduct and quotient cases follow analogously. Suppose that T^+ defines σ as a product sort with projections π_1 and π_2 of arity $\sigma \to \sigma_1$ and $\sigma \to \sigma_2$. A code ξ for the variables x and x_i must therefore be the formula

$$\pi_1(x) = y_1 \wedge \pi_2(x) = y_2 \wedge \pi_1(x_i) = y_{i1} \wedge \pi_2(x_i) = y_{i2}.$$

One defines the Σ-formula ϕ_t to be $y_1 = y_{i1} \wedge y_2 = y_{i2}$ and verifies that it satisfies the desired property. On the other hand, suppose that T^+ defines σ as a subsort with injection i of arity $\sigma \to \sigma_1$. A code ξ for the variables x and x_i is therefore the formula

$$i(x) = y \wedge i(x_i) = y_{i1}.$$

Let ϕ_t be the Σ-formula $y = y_{i1}$. The desired property again holds.

Second, suppose that t is the constant symbol c. Note that it must be the case that c is of sort $\sigma \in \Sigma$. If $c \in \Sigma$, then letting ϕ_t be the Σ-formula $x = c$ trivially yields the result. If $c \in \Sigma^+ \backslash \Sigma$, then there is some Σ-formula $\psi(x)$ that T^+ uses to explicitly define c. Letting $\phi_t = \psi$ yields the desired result.

For the third (and final) step of the induction, we suppose that t is a term of the form

$$f\big(t_1(x_1, \ldots, x_n, \overline{x}_1, \ldots, \overline{x}_m), \ldots, t_k(x_1, \ldots, x_n, \overline{x}_1, \ldots, \overline{x}_m)\big),$$

where $f \in \Sigma^+$ is a function symbol. We show that the result holds for t if it holds for all of the terms $t_1, \ldots, t_k$. There are three cases to consider. First, if $f \in \Sigma$, then it must be that f has arity $\sigma_1 \times \ldots \times \sigma_k \to \sigma$, where $\sigma, \sigma_1, \ldots, \sigma_k \in \Sigma$. Let ξ be a code for $x_1, \ldots, x_n$. We define ϕ_t to be the Σ-formula

$$\exists_{\sigma_1} z_1 \ldots \exists_{\sigma_k} z_k \big(\phi_{t_1}(z_1, \overline{x}_1, \ldots, y_{n2}) \wedge \ldots \wedge \phi_{t_k}(z_k, \overline{x}_1, \ldots y_{n2}) \wedge f(z_1, \ldots, z_k) = x\big),$$

where each of the ϕ_{t_i} exists by our inductive hypothesis. One can verify that ϕ_t satisfies the desired property. Second, if $f \in \Sigma^+ \backslash \Sigma$ is defined by a Σ-formula $\psi(z_1, \ldots, z_k, x)$, then one defines ϕ_t in an analogous manner to above. (Note that, in this case, the arity of f is again $\sigma_1 \times \ldots \times \sigma_k \to \sigma$ with $\sigma_1, \ldots, \sigma_k, \sigma \in \Sigma$.)

Third, we need to verify that the result holds if f is a function symbol that is used in the definition of a new sort. We discuss the cases where f is π_1 and where f is ϵ. Suppose that f is π_1 with arity $\sigma \to \sigma_1$. Then it must be that the term t_1 is a variable x_i of sort σ since there are no other Σ^+-terms of sort σ. So the term t is $\pi_1(x_i)$. Let $\xi(x_i, y_{i1}, y_{i2})$ be a code for x_i. It must be that ξ is the formula

$$\pi_1(x_i) = y_{i1} \wedge \pi_2(x_i) = y_{i2}.$$

Letting ϕ_t be the formula $y_{i1} = x$ yields the desired result. On the other hand, suppose that f is the function symbol ϵ of arity $\sigma_1 \to \sigma$, where σ is a quotient sort defined by the Σ-formula $\psi(z_1, z_2)$. The term t in this case is $\epsilon(t_1(x_1, \ldots, x_n, \overline{x}_1, \ldots, \overline{x}_m))$, and we assume that the result holds for the Σ^+-term t_1 of sort $\sigma_1 \in \Sigma$. Let ξ be a code for the variables $x, x_1, \ldots, x_n$. This code determines a code $\overline{\xi}$ for the variables $x_1, \ldots, x_n$ by "forgetting" the conjunct $\epsilon(y) = x$ that involves the variable x. We use the code $\overline{\xi}$ and the inductive hypothesis to obtain the formula ϕ_{t_1}. Then we define ϕ_t to be the Σ-formula

$$\exists_{\sigma_1} z\big(\phi_{t_1}(z,\overline{x}_1,\ldots,\overline{x}_m,y_{11},\ldots,y_{n2}) \wedge \psi(y,z)\big).$$

Considering the original code ξ, one verifies that the result holds for ϕ_{t_1}. □

We now turn to the proof of the main result.

Proof We induct on the complexity of ϕ. Suppose that ϕ is the formula $t(x_1,\ldots,x_n,\overline{x}_1,\ldots,\overline{x}_m) = s(x_1,\ldots,x_n,\overline{x}_1,\ldots,\overline{x}_m)$, where t and s are Σ^+-terms of sort σ. Let $\xi(x_1,\ldots,y_{n2})$ be a code for $x_1,\ldots,x_n$, and let x be a variable of sort σ. By Lemma 7.3.8, there are corresponding Σ-formulas $\phi_t(x,\overline{x}_1,\ldots,\overline{x}_m,y_{11},\ldots,y_{n2})$ and $\phi_s(x,\overline{x}_1,\ldots,\overline{x}_m,y_{11},\ldots,y_{n2})$. The Σ-formula ϕ^* is then defined to be

$$\exists_\sigma x\big(\phi_t(x,\overline{x}_1,\ldots,\overline{x}_m,y_{11},\ldots,y_{n2}) \wedge \phi_s(x,\overline{x}_1,\ldots,\overline{x}_m,y_{11},\ldots,y_{n2})\big).$$

One can verify that this definition of ϕ^* satisfies the desired result.

If t and s are of sort $\sigma \in \Sigma^+\backslash\Sigma$, then there are four cases to consider. We show that the result holds when T^+ defines σ as a product sort or a quotient sort. The coproduct and subsort cases follow analogously. If T^+ defines σ as a product sort with projections π_1 and π_2 of arity $\sigma \to \sigma_1$ and $\sigma \to \sigma_2$, then we define a code $\overline{\xi}(x,x_1,\ldots,y_{n2},v_1,v_2)$ for the variables $x,x_1,\ldots,x_n$ by

$$\xi(x_1,\ldots,y_{n2}) \wedge \pi_1(x) = v_1 \wedge \pi_2(x) = v_2.$$

Lemma 7.3.8 and the code $\overline{\xi}$ for the variables $x,x_1,\ldots,x_n$ generate the Σ-formulas $\phi_t(\overline{x}_1,\ldots,\overline{x}_m,y_{11},\ldots,y_{n2},v_1,v_2)$ and $\phi_s(\overline{x}_1,\ldots,\overline{x}_m,y_{11},\ldots,y_{n2},v_1,v_2)$. We then define the Σ-formula ϕ^* to be

$$\exists_{\sigma_1} v_1 \exists_{\sigma_2} v_2\big(\phi_t(\overline{x}_1,\ldots,\overline{x}_m,y_{11},\ldots,y_{n2},v_1,v_2) \\ \wedge \phi_s(\overline{x}_1,\ldots,\overline{x}_m,y_{11},\ldots,y_{n2},v_1,v_2)\big).$$

One can verify that ϕ^* again satisfies the desired result.

If T^+ defines σ as a quotient sort with projection ϵ of arity $\sigma_1 \to \sigma$, then we again define a new code $\overline{\xi}(x,x_1,\ldots,y_{n2},v)$ for the variables $x,x_1,\ldots,x_n$ by

$$\xi(x_1,\ldots,y_{n2}) \wedge \epsilon(v) = x,$$

Lemma 7.3.8 and the code $\overline{\xi}$ for the variables $x,x_1,\ldots,x_n$ again generate the Σ-formulas $\phi_t(\overline{x}_1,\ldots,\overline{x}_m,y_{11},\ldots,y_{n2},v)$ and $\phi_s(\overline{x}_1,\ldots,\overline{x}_m,y_{11},\ldots,y_{n2},v)$. We define the Σ-formula ϕ^* to be

$$\exists_{\sigma_1} v\big(\phi_t(\overline{x}_1,\ldots,\overline{x}_m,y_{11},\ldots,y_{n2},v) \wedge \phi_s(\overline{x}_1,\ldots,\overline{x}_m,y_{11},\ldots,y_{n2},v)\big).$$

One again verifies that this ϕ^* satisfies the desired property. So the result holds when ϕ is of the form $t = s$ for Σ^+-terms t and s.

Now suppose that $\phi(x_1,\ldots,x_n,\overline{x}_1,\ldots,\overline{x}_m)$ is a Σ^+-formula of the form

$$p(t_1(x_1,\ldots,x_n,\overline{x}_1,\ldots,\overline{x}_m),\ldots,t_k(x_1,\ldots,x_n,\overline{x}_1,\ldots,\overline{x}_m)),$$

where p has arity $\sigma_1 \times \ldots \times \sigma_k$. Note that it must be that $\sigma_1,\ldots,\sigma_k \in \Sigma$. Either $p \in \Sigma$ or $p \in \Sigma^+\backslash\Sigma$. We consider the second case. (The first is analogous.) Let $\psi(z_1,\ldots,z_k)$

be the Σ-formula that T^+ uses to explicitly define p and let $\xi(x_1, \dots, y_{n2})$ be a code for $x_1, \dots, x_n$. Lemma 7.3.8 and ξ generate the Σ-formulas $\phi_{t_i}(z_i, \overline{x}_1, \dots, \overline{x}_m, y_{11}, \dots, y_{n2})$ for each $i = 1, \dots, k$. We define ϕ^* to be the Σ-formula

$$\exists_{\sigma_1} z_1 \dots \exists_{\sigma_k} z_k \big(\phi_{t_1}(z_1, \overline{x}_1, \dots, \overline{x}_m, y_{11}, \dots, y_{n2}) \wedge \dots \\ \wedge \phi_{t_k}(z_k, \overline{x}_1, \dots, \overline{x}_m, y_{11}, \dots, y_{n2}) \wedge \psi(z_1, \dots, z_k)\big).$$

One can again verify that the result holds for this choice of ϕ^*.

We have covered the "base cases" for our induction. We now turn to the inductive step. We consider the cases of $\neg$, $\wedge$, and $\forall$. Suppose that the result holds for Σ^+-formulas ϕ_1 and ϕ_2. Then it trivially holds for $\neg\phi_1$ by letting $(\neg\phi)^*$ be $\neg(\phi^*)$. It also trivially holds for $\phi_1 \wedge \phi_2$ by letting $(\phi_1 \wedge \phi_2)^*$ be $\phi_1^* \wedge \phi_2^*$.

The $\forall_{\sigma_i}$ case requires more work. If x_i is a variable of sort $\sigma_i \in \Sigma$, we let $(\forall_{\sigma_i} x_i \phi_1)^*$ be $\forall_{\sigma_i} x_i (\phi_1^*)$. The only nontrivial part of the inductive step is when one quantifies over variables with sorts in $\Sigma^+ \backslash \Sigma$. Suppose that $\phi(x_1, \dots, x_n, \overline{x}_1, \dots, \overline{x}_m)$ is a Σ^+-formula and that the result holds for it. We let x_i be a variable of sort $\sigma_i \in \Sigma^+ \backslash \Sigma$, and we show that the result also holds for the Σ-formula $\forall_{\sigma_i} x_i \phi(x_1, \dots, x_n, \overline{x}_1, \dots, \overline{x}_m)$. There are again four cases. We show that the result holds when σ_i is a product sort and a coproduct sort. The cases of subsorts and quotient sorts follow analogously.

Suppose that T^+ defines σ_i as a product sort with projections π_1 and π_2 of arity $\sigma_i \to \sigma_{i1}$ and $\sigma_i \to \sigma_{i2}$. Quantifying over a variable x_i of product sort σ_i can be thought of as "quantifying over pairs of elements of sorts σ_{i1} and σ_{i2}." Indeed, let $\xi(x_1, \dots, y_{n2})$ be a code for the variables $x_1, \dots, x_{i-1}, x_{i+1}, \dots, x_n$ (these are all of the free variables in $\forall_{\sigma_i} x_i \phi$ with sorts in $\Sigma^+ \backslash \Sigma$). We define a code $\overline{\xi}$ for the variables $x_1, \dots, x_{i-1}, x_i, x_{i+1}, \dots, x_n$ by

$$\xi(x_1, \dots, y_{n2}) \wedge \pi_1(x_i) = v_1 \wedge \pi_2(x_i) = v_2.$$

One uses the code $\overline{\xi}$ and the inductive hypothesis to generate the Σ-formula $\phi^*(\overline{x}_1, \dots, \overline{x}_m, y_{11}, \dots, y_{n2}, v_1, v_2)$. We then define the Σ-formula $(\forall_{\sigma_i} x_i \phi)^*$ to be

$$\forall_{\sigma_{i1}} v_1 \forall_{\sigma_{i2}} v_2 \phi^*(\overline{x}_1, \dots, \overline{x}_m, y_{11}, \dots, y_{n2}, v_1, v_2).$$

And one verifies that the desired result holds for this choice of $(\forall_{\sigma_i} x_i \phi)^*$.

Suppose that T^+ defines σ_i as a coproduct sort with injections ρ_1 and ρ_2 of arity $\sigma_{i1} \to \sigma_i$ and $\sigma_{i2} \to \sigma_i$. Quantifying over a variable x_i of coproduct sort σ_i can be thought of as "quantifying over *both* elements of sort σ_{i1} and elements of sort σ_{i2}." Indeed, let $\xi(x_1, \dots, y_{n2})$ be a code for the variables $x_1, \dots, x_{i-1}, x_{i+1}, \dots, x_n$ (these are again all of the free variables in $\forall_{\sigma_i} x_i \phi$ with sorts in $\Sigma^+ \backslash \Sigma$). We define two different codes $\overline{\xi}$ for the variables $x_1, \dots, x_{i-1}, x_i, x_{i+1}, \dots, x_n$ by

$$\xi(x_1, \dots, y_{n2}) \wedge \rho_1(v_1) = x_i \\ \xi(x_1, \dots, y_{n2}) \wedge \rho_2(v_2) = x_i.$$

We will call the first code $\xi'(x_1, \dots, y_{n2}, v_1)$ and the second $\xi''(x_1, \dots, y_{n2}, v_2)$. We use these two codes and the inductive hypothesis to generate Σ-formulas $\phi^{*'}$ and $\phi^{*''}$. We then define the Σ-formula $(\forall_{\sigma_i} x_i \phi)^*$ to be

$$\forall_{\sigma_{i1}} v_1 \forall_{\sigma_{i2}} v_2 \big(\phi^{*'}(\overline{x}_1, \ldots, \overline{x}_m, y_{11}, \ldots, y_{n2}, v_2) \\ \wedge \phi^{*''}(\overline{x}_1, \ldots, \overline{x}_m, y_{11}, \ldots, y_{n2}, v_2)\big).$$

One can verify that the desired result holds again for this definition of $(\forall_{\sigma_i} x_i \phi)^*$. □

Theorem 7.3.7 has the following immediate corollary.

COROLLARY 7.3.9 *Let $\Sigma \subseteq \Sigma^+$ be signatures and T a Σ-theory. If T^+ is a Morita extension of T to Σ^+, then for every Σ^+-sentence ϕ there is a Σ-sentence ϕ^* such that $T^+ \vdash \phi \leftrightarrow \phi^*$.*

Proof Let ϕ be a Σ^+-sentence, and consider the empty code ξ. Theorem 7.3.7 implies that there is a Σ-sentence ϕ^* such that $T^+ \vdash \xi \rightarrow (\phi \leftrightarrow \phi^*)$. Since ξ is a tautology, we trivially have that $T^+ \vdash \phi \leftrightarrow \phi^*$. □

The theorems in this section capture different senses in which a Morita extension of a theory "says no more" than the original theory. In this way, Morita equivalence is analogous to definitional equivalence.

At first glance, Morita equivalence might strike one as different from definitional equivalence in an important way. To show that theories are Morita equivalent, one is allowed to take any finite number of Morita extensions of the theories. On the other hand, to show that two theories are definitionally equivalent, it appears that one is only allowed to take *one* definitional extension of each theory. One might worry that Morita equivalence is therefore not perfectly analogous to definitional equivalence.

Fortunately, this is not the case. Theorem 3.3 implies that if theories $T_1, \ldots, T_n$ are such that each T_{i+1} is a definitional extension of T_i, then T_n is, in fact, a definitional extension of T_1. (One can easily verify that this is not true of Morita extensions.) To show that two theories are definitionally equivalent, therefore, one actually *is* allowed to take any finite number of definitional extensions of each theory.

If two theories are definitionally equivalent, then they are trivially Morita equivalent. Unlike definitional equivalence, however, Morita equivalence is capable of capturing a sense in which theories with different sort symbols are equivalent. The following example demonstrates that Morita equivalence is a more liberal criterion for theoretical equivalence.

Example 7.3.10 Let $\Sigma_1 = \{\sigma_1, p, q\}$ and $\Sigma_2 = \{\sigma_2, \sigma_3\}$ be signatures with σ_i sort symbols, and p and q predicate symbols of arity σ_1. Let T_1 be the Σ_1-theory that says p and q are nonempty, mutually exclusive, and exhaustive. Let T_2 be the empty theory in Σ_2. Since the signatures Σ_1 and Σ_2 have different sort symbols, T_1 and T_2 can't possibly be definitionally equivalent. Nonetheless, it's easy to see that T_1 and T_2 are Morita equivalent. Let $\Sigma = \Sigma_1 \cup \Sigma_2 \cup \{i_2, i_3\}$ be a signature with i_2 and i_3 function symbols of arity $\sigma_2 \rightarrow \sigma_1$ and $\sigma_3 \rightarrow \sigma_1$. Consider the following Σ-sentences.

$$\forall_{\sigma_1} x\big(p(x) \leftrightarrow \exists_{\sigma_2} y(i_2(y) = x)\big) \wedge \forall_{\sigma_2} y_1 \forall_{\sigma_2} y_2\big(i_2(y_1) = i_2(y_2) \rightarrow y_1 = y_2\big) \qquad (\delta_{\sigma_2})$$

$$\forall_{\sigma_1} x\big(q(x) \leftrightarrow \exists_{\sigma_3} z(i_3(z) = x)\big) \wedge \forall_{\sigma_3} z_1 \forall_{\sigma_3} z_2\big(i_3(z_1) = i_3(z_2) \rightarrow z_1 = z_2\big) \qquad (\delta_{\sigma_3})$$

$$\forall_{\sigma_1} x\big(\exists_{\sigma_2=1} y(i_2(y) = x) \vee \exists_{\sigma_3=1} z(i_3(z) = x)\big) \wedge \forall_{\sigma_2} y \forall_{\sigma_3} z \neg\big(i_2(y) = i_3(z)\big) \qquad (\delta_{\sigma_1})$$

$$\forall_{\sigma_1} x\big(p(x) \leftrightarrow \exists_{\sigma_2} y(i_2(y) = x)\big) \qquad (\delta_p)$$

$$\forall_{\sigma_1} x\big(q(x) \leftrightarrow \exists_{\sigma_3} z(i_3(z) = x)\big) \qquad (\delta_q)$$

The Σ-theory $T_1^1 = T_1 \cup \{\delta_{\sigma_2}, \delta_{\sigma_3}\}$ is a Morita extension of T_1 to the signature Σ. It defines σ_2 to be the subsort of "elements that are p" and σ_3 to be the subsort of "elements that are q."

The theory $T_2^1 = T_2 \cup \{\delta_{\sigma_1}\}$ is a Morita extension of T_2 to the signature $\Sigma_2 \cup \{\sigma_1, i_2, i_3\}$. It defines σ_1 to be the coproduct sort of σ_2 and σ_3. Lastly, the Σ-theory $T_2^2 = T_2^1 \cup \{\delta_p, \delta_q\}$ is a Morita extension of T_2^1 to the signature Σ. It defines the predicates p and q to apply to elements in the "images" of i_2 and i_3, respectively. One can verify that T_1^1 and T_2^2 are logically equivalent, so T_1 and T_2 are Morita equivalent. ⌟

Morita equivalence captures a clear and robust sense in which theories might be equivalent, but it is a difficult criterion to apply outside of the framework of first-order logic. Indeed, without a formal language one does not have the resources to say what an explicit definition is. Questions of equivalence and inequivalence of theories, however, still come up outside of this framework. It is well known, for example, that there are different ways of formulating the theory of smooth manifolds (Nestruev, 2002). There are also different formulations of the theory of topological spaces (Kuratowski, 1966). None of these formulations are first-order theories. Physical theories, too, are rarely formulated in first-order logic, and there are many pairs of physical theories that have been considered to be equivalent. We list just a few examples.

- According to the standard view in physics, Heisenberg's matrix mechanics is equivalent to Schrödinger's wave mechanics – despite the fact that these theories use completely different formalisms, and neither is axiomatizable in first-order logic. Or, if you prefer to be more mathematically rigorous, quantum mechanics can be formulated either in terms of Hilbert spaces or in terms of C^*-algebras. There are good reasons, however, to think that these two formulations are equivalent.
- A model of Einstein's general theory of relativity (GTR) is typically taken to be a smooth manifold with a Lorentzian metric. However, we have a free choice: either we can use a metric of signature $(3, 1)$ or a metric of signature $(1, 3)$. These two formulations of GTR seem to be equivalent – but it's doubtful that we could explicate that equivalence in terms of some regimentation of these theories in first-order logic.

In fact, GTR can also be formulated with a completely different mathematical apparatus, viz. "Einstein algebras," and there is a precise sense in which this formulation is equivalent to the formulation in terms of manifolds (see Rosenstock et al., 2015; Weatherall, 2018).

- GTR seems to differ radically from classical Newtonian gravitation, since the latter posits a static spacetime structure. Some have claimed, in fact, that GTR has a special property, called "general covariance," that disguishes it from all previous spacetime theories. However, in the mid-twentieth century, Henri Cartan formulated a coordinate-free version of Newtonian gravitation on a curved spacetime. If this Newton–Cartan gravitational theory is equivalent to Newtonian gravity, then the latter is also generally covariant. For discussion of this example, see Glymour (1977); Knox (2014); Weatherall (2016a).
- In typical presentations of rigorous methods in classical physics, it is usually assumed (or even partially demonstrated) that the Lagrangian formalism is equivalent to the Hamiltonian formalism. However, North (2009) argues that these two theories have different structure, and hence are inequivalent. For further discussion, see Halvorson (2011); Swanson and Halvorson (2012); Curiel (2014); Barrett (2015).
- Most cutting-edge theories in physics make use of the so-called gauge formalism, and this raises many challenging interpretive issues (see Healey, 2007). Philosophers of science have recently entered into a dispute about whether gauge theories are better thought of in terms of the fiber bundle formalism, or in terms of the holonomy formalism. However, Rosenstock and Weatherall (2016) argue that the two formalisms are equivalent.

Since none of the theories admits a first-order formulation (at least not in any obvious sense), Morita equivalence is incapable of validating these claims of equivalence. Philosophers of science are left with two options: either claim or deny equivalence without a precise account of the standards or propose a more broadly applicable explication of equivalence. We pursue the second option here.

Among the many ways we could explicate theoretical equivalence, we find it most promising to look for hints from contemporary mathematics. In other words, we look to which ideas are working well in contemporary mathematics, and we try to put them to work in the service of philosophy of science. One such fruitful ideas is the notion of **categorical equivalence**, which we first mentioned in Chapter 3. Historically speaking, categorical equivalence was first defined by Eilenberg and Mac Lane (1942, 1945), made a brief appearance in some earlier work in philosophy of science (Pearce, 1985), and has recently by reintroduced in philosophical discussion by Halvorson (2012, 2016); Weatherall (2016a). In the remainder of this section, we review this notion, and prove a few results that relate categorical and Morita equivalence. In summary, for theories with a first-order formulation, Morita equivalence implies categorical equivalence, but not vice versa.

Categorical equivalence is motivated by the following simple observation: First-order theories have categories of models. If T is a Σ-theory, we will use the notation $\mathrm{Mod}(T)$

to denote the **category of models** of T. The objects of Mod(T) are models of T. For the arrows of Mod(T), we have a couple of salient choices. On the one hand, we could choose arrows to be homomorphisms – i.e., $f : M \to N$ is a function (or family of functions) that preserves the extensions of the terms in the signature Σ. On the other hand, we could choose arrows to be elementary embeddings – i.e., $f : M \to N$ is an injective function (or family of functions) that preserves the extensions of all Σ formulas.

Let Mod(T) denote the category with elementary embeddings as arrows, and let $\text{Mod}_h(T)$ denote the category with homomorphisms as arrows. But which of these two categories, Mod(T) or $\text{Mod}_h(T)$, should we think of as representing the theory T? We will choose the category Mod(T), with elementary embeddings as arrows, for the following reasons. First, the image of a model of T under a homomorphism f is not necessarily a model of T. For example, let T be the theory (in a single-sorted signature) that says there are exactly two things. Then a model M of T is a set with two elements. However, the mapping $f : M \to M$ that takes both elements to a single element is a homomorphism, and its image $f(M)$ is not a model of T. Such a situation is not necessarily a disaster, but it shows that homomorphisms do not mesh well with full first-order logic. Second, $\text{Mod}_h(\cdot)$ does not even preserve definitional equivalence – i.e., there are definitionally equivalent theories T_1 and T_2 such that $\text{Mod}_h(T_1)$ is not categorically equivalent to $\text{Mod}_h(T_2)$.

Example 7.3.11 Let $\Sigma_1 = \{\sigma\}$, where σ is a sort symbol, and let T_1 be the theory in Σ_1 that says there are exactly two things. Let $\Sigma_2 = \{\sigma, \theta\}$ where θ is a relation of arity $\sigma \times \sigma$, and let T_2 be the theory in Σ_2 that says there are exactly two things, and $T_2 \vDash \theta(x, y) \leftrightarrow (x \neq y)$. Obviously T_2 is a definitional extension of T_1. Now, every arrow of $\text{Mod}_h(T_2)$ is an injection, since it preserves θ and hence $\neq$. But arrows of $\text{Mod}_h(T_1)$ need not be injections. Therefore, $\text{Mod}_h(T_1)$ and $\text{Mod}_h(T_2)$ are not categorically equivalent. ⌟

Because of these issues with homomorphisms, we will continue to associate a theory T with the category Mod(T) whose objects are models of T and whose arrows are elementary embeddings between these models. We recall now the definition of an equivalence of categories.

DEFINITION 7.3.12 A functor $F : \mathbf{C} \to \mathbf{D}$ is called an **equivalence of categories** just in case there is a functor $G : \mathbf{D} \to \mathbf{C}$, and natural isomorphisms $\eta : GF \Rightarrow 1_{\mathbf{C}}$ and $\varepsilon : FG \Rightarrow 1_{\mathbf{D}}$.

We will also need the following fact, a standard result of category theory (see Mac Lane, 1971, p. 93).

PROPOSITION 7.3.13 *A functor $F : \mathbf{C} \to \mathbf{D}$ is equivalence of categories iff F is full, faithful, and essentially surjective.*

While each first-order theory T defines a category Mod(T), this structure is not particular to first-order theories. Indeed, one can easily define categories of models for the different formulations of the theory of smooth manifolds and for the different formulations of the theory of topological spaces. The arrows in these categories are simply the structure-preserving maps between the objects in the categories. One can also define categories of models for physical theories; see, for example, Barrett (2015); Rosenstock et al. (2015); Weatherall (2016a,c, 2018). This means that the following criterion for theoretical equivalence is applicable in a more general setting than definitional equivalence and Morita equivalence. In particular, it can be applied outside of the framework of first-order logic.

DEFINITION 7.3.14 Theories T_1 and T_2 are **categorically equivalent** if their categories of models $\text{Mod}(T_1)$ and $\text{Mod}(T_2)$ are equivalent.

Categorical equivalence captures a sense in which theories have "isomorphic semantic structure." If T_1 and T_2 are categorically equivalent, then the relationships that models of T_1 bear to one another are "isomorphic" to the relationships that models of T_2 bear to one another.

In order to show how categorical equivalence relates to Morita equivalence, we focus on first-order theories. We will show that categorical equivalence is a strictly weaker criterion for theoretical equivalence than Morita equivalence is. We first need some preliminaries about the category of models $\text{Mod}(T)$ for a first-order theory T. Suppose that $\Sigma \subseteq \Sigma^+$ are signatures and that the Σ^+-theory T^+ is an extension of the Σ-theory T. There is a natural "projection" functor $\Pi : \text{Mod}(T^+) \to \text{Mod}(T)$ from the category of models of T^+ to the category of models of T. The functor Π is defined as follows.

- $\Pi(M) = M|_\Sigma$ for every object M in $\text{Mod}(T^+)$.
- $\Pi(h) = h|_\Sigma$ for every arrow $h : M \to N$ in $\text{Mod}(T^+)$, where the family of maps $h|_\Sigma$ is defined to be $h|_\Sigma = \{h_\sigma : M_\sigma \to N_\sigma$ such that $\sigma \in \Sigma\}$.

Since T^+ is an extension of T, the Σ-structure $\Pi(M)$ is guaranteed to be a model of T. Likewise, the map $\Pi(h) : M|_\Sigma \to N|_\Sigma$ is guaranteed to be an elementary embedding. One can easily verify that $\Pi : \text{Mod}(T^+) \to \text{Mod}(T)$ is a functor.

The following three propositions will together establish the relationship between $\text{Mod}(T^+)$ and $\text{Mod}(T)$ when T^+ is a Morita extension of T. They imply that when T^+ is a Morita extension of T, the functor $\Pi : \text{Mod}(T^+) \to \text{Mod}(T)$ is full, faithful, and essentially surjective. The categories $\text{Mod}(T^+)$ and $\text{Mod}(T)$ are therefore equivalent.

PROPOSITION 7.3.15 *Let $\Sigma \subseteq \Sigma^+$ be signatures and T a Σ-theory. If T^+ is a Morita extension of T to Σ^+, then Π is essentially surjective.*

Proof If M is a model of T, then Theorem 7.3.1 implies that there is a model M^+ of T^+ that is an expansion of M. Since $\Pi(M^+) = M^+|_\Sigma = M$ the functor Π is essentially surjective. □

PROPOSITION 7.3.16 *Let $\Sigma \subseteq \Sigma^+$ be signatures and T a Σ-theory. If T^+ is a Morita extension of T to Σ^+, then Π is faithful.*

Proof Let $h : M \to N$ and $g : M \to N$ be arrows in $\text{Mod}(T^+)$, and suppose that $\Pi(h) = \Pi(g)$. We show that $h = g$. By assumption, $h_\sigma = g_\sigma$ for every sort symbol

$\sigma \in \Sigma$. We show that $h_\sigma = g_\sigma$ also for $\sigma \in \Sigma^+\backslash\Sigma$. We consider the cases where T^+ defines σ as a product sort or a subsort. The coproduct and quotient sort cases follow analogously.

Suppose that T^+ defines σ as a product sort with projections π_1 and π_2 of arity $\sigma \to \sigma_1$ and $\sigma \to \sigma_2$. Then the following equalities hold.

$$\pi_1^N \circ h_\sigma = h_{\sigma_1} \circ \pi_1^M = g_{\sigma_1} \circ \pi_1^M = \pi_1^N \circ g_\sigma$$

The first and third equalities hold since h and g are elementary embeddings, and the second since $h_{\sigma_1} = g_{\sigma_1}$. One can verify in the same manner that $\pi_2^N \circ h_\sigma = \pi_2^N \circ g_\sigma$. Since N is a model of T^+ and T^+ defines σ as a product sort, we know that $N \vDash \forall_{\sigma_1} x \forall_{\sigma_2} y \exists_{\sigma=1} z(\pi_1(z) = x \wedge \pi_2(z) = y)$. This implies that $h_\sigma = g_\sigma$.

On the other hand, if T^+ defines σ as a subsort with injection i of arity $\sigma \to \sigma_1$, then the following equalities hold:

$$i^N \circ h_\sigma = h_{\sigma_1} \circ i^M = g_{\sigma_1} \circ i^M = i^N \circ g_\sigma.$$

These equalities follow in the same manner as previously. Since i^N is an injection it must be that $h_\sigma = g_\sigma$. □

Before proving that Π is full, we need the following simple lemma.

LEMMA 7.3.17 *Let M be a model of T^+ with $a_1, \ldots, a_n$ elements of M of sorts $\sigma_1, \ldots, \sigma_n \in \Sigma^+\backslash\Sigma$. If $x_1, \ldots, x_n$ are variables sorts $\sigma_1, \ldots, \sigma_n$, then there is a code $\xi(x_1, \ldots, x_n, y_{11}, \ldots, y_{n2})$ and elements $b_{11}, \ldots, b_{n2}$ of M such that $M \vDash \xi[a_1, \ldots, a_n, b_{11}, \ldots, b_{n2}]$.*

Proof We define the code $\xi(x_1, \ldots, y_{n2})$. If T^+ defines σ_i as a product sort, quotient sort, or subsort, then we have no choice about what the conjunct $\xi_i(x_i, y_{i1}, y_{i2})$ is. If T^+ defines σ_i as a coproduct sort, then we know that either there is an element b_{i1} of M such that $\rho_1(b_{i1}) = a_i$ or there is an element b_{i2} of M such that $\rho_2(b_{i2}) = a_i$. If the former, we let ξ_i be $\rho_1(y_{i1}) = x_i$, and if the latter, we let ξ_i be $\rho_2(y_{i2}) = x_i$. One defines the elements $b_{11}, \ldots, b_{n2}$ in the obvious way. For example, if σ_i is a product sort, then we let $b_{i1} = \pi_1^M(a_i)$ and $b_{i2} = \pi_2^M(a_i)$. By construction, we have that $M \vDash \xi[a_1, \ldots, a_n, b_{11}, \ldots, b_{n2}]$. □

We now use this lemma to show that Π is full.

PROPOSITION 7.3.18 *Let $\Sigma \subseteq \Sigma^+$ be signatures and T a Σ-theory. If T^+ is a Morita extension of T to Σ^+, then Π is full.*

Proof Let M and N be models of T^+ with $h : \Pi(M) \to \Pi(N)$ an arrow in Mod(T). This means that $h : M|_\Sigma \to N|_\Sigma$ is an elementary embedding. We show that the map $h^+ : M \to N$ is an elementary embedding and therefore an arrow in Mod(T^+). Since $\Pi(h^+) = h$, this will imply that Π is full.

Let $\phi(x_1, \ldots, x_n, \overline{x}_1, \ldots, \overline{x}_m)$ be a Σ^+-formula, and let $a_1, \ldots, a_n, \overline{a}_1, \ldots, \overline{a}_m$ be elements of M of the same sorts as the variables $x_1, \ldots, x_n, \overline{x}_1, \ldots, \overline{x}_m$. Lemma 7.3.17 implies that there is a code $\xi(x_1, \ldots, x_n, y_{11}, \ldots, y_{n2})$ and elements $b_{11}, \ldots, b_{n2}$ of M such that $M \vDash \xi[a_1, \ldots, a_n, b_{11}, \ldots, b_{n2}]$. The definition of the map h^+ implies that

$N \vDash \xi[h^+(a_1, \ldots, a_n, b_{11}, \ldots, b_{n2})]$. We now show that $M \vDash \phi[a_1, \ldots, a_n, \overline{a}_1, \ldots, \overline{a}_m]$ if and only if $N \vDash \phi[h^+(a_1, \ldots, a_n, \overline{a}_1, \ldots, \overline{a}_m)]$. By Theorem 7.3.7, there is a Σ-formula $\phi^*(\overline{x}_1, \ldots, \overline{x}_m, y_{11}, \ldots, y_{n2})$ such that

$$T^+ \vDash \forall_{\sigma_1} x_1 \ldots \forall_{\sigma_n} x_n \forall_{\overline{\sigma}_1} \overline{x}_1 \ldots \forall_{\overline{\sigma}_m} \overline{x}_m \forall_{\sigma_{11}} y_{11} \ldots \forall_{\sigma_{n2}} y_{n2} \big(\xi(x_1, \ldots, y_{n2}) \rightarrow \\ \big(\phi(x_1, \ldots x_n, \overline{x}_1, \ldots, \overline{x}_m) \leftrightarrow \phi^*(\overline{x}_1, \ldots, \overline{x}_m, y_{11}, \ldots, y_{n2})\big)\big) \tag{7.1}$$

We then see that the following string of equivalences holds.

$$\begin{aligned} M \vDash \phi[a_1, \ldots, a_n, \overline{a}_1, \ldots, \overline{a}_m] &\Longleftrightarrow M \vDash \phi^*[\overline{a}_1, \ldots, \overline{a}_m, b_{11}, \ldots, b_{n2}] \\ &\Longleftrightarrow M|_\Sigma \vDash \phi^*[\overline{a}_1, \ldots, \overline{a}_m, b_{11}, \ldots, b_{n2}] \\ &\Longleftrightarrow N|_\Sigma \vDash \phi^*[h(\overline{a}_1, \ldots, \overline{a}_m, b_{11}, \ldots, b_{n2})] \\ &\Longleftrightarrow N \vDash \phi^*[h(\overline{a}_1, \ldots, \overline{a}_m, b_{11}, \ldots, b_{n2})] \\ &\Longleftrightarrow N \vDash \phi^*[h^+(\overline{a}_1, \ldots, \overline{a}_m, b_{11}, \ldots, b_{n2})] \\ &\Longleftrightarrow N \vDash \phi[h^+(a_1, \ldots, a_n, \overline{a}_1, \ldots, \overline{a}_m)] \end{aligned}$$

The first and sixth equivalences hold by (5) and the fact that M and N are models of T^+, the second and fourth hold since ϕ^* is a Σ-formula, the third since $h : M|_\Sigma \rightarrow N|_\Sigma$ is an elementary embedding, and the fifth by the definition of h^+ and the fact that the elements $\overline{a}_1, \ldots, \overline{a}_m, b_{11}, \ldots, b_{n2}$ have sorts in Σ. □

These three propositions provide us with the resources to show how categorical equivalence is related to Morita equivalence. Our first result follows as an immediate corollary.

THEOREM 7.3.19 (Barrett) *Morita equivalence entails categorical equivalence.*

Proof Suppose that T_1 and T_2 are Morita equivalent. Then there are theories $T_1^1, \ldots, T_1^n$ and $T_2^1, \ldots, T_2^m$ that satisfy the three conditions in the definition of Morita equivalence. Propositions 7.3.15, 7.3.16, and 7.3.18 imply that the Π functors between these theories, represented by the arrows in the following figure, are all equivalences.

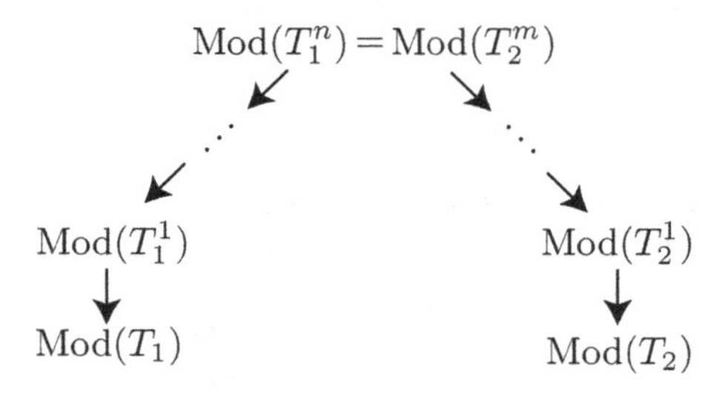

This implies that Mod(T_1) is equivalent to Mod(T_2), and so T_1 and T_2 are categorically equivalent. □

The converse to Theorem 7.3.19, however, does not hold. There are theories that are categorically equivalent but not Morita equivalent. In order to show this, we need one piece of terminology.

DEFINITION 7.3.20 A category **C** is **discrete** if it is equivalent to a category whose only arrows are identity arrows.

Note that discrete categories are essentially just sets. In other words, each discrete category is uniquely determined by its underlying set of objects.

THEOREM 7.3.21 *Categorical equivalence does not entail Morita equivalence.*

Proof Let $\Sigma_1 = \{\sigma_1, p_0, p_1, p_2, \ldots\}$ be a signature with a single sort symbol σ_1 and a countable infinity of predicate symbols p_i of arity σ_1. Let $\Sigma_2 = \{\sigma_2, q_0, q_1, q_2, \ldots\}$ be a signature with a single sort symbol σ_2 and a countable infinity of predicate symbols q_i of arity σ_2. Define the Σ_1-theory T_1 and Σ_2-theory T_2 as follows.

$$T_1 = \{\exists_{\sigma_1 = 1} x(x = x)\}$$
$$T_2 = \{\exists_{\sigma_2 = 1} y(y = y), \forall_{\sigma_2} y(q_0(y) \to q_1(y)), \forall_{\sigma_2} y(q_0(y) \to q_2(y)), \ldots\}$$

The theory T_2 has the sentence $\forall_{\sigma_2} y(q_0(y) \to q_i(y))$ as an axiom for each $i \in \mathbb{N}$.

We first show that T_1 and T_2 are categorically equivalent. It is easy to see that $\text{Mod}(T_1)$ and $\text{Mod}(T_2)$ both have $2^{\aleph_0}$ (non-isomorphic) objects. Furthermore, $\text{Mod}(T_1)$ and $\text{Mod}(T_2)$ are both discrete categories. We show here that $\text{Mod}(T_1)$ is discrete. Suppose that there is an elementary embedding $f : M \to N$ between models M and N of T_1. It must be that f maps the unique element $m \in M$ to the unique element $n \in N$. Furthermore, since f is an elementary embedding, $M \vDash p_i[m]$ if and only if $N \vDash p_i[n]$ for every predicate $p_i \in \Sigma_1$. This implies that $f : M \to N$ is actually an isomorphism. Every arrow $f : M \to N$ in $\text{Mod}(T_1)$ is therefore an isomorphism, and there is at most one arrow between any two objects of $\text{Mod}(T_1)$. This immediately implies that $\text{Mod}(T_1)$ is discrete. An analogous argument demonstrates that $\text{Mod}(T_2)$ is discrete. Any bijection between the objects of $\text{Mod}(T_1)$ and $\text{Mod}(T_2)$ is therefore an equivalence of categories.

But T_1 and T_2 are not Morita equivalent. Suppose, for contradiction, that T is a "common Morita extension" of T_1 and T_2. Corollary 7.3.9 implies that there is a Σ_1-sentence ϕ such that $T \vdash \forall y q_0(y) \leftrightarrow \phi$. One can verify using Theorem 7.3.1 and Corollary 7.3.9 that the sentence ϕ has the following property: If ψ is a Σ_1-sentence and $T_1 \vdash \psi \to \phi$, then either (i) $T_1 \vdash \neg\psi$ or (ii) $T_1 \vdash \phi \to \psi$. But ϕ cannot have this property. Consider the Σ_1-sentence

$$\psi := \phi \wedge \forall x p_i(x),$$

where p_i is a predicate symbol that does not occur in ϕ. We trivially see that $T_1 \vdash \psi \to \phi$, but neither (i) nor (ii) hold of ψ. This implies that T_1 and T_2 are not Morita equivalent. □

7.4 From Geometry to Conceptual Relativity

The twentieth century saw wide swings in prevailing philosophical opinion. In the 1920s, the logical positivists staked out a decidedly antirealist position, particularly in their rejection of the possibility of metaphysical knowledge. Only a few decades later, prevailing opinion had reached the opposite end of the spectrum. The great analytic philosophers of the 1970s and 1980s – Putnam, Lewis, Kripke, etc. – were unabashed proponents of scientific and metaphysical realism. Or perhaps it would be more accurate to say that these philosophers presupposed realism and built their philosophical programs on the assumption that there is a kind of knowledge that transcends the claims of the empirical sciences.

But the pendulum didn't rest there. By the end of the twentieth century, several analytic philosophers were giving arguments against realism, saying that it didn't mesh well with the way that the sciences actually work. For example, Putnam and Goodman pointed to the existence of different formulations of Euclidean geometry, some of which take points as primitives, and some of which take lines as primitives, saying that realists must render the incorrect verdict that these are inequivalent theories. We will call the invocation of this particular example the *argument from geometry against realism.*

According to the argument from geometry, certain situations can equally well be described using a theory that takes points as fundamental entities or, instead, using a theory that takes lines as fundamental entities. Someone who adopts the first theory is committed to the existence of points and not lines, while someone who adopts the second theory is committed to the existence of lines and not points. But points and lines are different kinds of things, and, in general, the number of points (according to the first theory) will be different from the number of lines (according to the second theory). Since both parties correctly describe the world but use different ontologies to do so, it's supposed to follow that there is no matter of fact about what the ontology of the world is – in direct contradiction with a fundamental tenet of metaphysical realism.

In responding to examples of this sort, metaphysical realists typically agree that the two theories in question involve incompatible ontological commitments (see Sider, 2009; van Inwagen, 2009). These realists then claim, however, that at most one of the two theories can be correct, at least in a fundamental sense. The upshot of this kind of response, of course, is that a realist ontology has been purchased at the price of an epistemic predicament: Only one of the theories is correct, but we will never know which one.

In this section, we propose another reply to arguments of this sort, and specifically to the argument from geometry. We show that geometries with points can naturally be considered equivalent to geometries with lines, and we argue that this equivalence does not in any way threaten the idea that there is an objective world. In other words, since these two theories are equivalent, there is a sense in which they involve exactly the same ontological commitments. The example of geometries with points and geometries with lines does not undermine metaphysical realism in the way that Putnam and Goodman suggested.

There are many ways to formulate a particular geometric theory, and these formulations often differ with respect to the kinds of objects that are taken as primitive. The most famous example of this phenomenon is Euclidean geometry. Tarski first formulated Euclidean geometry using open balls (Tarski, 1929), and later using points (Tarski, 1959). Schwabhäuser and Szczerba (1975) formulated Euclidean geometry using lines, and Hilbert (1930) used points, lines, planes, and angles. These formulations of Euclidean geometry all take different kinds of objects to be primitive, but despite this ostensible difference, they nonetheless manage to express the same geometric facts. Indeed, it is standard to recognize some sense in which all of these formulations of Euclidean geometry are *equivalent*. This sense of equivalence, however, is rarely made perfectly precise.

In fact, from a certain point of view, it might seem that these theories cannot be equivalent. Consider a simple example: Take six lines in the Euclidean plane, as in the following diagram.

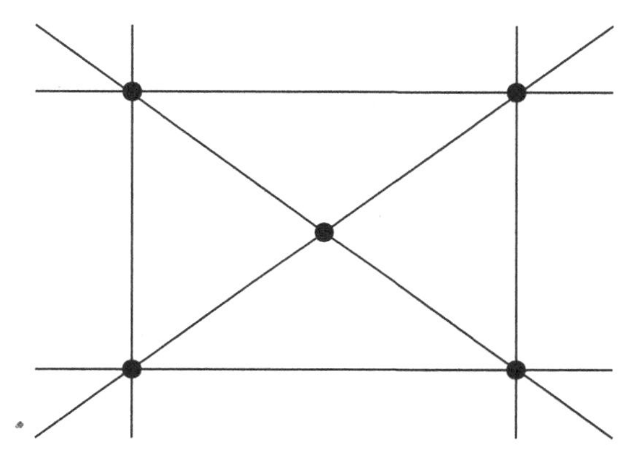

On the one hand, if this diagram were described in terms of the point-based version of Euclidean geometry (T_p), then we would say that there are exactly five things. On the other hand, if this diagram were described in terms of the line-based version of Euclidean geometry (T_ℓ), then we would say that there are exactly six things. The point-based and line-based descriptions therefore seem to disagree about a feature of the diagram – namely, how many things there are in the diagram.

Indeed, according to one natural notion of theoretical equivalence, the first description T_p is not equivalent to the second description T_ℓ. The notion we have in mind is **definitional equivalence**, which we introduced in Section 4.6, and which first entered into philosophy of science through the work of Glymour (1971, 1977, 1980). If two theories are definitionally equivalent, then the cardinalities of their respective domains will be equal. Since the domains of T_p and T_ℓ do not have the same cardinality, these descriptions cannot be definitionally equivalent.

This would be the end of the matter if definitional equivalence were the only legitimate notion of theoretical equivalence. But, as we now know, there is a better notion of theoretical equivalence that does not prejudge issues about the cardinality of domains.

All of the geometries that we will consider are formulated using (some subset of) the following vocabulary. Here we follow Schwabhäuser et al. (1983).

- The sort symbols σ_p and σ_ℓ will indicate the sort of points and the sort of lines, respectively. We will use letters from the beginning of the alphabet like a, b, c to denote variables of sort σ_p, and letters from the end of the alphabet like x, y, z to denote variables of sort σ_ℓ.

- The predicate symbol $r(a,x)$ of arity $\sigma_p \times \sigma_\ell$ indicates that the point a lies on the line x.
- The predicate symbol $s(a,b,c)$ of arity $\sigma_p \times \sigma_p \times \sigma_p$ indicates that the points $a, b,$ and c are colinear.
- The predicate symbol $p(x,y)$ of arity $\sigma_\ell \times \sigma_\ell$ indicates that the lines x and y intersect.
- Lastly, the predicate symbol $o(x,y,z)$ of arity $\sigma_\ell \times \sigma_\ell \times \sigma_\ell$ indicates that the lines $x, y,$ and z are compunctual – i.e., that they all intersect at a single point.

We now prove two theorems that capture the equivalence between geometries with points and geometries with lines. We then provide three examples that illustrate the generality of these results.

Suppose that we are given a formulation of geometry T that uses both of the sort symbols σ_p and σ_ℓ. The two theorems that we will prove in this section show that, given some natural assumptions, the theory T is Morita equivalent both to a theory T_p that only uses the sort σ_p and to a theory T_ℓ that only uses the sort σ_ℓ. In this sense, therefore, the geometry T can be formulated using only points, only lines, or both points and lines.

Our first theorem captures a sense in which the geometry T can be formulated using only points. In order to prove this theorem, we will need the following important result. The proof of this proposition is given by Schwabhäuser et al. (1983, Proposition 4.59).

PROPOSITION 7.4.1 (Elimination of line variables) *Let T be a theory formulated in the signature $\Sigma = \{\sigma_p, \sigma_\ell, r, s\}$, and suppose that T entails the following sentences:*

1. $(a \neq b) \rightarrow \exists_{=1} x\,(r(a,x) \wedge r(b,x))$
2. $\forall x \exists a \exists b\,(r(a,x) \wedge r(b,x) \wedge (a \neq b))$
3. $s(a,b,c) \leftrightarrow \exists x\,(r(a,x) \wedge r(b,x) \wedge r(c,x))$

Then for every Σ-formula ϕ without free variables of sort σ_l, there is a Σ-formula ϕ^*, whose free variables are included in those of ϕ, that contains no variables of sort σ_ℓ, and such that $T \vDash \forall \vec{a}(\phi(\vec{a}) \leftrightarrow \phi^*(\vec{a}))$.

We should take a moment here to unravel the intuition behind this proposition. The theory T can be thought of as a geometry that is formulated in terms of points and lines, using the basic notions of a point lying on a line and three points being colinear. Since the theory T is a geometry, the sentences 1, 2, and 3 are sentences that one should naturally expect T to satisfy. Given these assumptions on T, Proposition 7.4.1 simply guarantees that Σ-formulas ϕ can be "translated" into corresponding formulas ϕ^* that do not use the apparatus of lines. This translation eliminates the line variables from every Σ-formula in two steps. First, one uses the fact that every line is uniquely characterized by two nonidentical points lying on it to replace equalities between line variables with more complex expressions using the predicate r. Second, one replaces instances of the predicate $r(a,x)$ by using complex expressions involving the colinearity predicate $s(a,b,c)$. The reader is encouraged to consult Schwabhäuser et al. (1983, Proposition 4.59) for details.

With this proposition in hand, we have the following result.

THEOREM 7.4.2 (Barrett) *Let T be a theory that satisfies the hypotheses of Proposition 7.4.1. Then there is a theory T_p in the restricted signature $\Sigma_0 = \Sigma \backslash \{\sigma_\ell, r\}$ that is Morita equivalent to T.*

Theorem 7.4.2 captures a sense in which every geometry that is formulated with points and lines could be formulated equally well using only points. The idea behind the proof of Theorem 7.4.2 should be clear. Consider the Σ_0-theory defined by

$$T_p = \{\phi^* : T \vdash \phi\},$$

where the existence of the sentences ϕ^* is guaranteed by the fact that T satisfies the hypotheses of Proposition 7.4.1. The theory T_p can be thought of as a theory that "says the same thing as T" but uses only the apparatus of points. One proves Theorem 7.4.2 by showing that this theory T_p has the resources to define the sort σ_ℓ of lines. (Note that in the following proof we abuse our convention and occasionally use the variables x, y, z as variables that are not of sort σ_ℓ. But the sort of variables should always be clear from context.)

Proof of Theorem 7.4.2 It suffices to show that the theories T and T_p are Morita equivalent. The following figure illustrates the structure of our argument:

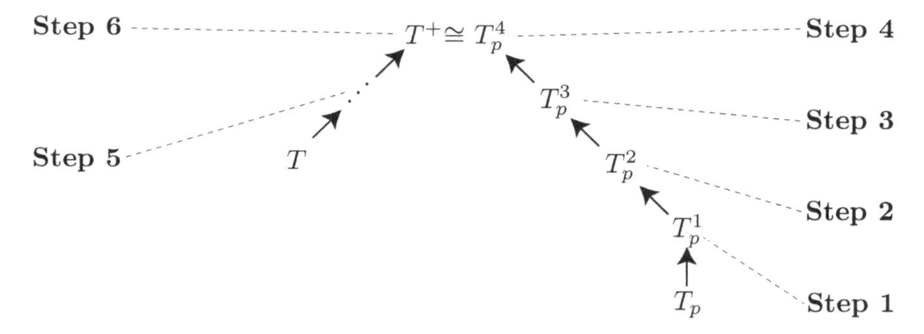

We begin on the right-hand side of the figure by building four theories T_p^1, T_p^2, T_p^3, and T_p^4. The purpose of these theories is to define, using the resources of the theory T_p, the symbols σ_ℓ and r.

Step 1: The theory T_p^1 is the Morita extension of T_p obtained by defining a new sort symbol $\sigma_p \times \sigma_p$ as a product sort (of the sort σ_p with itself). We can think of the elements of the sort $\sigma_p \times \sigma_p$ as pairs of points. The theory T_p^1 is a Morita extension of T_p to the signature $\Sigma_0 \cup \{\sigma_p \times \sigma_p, \pi_1, \pi_2\}$, where π_1 and π_2 are both function symbols of arity $\sigma_p \times \sigma_p \to \sigma_p$.

Step 2: The theory T_p^2 is the Morita extension of T_p^1 obtained by defining a new sort symbol σ_s as a subsort of $\sigma_p \times \sigma_p$. The elements of sort σ_s are the elements (a, b) of sort $\sigma_p \times \sigma_p$ such that $a \neq b$. One can easily write out the defining formula for the subsort σ_s to guarantee that this is the case. We can think of the elements of sort σ_s as the pairs of distinct points or, more intuitively, as the "line segments formed between distinct points." The theory T_p^2 is a Morita extension of T_p^1 to the signature $\Sigma_0 \cup \{\sigma_p \times \sigma_p, \pi_1, \pi_2, \sigma_s, i\}$, where i is a function symbol of arity $\sigma_s \to \sigma_p \times \sigma_p$.

Step 3: The theory T_p^2 employs a sort of "line segments," but we do not yet have a sort of lines. Indeed, we need to take care of the fact that some line segments determine the same line. We do this by considering the theory T_p^3, the Morita extension of T_p^2 obtained by defining the sort symbol σ_ℓ as a quotient sort of σ_s using the formula

$$s(\pi_1 \circ i(x), \pi_1 \circ i(y), \pi_2 \circ i(y)) \wedge s(\pi_2 \circ i(x), \pi_1 \circ i(y), \pi_2 \circ i(y)).$$

Using the fact that T is a conservative extension of T_p, one can easily verify that T_p^2 satisfies the admissibility conditions for this definition – i.e., the preceding formula is an equivalence relation according to T_p^2. The idea here is simple: two line segments (a_1, a_2) and (b_1, b_2) determine the same line just in case the points a_1, b_1, b_2 are colinear, and the points a_2, b_1, b_2 are, too. The theory T_p^3 simply identifies the line segments that determine the same line in this sense. We have now defined the sort σ_ℓ of lines. The theory T_p^3 is a Morita extension of T_p^2 to the signature $\Sigma_0 \cup \{\sigma_p \times \sigma_p, \pi_1, \pi_2, \sigma_s, i, \sigma_\ell, \epsilon\}$, where ϵ is a function symbol of sort $\sigma_s \to \sigma_\ell$.

Step 4: All that remains on the right-hand side of the figure is to define the predicate symbol r. The theory T_p^4 is the Morita extension of T_p^3 obtained by defining the predicate $r(a, z)$ using the formula

$$\exists_{\sigma_p \times \sigma_p} x \exists_{\sigma_s} y (\pi_1(x) = a \wedge i(y) = x \wedge \epsilon(y) = z).$$

The idea here is again intuitive. A point a is on a line z just in case there is another point b such that the pair of points (a, b) determines the line l. (In the preceding formula, one can think of the variable x as playing the role of this pair (a, b).) The theory T_p^4 is a Morita extension of T_p^3 to the signature $\Sigma_0 \cup \{\sigma_p \times \sigma_p, \pi_1, \pi_2, \sigma_s, i, \sigma_\ell, \epsilon, r\}$.

Step 5: We now turn to the left-hand side of our organizational figure. The theory T is formulated in the signature Σ, so it needs to define all of the new symbols that we added to the theory T_p in the course of defining σ_p and r. The theory T defines the symbols $\sigma_p \times \sigma_p, \pi_1, \pi_2, \sigma_s, i$ in the obvious manner. For example, it defines $\sigma_p \times \sigma_p$ as the product sort (of σ_p with itself) with the projections π_1 and π_2.

We still need, however, to define the function symbol ϵ. The function ϵ intuitively maps a pair of distinct points to the line that they determine. This suggests that we define $\epsilon(x) = y$ using the formula

$$r(\pi_1 \circ i(x), y) \wedge r(\pi_2 \circ i(x), y).$$

Intuitively, this formula is saying that a pair of points $x = (x_1, x_2)$ determines a line y just in case x_1 is on y and x_2 is on y. We call the theory that results from defining all of these symbols T^+.

Step 6: All that remains now is to show that the theory T_p^4 is logically equivalent to the theory T^+. This argument is mainly a tedious verification. The only nontrivial part of the argument is the following: one needs to show that $T_p^4 \vDash \phi$ for every sentence ϕ such that $T \vDash \phi$. One does this by verifying that T_p^4 itself entails the three sentences 1, 2, and 3 in the statement of Proposition 1. This means that T_p^4 entails the sentences

$\phi \leftrightarrow \phi^*$ for every Σ-sentence ϕ. In conjunction with the fact that $T_p^4 \vDash \phi^*$ for every consequence ϕ of T, this implies that $T_p^4 \vDash \phi$. The theories T_p^4 and T^+ are logically equivalent, so T_p and T must be Morita equivalent. □

Our second theorem is perfectly analogous to Theorem 7.4.2. It captures a sense in which a geometry T can be formulated using only lines. As with Theorem 7.4.2, we will need a preliminary result. The proof of the following proposition is given by Schwabhäuser et al. (1983, Proposition 4.89).

PROPOSITION 7.4.3 (Elimination of point variables) *Let T be a theory formulated in the signature $\Sigma = \{\sigma_p, \sigma_\ell, r, p, o\}$, and suppose that T implies the following sentences.*

1. $(x \neq y) \rightarrow \exists_{\leq 1} a(r(a,x) \wedge r(a,y))$
2. $\forall a \exists x \exists y((x \neq y) \wedge r(a,x) \wedge r(a,y))$
3. $o(x,y,z) \leftrightarrow \exists a(r(a,x) \wedge r(a,y) \wedge r(a,z))$
4. $p(x,y) \leftrightarrow ((x \neq y) \wedge s(x,y,y))$
5. $p(x,y) \leftrightarrow ((x \neq y) \wedge \exists a(r(a,x) \wedge r(a,y)))$

Then for every Σ-formula ϕ without free variables of sort σ_p, there is a Σ-formula ϕ^*, whose free variables are included in those of ϕ, that contains no variables of sort σ_p, and such that $T \vDash \forall \vec{x}(\phi(\vec{x}) \leftrightarrow \phi^*(\vec{x}))$.

Proposition 7.4.3 is perfectly analogous to Proposition 7.4.1. One again thinks of the theory T as a geometry, and so sentences 1–5 are sentences that one naturally expects T to satisfy. Proposition 7.4.3 guarantees that Σ-formulas can be "translated" into formulas ϕ^* that do not use the apparatus of points. Analogous to Proposition 7.4.1, one proves this proposition by showing that variables of sort σ_p can be eliminated in the following manner. One first replaces equalities between these variables and then interprets $r(a,x)$ in terms of $o(y,z,x)$, where y and z have a as their intersection point. The reader is invited to consult Schwabhäuser et al. (1983, Proposition 4.89) for further details.

With Proposition 7.4.3 in hand, we have the following result.

THEOREM 7.4.4 (Barrett) *Let T be a theory that satisfies the hypotheses of Proposition 7.4.3. There is a theory T_ℓ in the restricted signature $\Sigma_0 = \Sigma \backslash \{\sigma_p, r\}$ that is Morita equivalent to T.*

Proof The proof is analogous to the proof of Theorem 7.4.2, so we will not go into as much detail. Consider the Σ_0-theory T_ℓ defined by $T_\ell = \{\phi^* : T \vdash \phi\}$, where the existence of the sentences ϕ^* is guaranteed since T satisfies the hypotheses of Proposition 7.4.3. One shows that the theory T_ℓ is Morita equivalent to T. The theory T_ℓ needs to define the sort symbol σ_p. It does this by first defining a product sort of "pairs of lines," and then a subsort of "pairs of intersecting lines." The sort of points is then the quotient sort that results from identifying two pairs of intersecting lines (w,x) and (y,z) just in case both w,x,y and w,x,z are compunctual. The theory T_ℓ also needs to define the symbol r. It does this simply by requiring that $r(a,x)$ holds of a point a and a line x

just in case there is another line y such that the pair of lines (x, y) intersect at the point a. As in the proof of Theorem 7.4.2, T defines the symbols of T_ℓ in the natural way. □

Theorem 7.4.2 shows that every geometry formulated using points and lines could be formulated equally well using only points; Theorem 7.4.4 shows that it could be formulated equally well using only lines. These two results together capture a robust sense in which geometries with points and geometries with lines are equivalent theories.

Theorems 7.4.2 and 7.4.4 are quite general. Indeed, one can verify that many of the theories that we usually think of as geometries satisfy the hypotheses of the two theorems. We provide three examples here. We begin by revisiting a simple geometric theory that we considered earlier.

Example 7.4.5 Recall the earlier diagram of six lines and five points in the Euclidean plane. By interpreting the symbols $\sigma_p, \sigma_\ell, r, s, p,$ and o in the natural way, one can easily convert this diagram into a $\{\sigma_p, \sigma_\ell, r, s, p, o\}$-structure M. We now consider the geometric theory $\text{Th}(M) = \{\phi : M \vdash \phi\}$. One can verify by inspection that $\text{Th}(M)$ satisfies the hypotheses of both Theorems 7.4.2 and 7.4.4. Theorem 7.4.2 implies that this diagram can be fully described using only the apparatus of points (using the theory $\text{Th}(M)_p$), while Theorem 7.4.4 implies that it can be fully described using only the apparatus of lines (using the theory $\text{Th}(M)_\ell$), and all three of these theories are Morita equivalent. ⌟

In our next two examples, we consider more general geometric theories: projective geometry and affine geometry.

Example 7.4.6 (Projective geometry) Projective geometry is a theory T_{proj} formulated in the signature $\{\sigma_p, \sigma_\ell, r\}$, where all of these symbols are understood exactly as they were earlier. The theory T_{proj} has the following three axioms (Barnes and Mack, 1975).

- $a \neq b \to \exists_{=1} x(r(a,x) \wedge r(b,x))$
- $x \neq y \to \exists_{=1} a(r(a,x) \wedge r(a,y))$
- There are at least four points, no three of which lie on the same line.

(One can easily express the third axiom as a sentence of first-order logic, but we here refrain for the sake of clarity.)

Projective geometry satisfies the hypotheses of both Theorems 7.4.2 and 7.4.4. We consider Theorem 7.4.4. In order to apply this result, we need to add the following two axioms that define the symbols p and o:

$$p(x, y) \leftrightarrow (x \neq y \wedge \exists a(r(a,x) \wedge r(a,y))) \qquad (\theta_p)$$

$$o(x, y, z) \leftrightarrow \exists a(r(a,x) \wedge r(a,y) \wedge r(a,z)) \qquad (\theta_o)$$

One can easily verify that the $\{\sigma_\ell, \sigma_p, r, p, o\}$-theory T^+_{proj} obtained by adding the definitions θ_p and θ_o to the axioms of T_{proj} satisfies sentences 1–5 of Proposition 7.4.3. Theorem 7.4.4 then implies that there is a theory in the restricted signature $\{\sigma_\ell, p, c\}$ that is Morita equivalent to T^+_{proj}. Projective geometry can therefore be formulated using only the apparatus of lines. One argues in a perfectly analogous manner to show that Theorem 7.4.2 also applies to projective geometry, so it can also be formulated using only the apparatus of points. ┘

Example 7.4.7 (Affine geometry) Affine geometry is a theory T_{aff} formulated in the signature $\{\sigma_\ell, \sigma_p, r\}$, where all of these symbols are again understood exactly as earlier. The theory T_{aff} has the following five axioms (Veblen and Young, 1918, 118).

- $a \neq b \to \exists x(r(a,x) \wedge r(b,x))$
- $\neg r(a,x) \to \exists_{=1} y(r(a,y) \wedge \forall b(r(b,y) \to \neg r(b,x)))$
- $\forall x \exists a \exists b(a \neq b \wedge r(a,x) \wedge r(b,x))$
- $\exists a \exists b \exists c(a \neq b \wedge a \neq c \wedge b \neq c \wedge \neg \exists x(r(a,x) \wedge r(b,x) \wedge r(c,x)))$
- Pappus' theorem (Veblen and Young, 1918, p. 103 and Figure 40).

The fifth axiom can easily be written as a first-order sentence in the signature $\{\sigma_\ell, \sigma_p, r\}$, but since this axiom is not used in the following argument, we leave its translation to the reader. (Indeed, one only needs the first, third, and fourth axioms of T_{aff} to complete all of the following verifications.)

Affine geometry satisfies the hypotheses of both Theorems 7.4.2 and 7.4.4. We consider Theorem 7.4.2. In order to apply this result, we need to add one additional axiom to T_{aff} that defines the symbol s as follows:

$$s(a,b,c) \leftrightarrow \exists x(r(a,x) \wedge r(b,x) \wedge r(c,x)). \qquad (\theta_s)$$

It is now trivial to verify that sentences 1–3 of Proposition 7.4.1 are satisfied by the $\{\sigma_\ell, \sigma_p, r, s\}$-theory T^+_{aff} that is obtained by adding the sentence θ_s to the axioms of T_{aff}. Theorem 7.4.2 therefore implies that there is a theory in the restricted signature $\{\sigma_p, s\}$ that is Morita equivalent to T^+_{aff}, capturing a sense in which affine geometry can be formulated using only the apparatus of points. In a perfectly analogous manner, one can apply the Theorem 7.4.4 to the case of affine geometry. This captures a sense in which affine geometry can also be formulated using only lines. ┘

The previous example is more general than it might initially appear. Indeed, affine geometry serves as the foundation for many of our most familiar geometries. For example, by supplementing the affine geometry with the proper notion of orthogonality, one can obtain two dimensional Euclidean geometry or two dimensional Minkowski geometry. (See Coxeter [1955], Szczerba and Tarski [1979], Szczerba [1986, p. 910], or Goldblatt [1987] for details.) Theorems 7.4.2 and 7.4.4 therefore capture a sense in which both Euclidean geometry and Minkowski geometry can be formulated using either points or lines.

7.5 Morita Equivalence Is Intertranslatability

The results of the previous section might seem to validate Putnam's arguments against metaphysical realism. After all, we proved that line-based geometries are (Morita) equivalent to point-based geometries. However, in order to make a case against realism, one would need say something more about why Morita equivalence is the right notion of equivalence. Perhaps, you might worry that the inventors of Morita equivalence cooked it up precisely to deliver this kind of antirealistic verdict. Indeed, the definition of Morita equivalence seems to include several arbitrary choices. Why, for example, allow the construction of just these kinds of sorts (products, coproducts, subsorts, quotient sorts) and not others (such as exponential sorts)?

In this section, we provide independent motivation for Morita equivalence. In particular, we show that Morita equivalence corresponds to the notion of intertranslatability described in Section 5.4. The coincidence between these two notions is remarkable, as they were developed independently of each other. On the one hand, Morita equivalence was proposed by Barrett and Halvorson (2016b) and was motivated by results in topos theory (see Johnstone, 2003). On the other hand, many-sorted intertranslatability was being used already in model theory in the 1970s. It was given a precise formulation by van Benthem and Pearce (1984), and has been further articulated by Visser (2006). The coincidence of these two notions – Morita equivalence and intertranslatability – suggests that there is something natural about them, at least from a mathematical point of view.

We established previously that definitional equivalence of single-sorted theories corresponds to strong intertranslatability (4.6.17 and 6.6.21). We now generalize this result as follows: for many-sorted theories without trivial sorts (i.e., sorts that are restricted to only one thing), Morita equivalence corresponds to weak intertranslatability. Our argument proceeds as follows: first we show that if T^+ is a Morita extension of T, then there is a reduction $R : T^+ \to T$ that is inverse (up to homotopy) to the inclusion $I : T \to T^+$. Intuitively speaking, R expands each definiendum in Σ^+ into its definiens in Σ. The trick here is figuring out how an equality relation $x =_\sigma y$, with $\sigma \in \Sigma^+$, can be reconstrued in terms of Σ-formulas. For product sorts in Σ^+, the answer is simple: $x =_{\sigma_1 \times \sigma_2} y$ can be reconstrued as $(x_1 =_{\sigma_1} y_1) \wedge (x_2 =_{\sigma_2} y_2)$. For coproduct sorts in Σ^+, the answer is more complicated. The problem here is that an equality $x =_{(\sigma_1 + \sigma_2)} y$ defines an equivalence relation, but $(x_1 =_{\sigma_1} y_1) \vee (x_2 =_{\sigma_2} y_2)$ does not define an equivalence relation; hence the former cannot be reconstrued as the latter. Thus, to reconstrue equality statements over a coproduct sort, we will need a more roundabout construction. To this end, we borrow the following definition from Harnik (2011).

DEFINITION 7.5.1 Let T be a theory in signature Σ. We say that T is **proper** just in case there is a Σ-formula $\phi(z)$ such that $T \vdash \exists z \phi(z)$ and $T \vdash \exists z \neg \phi(z)$. Here we allow z also to be a sequence of variables, possibly of various sorts.

NOTE 7.5.2 Suppose that Σ has a sort symbol σ and that $T \vdash \exists x \exists y (x \neq_\sigma y)$. Then T is proper, as witnessed by the formula $\phi(x, y) \equiv (x =_\sigma y)$.

THEOREM 7.5.3 (Washington) *Let T be a proper theory, and let T^+ be a Morita extension of T. Then there is a translation $R : T^+ \to T$ that is inverse to the inclusion $I : T \to T^+$.*

Before we give a proof of this result, we give an example to show why it's necessary to restrict to proper theories.

Example 7.5.4 Let T be the theory of equality over a single sort σ. Let T^+ be the Morita extension of T to the signature $\{\sigma, \sigma', \rho_1, \rho_2\}$, where T^+ defines σ' as a coproduct with coprojections $\rho_1 : \sigma \to \sigma'$ and $\rho_2 : \sigma \to \sigma'$. In this case, $T^+ \vdash \rho_1(x_1) \neq \rho_2(x_2)$; hence $T^+ \vdash \exists y_1 \exists y_2 (y_1 \neq y_2)$, with y_1, y_2 variables of sort σ'.

If there were a translation $R : T^+ \to T$, then we would have a corresponding model functor $R^* : \text{Mod}(T) \to \text{Mod}(T^+)$. But consider the model M of T with $M(\sigma)$ a singleton set. In that case, $(R^*M)(\sigma')$ would be a quotient of a subset of $M(\sigma) \times \cdots \times M(\sigma)$, which is again a singleton set. This contradicts the fact that $T^+ \vdash \exists y_1 \exists y_2 (y_1 \neq y_2)$. Therefore, there is no translation (in the sense of 5.4.2) from T^+ to T. ⌟

Proof of 7.5.3 Since T is proper, there is a sort σ_* of Σ and a formula $\phi(z)$ with $z : \sigma_*$ such that $T \vdash \exists z \phi(z)$ and $T \vdash \exists z \neg \phi(z)$. We first define $R : S \to (S^+)^*$. All cases are straightforward, except for coproduct sorts, which require a special treatment.

- Suppose that T^+ defines σ as a product with projections $\pi_1 : \sigma \to \sigma_1$ and $\pi_2 : \sigma \to \sigma_2$. Then we define $R(\sigma) = \sigma_1, \sigma_2$.
- Suppose that T^+ defines σ as a coproduct with coprojections $\rho_1 : \sigma_1 \to \sigma$ and $\rho_2 : \sigma_2 \to \sigma$. Then we define $R(\sigma) = \sigma_1, \sigma_2, \sigma_*$. Here the final sort σ_* plays an auxiliary role that permits us to define a coproduct of two sorts as a quotient of a product of sorts.
- Suppose that T^+ defines σ as a subsort with injection $i : \sigma \to \sigma'$. Then we define $R(\sigma) = \sigma'$.
- Suppose that T^+ defines σ as a quotient sort with projection $p : \sigma' \to \sigma$. Then we define $R(\sigma) = \sigma'$.
- Finally, if $\sigma \in \Sigma$, we define $R(\sigma) = \sigma$.

We now define the formulas E_σ for each sort symbol $\sigma \in \Sigma^+$.

- If σ is defined as a product sort $\sigma_1 \times \sigma_2$, then we set

$$E(x_1, x_2, y_1, y_2) \equiv (x_1 = y_1) \wedge (x_2 = y_2).$$

- Suppose that T^+ defines σ as a coproduct sort $\sigma_1 + \sigma_2$, in which case $R(\sigma) = \sigma_1, \sigma_2, \sigma_*$. Intuitively speaking, we will use a triple x, y, z to represent a variable of sort $\sigma_1 + \sigma_2$. We will think of the triples satisfying $\phi(z)$ as ranging over σ_1 (with y and z as dummy variables), and we will think of the triples satisfying $\neg\phi(z)$ as ranging over σ_2 (with x and z as dummy variables). Since $\vdash \phi(z) \vee \neg\phi(z)$, any triple x, y, z satisfies exactly one of these two conditions. We can then explicitly define the relevant formula $E(x_1, x_2, z; x_1', x_2', z')$ as

$$(\phi(z) \wedge \phi(z') \wedge (x_1 = x_1')) \vee (\neg\phi(z) \wedge \neg\phi(z') \wedge (x_2 = x_2')).$$

- If T^+ defines σ as a quotient sort in terms of a Σ-formula ϕ, then define $E(x,y) \equiv \phi(x,y)$.
- If σ is defined as a subsort in terms of a Σ-formula ϕ, then define $E(x,y) \equiv \phi(x) \wedge \phi(y) \wedge (x = y)$.

To complete the definition of the reconstrual, we need to give the mapping from predicate symbols and function symbols of Σ^+ to Σ.

- If $p \in \Sigma^+\backslash\Sigma$ is a predicate symbol with explicit definition $p \leftrightarrow \psi_p$, then we define $R(p)(x_1,\dots,x_n)$ as $\psi_p(x_1,\dots,x_n)$. If $p \in \Sigma$, then we define the image to be $p(x_1,\dots,x_n)$.
- If $f \in \Sigma^+\backslash\Sigma$ is a function symbol that is not used in an explicit definition of a sort symbol, and if f has explicit definition $(f(\vec{x}) =_\sigma y) \leftrightarrow \psi_f(\vec{x},y)$, then we define $R(f)(\vec{x},y)$ as $\psi_f(\vec{x},y)$. If $f \in \Sigma$, then we define the image to be $f(\vec{x}) =_\sigma y$.
- For function symbols $\pi_i : \sigma \to \sigma_i$ that define a product sort, we define $R(\pi_1)(x_1,x_2,y_1) \equiv (x_1 =_{\sigma_1} y_1)$ and $R(\pi_2)(x_1,x_2,y_2) \equiv (x_2 =_{\sigma_2} y_2)$.
- For function symbols $\rho_i : \sigma_i \to \sigma$ that define a coproduct sort, we define $R(\rho_1)(v_1,x_1,x_2,z) \equiv (v_1 =_{\sigma_1} x_1)$ and $R(\rho_1)(v_2,x_1,x_2,z) \equiv (v_2 =_{\sigma_2} x_2)$.
- For a function symbol $\epsilon : \sigma' \to \sigma$ that defines a quotient sort, we define $R(\epsilon)(x,y) \equiv \phi(x,y)$.
- For a function symbol $i : \sigma \to \sigma'$ that defines a subsort, we define $R(i)(x,y) \equiv \phi(x) \wedge \phi(y) \wedge (x = y)$.

We now show that $RI \simeq 1_T$ and $IR \simeq 1_{T^+}$. The former case is trivial: since R acts as the identity on elements of Σ, it follows that $RI = 1_T$. For the proof that $IR \simeq 1_{T+}$, we will define a t-map $\chi : IR \Rightarrow 1_{T^+}$, and we will show that χ is a homotopy.

Recall that a homotopy is a family of formulas, one for each sort symbol $\sigma \in \Sigma^+$. We will treat only the case where T^+ defines σ as a coproduct over $\rho_1 : \sigma_1 \to \sigma$ and $\rho_2 : \sigma_2 \to \sigma$. We need to define a Σ^+-formula χ whose free variables are of sorts $R(\sigma)$ and σ. Intuitively speaking, χ should establish a bijection between elements of sort $(\sigma_1,\sigma_2,\sigma_*)/E$ and elements of sort σ. We define

$$\chi(x_1,x_2,z,x) \equiv (\phi(z) \wedge (\rho_1(x_1) = x)) \vee (\neg\phi(z) \wedge (\rho_2(x_2) = x)).$$

We sketch the argument for the various conditions in the definition of a t-map (5.4.11). Throughout, we argue internally to the theory T^+.

- We show that χ is well defined relative to the equivalence relation E on $\sigma_1,\sigma_2,\sigma_*$. That is,

  $$E(x_1,x_2,z;x_1',x_2',z') \wedge \chi(x_1,x_2,z,x) \to \chi(x_1',x_2',z',x).$$

  Indeed, if $E(x_1,x_2,z;x_1',x_2',z')$, then there are two cases: either $\phi(z) \wedge \phi(z')$ or $\neg\phi(z) \wedge \neg\phi(z')$. In the former case, we have both $x_1 = x_1'$ and $\chi(x_1,x_2,z,x) \leftrightarrow (\rho(x_1) = x)$. Hence $\chi(x_1',x_2',z',x)$. The second case is similar.

- The "exists" property – i.e., $\exists x \chi(x_1,x_2,z,x)$ – follows immediately from the fact that $\phi(z) \vee \neg\phi(z)$ and the fact that ρ_1,ρ_2 are functions.
- We show now that χ is one-to-one (relative to the equivalence relation E on $\sigma_1,\sigma_2,\sigma_1,\sigma_1$); that is,

$$\chi(x_1,x_2,z,x) \wedge \chi(x_1',x_2',z',x) \to E(x_1,x_2,z;x_1',x_2',z').$$

Assume that $\chi(x_1,x_2,z,x) \wedge \chi(x_1',x_2',z',x)$, which expands to

$$\begin{aligned}&\left[(\phi(z) \wedge \rho_1(x_1) = x) \vee (\neg\phi(z) \wedge \rho_2(x_2) = x)\right]\\ \wedge &\left[(\phi(z') \wedge \rho_1(x_1') = x) \vee (\neg\phi(z') \wedge \rho_2(x_2') = x)\right].\end{aligned}$$

Since $\rho_1(y_1) \neq \rho_2(y_2)$, the first conjunct is inconsistent with the fourth, and the second is inconsistent with the third. Since ρ_1 and ρ_2 are injective, that formula is equivalent to

$$(\phi(z) \wedge \phi(z') \wedge (x_1 = x_1')) \vee (\neg\phi(z) \wedge \neg\phi(z') \wedge (x_2 = x_2')),$$

which, of course, is $E(x_1,x_2,z;x_1',x_2',z')$.
- Finally, we show that χ is onto, i.e., $\exists z\exists x_1\exists x_2\, \chi(x_1,x_2,z,x)$. Fix x, in which case, we have $\exists x_1(\rho_1(x_1) = x) \vee \exists x_2(\rho_2(x_2) = x)$. Since T^+ is proper, $\exists z\phi(z)$. Hence, in the case that $\exists x_1(\rho_1(x_1) = x)$, we have

$$\exists z\exists x_1(\phi(z) \wedge (\rho_1(x_1) = x)),$$

from which it follows that

$$\exists z\exists x_1\exists x_2 \left[(\phi(z) \wedge \rho_1(x_1) = x) \vee (\neg\phi(z) \wedge \rho_2(x_2) = x)\right].$$

Again, since T is proper, $\exists z\neg\phi(z)$, hence the same holds in the case that $\exists x_2(\rho_2(x_2) = x)$. In either case, $\exists z\exists x_1\exists x_2\, \chi(x_1,x_2,z,x)$, as we needed to prove.

Thus, we have shown how to define the component of $\chi : IR \Rightarrow 1_{T^+}$ where $\sigma \in \Sigma^+$ is defined to be a coproduct sort. The other cases are simpler, and we leave them to the reader. □

This completes the proof that Morita equivalence implies weak intertranslatability. We now turn to the converse implication.

THEOREM 7.5.5 (Washington) *If T_1 and T_2 are weakly intertranslatable, then T_1 and T_2 are Morita equivalent.*

While this result is not surprising, it turns out that the proof is extremely complicated because of needing to keep track of all the newly defined symbols. Thus, before we descend into the details of the proof, we discuss the intuition behind it.

A weak translation $F : T_1 \to T_2$ doesn't necessarily map a sort symbol σ of T_1 to a sort symbol of T_2. Nor does it exactly map a sort symbol σ of T_1 to a "product" $\sigma_1 \times \cdots \times \sigma_n$ of sort symbols of T_2, because the domain formula D_F restricts to a "subsort" $F_\bullet(\sigma)$ of $\sigma_1 \times \cdots \times \sigma_n$. What's more, the equality relation $=_\sigma$ is translated to the equivalence relation E_σ, which means that σ is really translated into something like the "quotient sort" of $F_\bullet(\sigma)$ modulo E_σ. In what follows, we will frequently write $F(=_\sigma)$ instead of E_σ in order to explicitly indicate the reconstrual F.

Now, notice that each of the constructions we mentioned earlier is permitted in taking a Morita extension of T_2. Intuitively, then, T_2 has a Morita extension T_2^+ that has enough sorts so that the translation $F : T_1 \to T_2$ can be extended to a one-dimensional translation $\hat{F} : T_1 \to T_2^+$, i.e., such that $\hat{F}(\sigma)$ is a single sort symbol of T_2^+. Intuitively, then, this extended translation $\hat{F}$ should be one-half of a homotopy equivalence in the strict sense.

One can then repeat this process to define a one-dimensional translation $\hat{G} : T_2 \to T_2^+$. Then, using the reductions $R_i : T_i^+ \to T_i$, one hopes to show that T_1^+ and T_2^+ are intertranslatable in the strict (one-dimensional) sense, which entails that they have a common definitional extension.

In practice, there are many complications in working out this idea. Thus, in the following proof, it will be convenient to allow ourselves a liberalized notion of a Morita extension where we can, in one step, add subsorts of product sorts. Suppose that Σ has sort symbols $\sigma_1, \dots, \sigma_n$, and a formula $\phi(\vec{x})$, with $x_i : \sigma_i$, and such that $T \vdash \exists \vec{x}\phi(\vec{x})$. Then we may take

$$\Sigma^+ = \Sigma \cup \{\sigma\} \cup \{\pi_1, \dots, \pi_n\},$$

where $\pi_i : \sigma \to \sigma_i$, and we may add explicit definitions that specify σ as the subsort of $\sigma_1 \times \cdots \times \sigma_n$ determined by the formula $\phi(\vec{x})$:

1. The projections π_i are jointly injective, i.e.,

$$\bigwedge_{i=1}^{n} (\pi_i(x) = \pi_i(y)) \to (x = y).$$

2. The projections π_i are jointly surjective, with image in $\phi(\vec{x})$, i.e.,

$$\phi(x_1, \dots, x_n) \leftrightarrow \exists x : \sigma \bigwedge_{i=1}^{n} (\pi_i(x) = x_i).$$

This liberalized notion of Morita equivalence is clearly equivalent to the original. So, there is no harm in allowing the direct construction of subsorts $\sigma \rightarrowtail \sigma_1 \times \cdots \times \sigma_n$, given that there is an appropriate formula $\phi(\vec{x})$.

Proof Let T_1 be a Σ_1-theory and T_2 a Σ_2 theory that are intertranslatable by the translations $F : T_1 \to T_2$ and $G : T_2 \to T_1$, and homotopies $\chi : GF \cong 1_{T_1}$ and $\chi' : FG \cong 1_{T_2}$. We will create Morita extensions of T_1 and T_2 in several stages, first defining new sort symbols and then defining new relation and function symbols.

Step 1: Suppose that $\sigma \in \Sigma_1$ is a sort symbol and that $F(\sigma) = F(\sigma)_1, \dots, F(\sigma)_n$. Let $F_\bullet(\sigma)$ be a new sort symbol, and let

$$\Sigma_2^1 = \Sigma_2 \cup \{F_\bullet(\sigma) \mid \sigma \in S_1\} \cup \{\pi_{F(\sigma)_i} \mid \sigma \in S_1\},$$

where $\pi_{F(\sigma)_i}$ is a function symbol of sort $F_\bullet(\sigma) \to F(\sigma)_i$. Let T_2^1 be the Morita extension of T_2 that defines $F_\bullet(\sigma) \rightarrowtail F(\sigma)_1 \times \cdots \times F(\sigma)_n$, with projections $\pi_{F(\sigma)_i}$, using the domain formula $D_F(\vec{x})$.

Similarly, let

$$\Sigma_1^1 = \Sigma_1 \cup \{G_\bullet(\sigma) \mid \sigma \in S_2\} \cup \{\pi_{G(\sigma)_i} \mid \sigma \in S_2\},$$

and let T_1^1 be the Morita extension of T_1 that defines each such $G_\bullet(\sigma)$ as a product of $G(\sigma)_1, \ldots, G(\sigma)_m$, with projections $\pi_{G(\sigma)_i}$.

Before proceeding to the next step, recall that $G(=_\sigma)$ is a T_1-provable equivalence relation on the domain $D_G(\vec{x}) \rightarrowtail G(\sigma)_1, \ldots, G(\sigma)_n$. Thus, we can use the projections $\pi_i \equiv \pi_{G(\sigma)_i}$ to define an equivalence relation $G_\bullet(=_\sigma)(x, y)$ on $G_\bullet(\sigma)$:

$$G_\bullet(=_\sigma)(x, y) \equiv G(=_\sigma)(\pi_1(x), \ldots, \pi_n(x); \pi_1(y), \ldots, \pi_n(y)).$$

Step 2: For $\sigma \in S_2$, we use T_1^1 to define σ as the quotient of $G_\bullet(\sigma)$ modulo $G_\bullet(=_\sigma)$. Let

$$\Sigma_1^2 = \Sigma_1^1 \cup \{\sigma \mid \sigma \in S_2\} \cup \{\epsilon_\sigma \mid \sigma \in S_2\},$$

where ϵ_σ is a new function symbol of sort $G_\bullet(\sigma) \to \sigma$. Let δ_σ be the explicit definition

$$\delta_\sigma \equiv ((\epsilon_\sigma(x) = \epsilon_\sigma(y)) \leftrightarrow G_\bullet(=_\sigma)(x, y)) \wedge \forall y \exists x (\epsilon_\sigma(x) = y). \tag{7.2}$$

We then define a Morita extension

$$T_1^2 = T_1^1 \cup \{\delta_\sigma \mid \sigma \in S_2\}.$$

Similarly, let

$$\Sigma_2^2 = \Sigma_2^1 \cup \{\sigma \mid \sigma \in S_1\} \cup \{\epsilon_\sigma \mid \sigma \in S_1\},$$

where $\epsilon_\sigma : F_\bullet(\sigma) \to \sigma$, and let T_2^2 be the Morita extension of T_2^1 that defines each $\sigma \in S_1$ as a quotient sort.

Before proceeding to the next step, we show that T_1^2 defines a functional relation ξ from the domain D_{GF} to $G_\bullet(F(\sigma)_1), \ldots, G_\bullet(F(\sigma)_n)$ or, more precisely, to the image of the latter in $GF(\sigma)$. Recall that the domain formulas of the composite GF are given by the general recipe $D_{GF} = G(D_F)$; and that G is defined so that

$$G(\phi)(\vec{x}_1, \ldots, \vec{x}_n) \vdash D_G(\vec{x}_i),$$

for any Σ_2-formula ϕ. Thus, $D_{GF}(\vec{x}_1, \ldots, \vec{x}_n) \vdash D_G(\vec{x}_i)$. Furthermore, $G_\bullet(F(\sigma)_i)$ is defined as a subsort of $G(F(\sigma)_i)_1, \ldots, G(F(\sigma)_i)_m$ via the formula $D_G(\vec{x}_i)$.

$$\begin{array}{ccc} & & D_{GF}(\vec{x}_1, \ldots, \vec{x}_n) \\ & & \big\downarrow \xi \\ G_\bullet(F(\sigma)_1), \ldots, G_\bullet(F(\sigma)_n) & \xrightarrow{\zeta} & D_G(\vec{x}_1) \wedge \cdots \wedge D_G(\vec{x}_n) \\ & & \big\downarrow \\ & & GF(\sigma) \end{array}$$

Step 3: In Step 1, we equipped T_2^1 with subsorts $F_\bullet(\sigma) \rightarrowtail F(\sigma)_1, \dots, F(\sigma)_n$. Now we add these sorts to T_1^2 as well. Given $\sigma \in \Sigma_1$, each $F(\sigma)_i$ is a sort in Σ_2, hence by Step 2, also in Σ_1^2. Now let

$$\Sigma_1^3 \;=\; \Sigma_1^2 \cup \{F_\bullet(\sigma) \mid \sigma \in S_1\} \cup \{\pi_{F(\sigma)_i} \mid \sigma \in S_1\},$$

where $\pi_{F(\sigma)_i}$ is the Σ_2^1 function symbol of sort $F_\bullet(\sigma) \to F(\sigma)_i$. In order to define $F_\bullet(\sigma)$, we need an appropriate formula $U(x_1, \dots, x_n) \rightarrowtail F(\sigma)_1, \dots, F(\sigma)_n$. We choose the image of D_{GF} under the function $\rho \equiv \epsilon_{F(\sigma)_1} \wedge \cdots \wedge \epsilon_{F(\sigma)_n}$.

$$\begin{array}{ccc}
D_{GF}(\vec{x}_1, \dots, \vec{x}_n) & \overset{\rho}{\dashrightarrow} & U(x_1, \dots, x_n) \\
\downarrow & & \downarrow \\
D_G(\vec{x}_1) \wedge \cdots \wedge D_G(\vec{x}_n) & \xrightarrow{\;\rho\;} & F(\sigma)_1, \dots, F(\sigma)_n \\
\downarrow & & \\
GF(\sigma) & &
\end{array}$$

That is,

$$U(x_1, \dots, x_n) \;\equiv\; \exists \vec{x}_1 \dots \exists \vec{x}_n \left(D_{GF}(\vec{x}_1, \dots, \vec{x}_n) \wedge \bigwedge_{i=1}^{n} (\epsilon_{F(\sigma)_i}(\vec{x}_i) = x_i) \right).$$

Since $T_1 \vdash \exists X\, D_{GF}(X)$, it follows that $T_1^2 \vdash \exists \vec{x}\, U(\vec{x})$. Thus, we can use $U(\vec{x})$ to define $F_\bullet(\sigma)$ as a subsort of $F(\sigma)_1, \dots, F(\sigma)_n$, and we let T_1^3 denote the resulting Morita extension of T_1^2.

Similarly, let

$$\Sigma_2^3 \;=\; \Sigma_2^2 \cup \{G_\bullet(\sigma) \mid \sigma \in S_2\} \cup \{\pi_{G(\sigma)_i} \mid \sigma \in S_2\},$$

and let T_2^3 be the Morita extension of T_2^2 that defines each $G_\bullet(\sigma)$ as a subsort of $G(\sigma)_1, \dots, G(\sigma)_n$.

Step 4: Let Σ_1^4 be the union of Σ_1^3 with all relation and function symbols from Σ_2. We extend T_1^3 to T_1^4 by adding explicit definitions for all the new symbols. For notational simplicity, we treat only the case of a predicate symbol $p \in \Sigma_2$ of sort $\sigma \in \Sigma_2$. We leave the other cases to the reader. Recall that T_1^3 defines $\epsilon_\sigma : G_\bullet(\sigma) \to \sigma$ as a quotient, and also the projections $\pi_{G(\sigma)_i} : G_\bullet(\sigma) \to G(\sigma)_i$ can be conjoined to give a bijection θ between $G_\bullet(\sigma)$ and $D_G(\vec{x})$.

$$\begin{array}{ccccc}
G(p)(\vec{x}) & \leftarrow\!\dashrightarrow & G_\bullet(p) & \dashrightarrow & \phi_p(x) \\
\downarrow & & \downarrow & & \downarrow \\
D_G(\vec{x}) & \overset{\theta}{\longleftrightarrow} & G_\bullet(\sigma) & \xrightarrow{\;\epsilon_\sigma\;} & \sigma
\end{array}$$

To define ϕ_p, first pull $G(p)(\vec{x})$ back along π to obtain $G_\bullet(p)$; then take the image of $G_\bullet(p)$ under ϵ_σ. That is,

$$\phi_p(x) \;\equiv\; \exists y \; (G(p)(\pi_1(y), \dots, \pi_n(y)) \wedge (\epsilon_\sigma(y) = x)) \,.$$

Recall that

$$G(=_\sigma)(\pi_1(y),\ldots,\pi_n(y);\pi_1(z),\ldots,\pi_n(z)) \vdash \epsilon_\sigma(y)=\epsilon_\sigma(z),$$

and also that

$$G(p)(\vec{x}), G(=_\sigma)(\vec{x},\vec{y}) \vdash G(p)(\vec{y}).$$

Hence the preceding diagram defines a functional relation from $G(p)(\vec{x})$ to $\phi_p(x)$, relative to the notion of equality given by $G(=_\sigma)$.

We now add explicit definitions $\delta_p \equiv p(x) \leftrightarrow \phi_p(x)$ for each relation symbol $p \in \Sigma_2$, creating a Morita extension T_1^4 of T_1^3. We perform the analogous construction to obtain extensions $\Sigma_2^4 \supseteq \Sigma_2^3$ and $T_2^4 \supseteq T_2^3$.

Before proceeding, we note that at this stage, the expanded signature Σ_2^4 has copies of the Σ_1-formulas D_{GF} and χ that define the homotopy $\chi : GF \Rightarrow 1_T$ for T.

$$\begin{array}{ccc} & & F_\bullet(\sigma) \\ & & \downarrow{\scriptstyle \epsilon_\sigma} \\ D_{GF}(\vec{x}_1,\ldots,\vec{x}_n) & \xrightarrow{\chi} & \sigma \end{array}$$

Step 5: In Step 3, we equipped T_2^3 with function symbols $\epsilon_\sigma : F_\bullet(\sigma) \to \sigma$, for $\sigma \in S_1$. We now add these function symbols to T_1^4 as well. Let

$$\Sigma_1^5 = \Sigma_1^4 \cup \{\epsilon_\sigma \mid \sigma \in S_1\}.$$

We need to find a Σ_1^4-formula that can serve as a suitable definiens for ϵ_σ. We construct a span of relations.

$$\begin{array}{ccccc} D_G(\vec{x}_1)\wedge\cdots\wedge D_G(\vec{x}_n) & \longleftarrow & D_{GF}(\vec{x}_1,\ldots,\vec{x}_n) & \xrightarrow{\chi} & \sigma \\ \downarrow{\scriptstyle \rho} & & \downarrow{\scriptstyle \rho} & & \\ F(\sigma)_1,\ldots,F(\sigma)_n & \longleftarrow & U(x_1,\ldots,x_n) & & \\ & & \downarrow{\scriptstyle \theta} & & \\ & & F_\bullet(\sigma) & & \end{array}$$

Here $D_G(\vec{x}_i)$ is the domain formula corresponding to the assignment $F(\sigma)_i \mapsto G(F(\sigma)_i)$; and $\rho \equiv \epsilon_{F(\sigma)_1} \wedge \cdots \wedge \epsilon_{F(\sigma)_n}$, where $\epsilon_{F(\sigma)_i} : G(F(\sigma)_i) \to F(\sigma)_i$ defines $F(\sigma)_i$ as a quotient sort via the equivalence relation $G(=_{F(\sigma)_i})$; and θ is given by

$$\theta(x_1,\ldots,x_n;y) \equiv U(x_1,\ldots,x_n) \wedge \bigwedge_{i=1}^{n}(x_i = \pi_i(y)),$$

for $U(x_1,\ldots,x_n)$, as defined in Step 3. Here θ is a bijection, so we ignore it. We show that the span of $\rho : D_{GF} \to U$ and $\chi : D_{GF} \to \sigma$ defines a functional relation from U to σ.

Since the homotopy formula χ is well defined relative to the equivalence relation $GF(=_\sigma)$, and surjective onto σ, we have

$$GF(=_\sigma)(Y,Z) \vdash \exists_{\sigma=1} x\,(\chi(Y,x) \wedge \chi(Z,x)).$$

Here we have used $Y = \vec{y}_1, \ldots, \vec{y}_n$ and $Z = \vec{z}_1, \ldots, \vec{z}_n$ for sequences of variables of sort $GF(\sigma)$. It will suffice then to show that

$$\rho(Y; x_1, \ldots, x_n), \rho(Z; x_1, \ldots, x_n) \vdash GF(=_\sigma)(Y, Z). \tag{7.3}$$

Now, the definition of ρ yields

$$\rho(Y; x_1, \ldots, x_n), \rho(Z; x_1, \ldots, x_n) \vdash \bigwedge_{i=1}^{n} G(=_{F(\sigma)_i})(\vec{y}_i, \vec{z}_i). \tag{7.4}$$

Moreover, since F is a translation, T_2 entails that $F(=_\sigma)$ is an equivalence relation on $F(\sigma)_1, \ldots F(\sigma)_n$. Hence, by reflexivity,

$$\bigwedge_{i=1}^{n}(y =_{F(\sigma)_i} z) \vdash F(=_\sigma)(y_1, \ldots, y_n; z_1, \ldots, z_n).$$

Since $G : T_2 \to T_1$ is a translation, the substitution theorem gives

$$\bigwedge_{i=1}^{n} G(=_{F(\sigma)_i})(\vec{y}_i, \vec{z}_i) \vdash GF(=_\sigma)(Y, Z). \tag{7.5}$$

The implications (7.4) and (7.5) together show that $\chi \circ \rho^{-1}$ is a functional relation from U to σ, where equality on the former is given by $GF(=_\sigma)$. Thus, $\chi \circ (\theta \circ \rho)^{-1}$ is a functional relation from $F_\bullet(\sigma)$ to σ. Using ψ to denote this relation, we introduce the explicit definition

$$\delta_{\epsilon_\sigma} \equiv (\epsilon_\sigma(x) = y) \leftrightarrow \psi(x, y), \tag{7.6}$$

and we define a Morita extension

$$T_1^5 = T_1^4 \cup \{\delta_{\epsilon_\sigma} \mid \sigma \in S_1\}.$$

We define a Morita extension T_2^5 of T_2^4 in an analogous fashion. Therefore, $\Sigma_1^5 = \Sigma_2^5$. This completes our construction of the Morita extensions T_1^5 of T_1, and T_2^5 of T_2.

We will now show that T_1^5 and T_2^5 are logically equivalent, thereby establishing the Morita equivalence of T_1 and T_2. To this end, note first that since T_1^5 is a Morita extension of T_1, the two theories are intertranslatable, and similarly for T_2^5 and T_2. (Note that the construction does not use coproduct sorts. Hence, the result holds even when T_1 and T_2 are not proper theories.) Composing these translations gives translations $F : T_1^5 \to T_2^5$ and $G : T_2^5 \to T_1^5$ that extend the original translations $F : T_1 \to T_2$ and $G : T_2 \to T_1$. We will use these translations to show that T_1^5 and T_2^5 have the same models in their shared signature $\Sigma_1^5 = \Sigma_2^5$. The intuition behind the result is clear: since T_1^5 is a Morita extension of T_1, each model of T_1 uniquely expands to a model of T_1^5, and similarly for T_2 and T_2^5. Since the original model functor $F^* : \text{Mod}(T_2) \to \text{Mod}(T_1)$ is an equivalence of categories, the lifted model functor $F^* : \text{Mod}(T_2^5) \to \text{Mod}(T_1^5)$ is also an equivalence of categories. We proceed now to the details of the argument.

Recall that we defined a reconstrual $\tilde{F} : \Sigma_1 \to \Sigma_2^3$ that is constant on sorts. (Hence, we may treat $\tilde{F}$ as a reconstrual in the more narrow sense.) We extend $\tilde{F}$ as usual to a map from Σ_1- formulas to Σ_2^3 formulas. Since F is a translation, $\tilde{F}$ is also a translation. Thus, the corresponding model map $\tilde{F}^*$ has the feature that

$$(\tilde{F}^*M)(\phi) = M(\tilde{F}(\phi)),$$

for each Σ_1-formula ϕ. In particular, $\tilde{F}^*M \models \phi$ iff $M \models \tilde{F}(\phi)$. The translation $\tilde{F} : T_1 \to T_2^3$ also has the feature that $T_2^5 \vdash F(\phi) \leftrightarrow \tilde{F}(\phi)$, for any sentence ϕ of T_1.

Now let M be a model of T_2^5. First we show that $M \models \phi$ for any Σ_1-sentence ϕ such that $T_1 \vdash \phi$. Since M satisfies the explicit definitions we gave for all the symbols in Σ_1, it follows (by induction) that $M(\tilde{F}(\phi)) = M(\phi)$ for any Σ_1-formula ϕ. Since M is a model of T_2^3, F^*M is a model of T_1, and $\tilde{F}^*M \models \phi$. By the previous paragraph, $M \models \tilde{F}(\phi)$, hence $M \models F(\phi)$, and, therefore, $M \models \phi$.

We now show that M satisfies the explicit definitions we added to T_1^5 in Steps 1–5. In Step 1, we added the definition of $G(\sigma)$ as a product sort. However, we added the same definition to T_2^3 in Step 3. Thus, since M is a model of T_2^3, these definitions are satisfied by M.

In Step 2, we expand Σ_2^1 to Σ_2^2 by adding sort symbols $\sigma \in S_1$ and function symbols $\epsilon_\sigma : F_\bullet(\sigma) \to \sigma$, and we let T_2^2 define $\epsilon_\sigma : F_\bullet(\sigma) \to \sigma$ as a quotient map corresponding to the equivalence relation $F_\bullet(=_\sigma)$. Hence, in any model M of T_2^5, we have

$$\epsilon_\sigma(a) = \epsilon_\sigma(b) \qquad \text{iff} \qquad F_\bullet(=_\sigma)(a,b),$$

for $a,b \in M_{F_\bullet(\sigma)}$. Recall also that T_2^5 explicitly defines $F_\bullet(\sigma)$ as a subsort of $F(\sigma)_1, \dots, F(\sigma)_n$, and that

$$F_\bullet(=_\sigma)(a,b) \qquad \text{iff} \qquad F(=_\sigma)(\vec{a},\vec{b}).$$

In Step 5, we stipulate that $T_1^5 \vdash \delta_{\epsilon_\sigma}$, where δ_{ϵ_σ} is the explicit definition:

$$\delta_{\epsilon_\sigma} \;\equiv\; (\epsilon_\sigma(x) = z) \leftrightarrow (\chi \circ \rho^{-1})(x,z).$$

We need to show that $T_2^5 \vdash \delta_{\epsilon_\sigma}$, and for this, we need to see how T_2^5 defines the symbols χ and ρ. First, $\chi : D_{GF} \to \sigma$ is the homotopy map, which is originally a Σ_1-formula. Thus, the symbols in χ are explicitly defined by T_2^4 in Step 4.

Next, $\rho \equiv \epsilon_{F(\sigma)_1} \wedge \cdots \wedge \epsilon_{F(\sigma)_n}$, where $F(\sigma)_i$ is a Σ_2 sort symbol, and $\epsilon_{F(\sigma)_i} : G_\bullet(F(\sigma)_i) \to F(\sigma)_i$ is a function symbol. In Step 1, we have T_1^1 define $G_\bullet(F(\sigma)_i)$ as a sub-product sort of $G(F(\sigma)_i)_1, \dots, G(F(\sigma)_i)_m$. In Step 5, we have T_2^5 define the function symbol $\epsilon_{F(\sigma)_i}$ in terms of the Σ_2 homotopy map χ'.

We need to show now that $M \models \delta_{\epsilon_\sigma}$ or, in other words, that $\epsilon_\sigma(x) = z$ and $\psi(x,z)$ define the same relation in M. We can show that the following diagram commutes (where the objects are meant to be domains of the sort symbols in the model M).

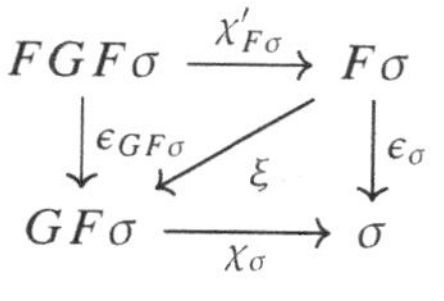

$$\begin{array}{ccc}
FGF\sigma & \xrightarrow{\chi'_{F\sigma}} & F\sigma \\
\downarrow{\scriptstyle \epsilon_{GF\sigma}} & \swarrow{\scriptstyle \xi} & \downarrow{\scriptstyle \epsilon_\sigma} \\
GF\sigma & \xrightarrow[\chi_\sigma]{} & \sigma
\end{array}$$

We can thus characterize χ_σ as the map that makes the preceding diagram commute. A key observation is that $F(\chi_\sigma) = \chi'_{F(\sigma)}$ for each sort $\sigma \in S_1$. □

TECHNICAL ASIDE 7.5.6 The sheer complexity of the previous proof shows one reason why it can be convenient to move to the context of categorical logic, where theories are treated as certain kinds of categories. We conjecture that a more intuitive (but conceptually laden) proof of this result could be obtained as follows.

Each first-order theory T has a unique classifying (Boolean) pretopos in the sense of Makkai (1987). Intuitively speaking, T and T' should have the same classifying pretopos iff T and T' are weakly intertranslatable in the sense we have described here. Furthermore, Tsementzis (2017b) shows that T and T' have the same classifying pretopos iff T and T' are Morita equivalent.

Having completed this result, we now have a much clearer picture of the various options for a precise notion of theoretical equivalence. We have placed the most salient options in the following chart.

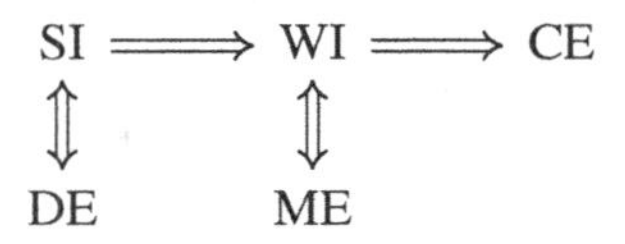

Here "I" represents the intertranslatability notions (strong and weak), and "E" represents the equivalence notions (definitional, Morita, and categorical). In this chart, the further to the right, the more liberal the notion of theoretical equivalence, and the fewer the invariants of equivalence. For example, if $F : T \to T'$ is a strong (one-dimensional) translation, then the dual functor $F^* : \text{Mod}(T') \to \text{Mod}(T)$ preserves the size of the underlying domains of models, which isn't necessarily the case for Morita equivalent theories. Similarly, if $F : T \to T'$ is a weak translation, then $F^* : \text{Mod}(T') \to \text{Mod}(T)$ preserves ultraproducts, which isn't necessarily the case for an arbitrary categorical equivalence between $\text{Mod}(T')$ and $\text{Mod}(T)$.

7.6 Open Questions

We do not mean to give the impression that we have answered all of the interesting questions that could be raised about theories and the relations between them. Quite to the contrary, we hope that our investigations serve to reinvigorate the sort of "exact philosophy" that Rudolf Carnap envisioned. We conclude this section, then, with a list of some open questions and lines of investigation that might be pursued.

1. We encourage philosophers of science to return to previous discussions of specific scientific theories, where claims of equivalence (or inequivalence) play a central role, but where the relevant notion of equivalence was not explicated. Can the tools we have developed here help clarify the commitments that led to certain judgments of equivalence or inequivalence?

2. It would be interesting to look again at the possibilities for providing perspicuous first-order formalizations of interesting scientific theories. Some work in this direction continues, e.g., with the Budapest group working on axiomatizations of relativity theory (see Andréka and Németi, 2014).
3. Some theories are so strong that new sorts (e.g., product sorts) seem to be encoded already into the original sorts. For example, in Peano arithmetic, n-tuples of natural numbers can be encoded as individual natural numbers. This encoding could perhaps be represented as an isomorphism $f : (\sigma \times \sigma) \to \sigma$ in a Morita extension T^+ of T. One might conjecture that for theories like Peano arithmetic, strict (one-dimensional) intertranslatability is equivalent to weak (many-dimensional) intertranslatability.
4. It's tempting to think that one could resort to "ontological maximalism" in the following sense: for a model M of theory T, the ontology for M consists of *all* the objects in every set that can be *constructed from* the domain M or, if the theory is many-sorted, from the domains $M(\sigma_1), \dots, M(\sigma_n)$. (This idea is in the spirit of the suggestion of Hawthorne [2006].)

   There are three immediate difficulties with this proposal. First, this proposal would make the ontology of every nontrivial theory infinite. In particular, infinitely many distinct elements occur in the tower of Cartesian products: $M, M \times M, M \times M \times M, \dots$ And that's even before we construct equivalence classes and coproducts from these sets. Second, it's not clear which constructions should be permitted. Should we allow the constructions from a Morita extension, or should we also allow, say, the construction of powersets? Third and finally, ontological maximalism runs contrary to the spirit of Ockham's razor.
5. One might worry that the definition of Morita equivalence is *arbitrary*. Why do we allow the particular definitions we do, and not others? Is there any intrinsic motivation for this choice? There is an intuition that the definitions permitted in a Morita extension are precisely those definitions that can be expressed in first-order language. How can we make that intuition precise?

7.7 Notes

- The notion of a dual functor $F^* : \mathrm{Mod}(T') \to \mathrm{Mod}(T)$ makes an appearance already in Makkai and Reyes (1977), who explore the correspondence between properties of F and properties of F^*. This exploration is part of their proof of the "conceptual completeness" of coherent logic.

  However, for Makkai and Reyes, first-order theories are replaced by coherent theories, and the latter are replaced by their corresponding pretoposes – all of which make their discussion a bit inaccessible for most philosophers. For an even more sophisticated investigation in this direction, see Breiner (2014). The dual functor makes an appearance in classical first-order logic in Gajda et al. (1987). The dual functor also seems to be quite closely related to the notion of a "model mapping" due to Gaifman (see Myers, 1997).

In later work, Makkai (1991) explores the question of which functors $G : \text{Mod}(T') \to \text{Mod}(T)$ are duals of translations. He makes some progress by assuming that Mod(T) and Mod(T') are not just categories, but *ultracategories*, i.e., categories with ultraproduct structure. In this case, the dual functors are those that preserve the ultraproduct structure.

- The proof that Morita equivalence implies categorical equivalence is from Barrett and Halvorson (2016b). In one sense, the result was no surprise all: the notion of Morita equivalence for first-order theories was modelled after the notion of Morita equivalence in categorical logic, i.e., when two theories T and T' have equivalent classifying toposes $\mathscr{E}_T$ and $\mathscr{E}_{T'}$. And when $\mathscr{E}_T \simeq \mathscr{E}_{T'}$, standard topos-theoretic methods show that $\text{Mod}(T) \simeq \text{Mod}(T')$ (see Johnstone, 2003, D1.4.13). Tsementzis (2017b) calls the notion we use here "T-Morita equivalence," and he gives a precise description of the relation between it at the topos-theoretic notion.
- The Morita equivalence of point and line geometries was demonstrated by Barrett and Halvorson (2017a). The arguments about geometry are novel, but not without precedent. Beth and Tarski (1956), Scott (1956), Tarski (1956), Robinson (1959), and Royden (1959) focus on the relationships between formulations of geometry that use different primitive *predicate* symbols, but not different primitive *sort* symbols. Szczerba (1977) and Schwabhäuser et al. (1983) take crucial steps toward capturing the relationships between geometries with different sorts but do not explicitly prove their equivalence. Andréka et al. (2008) and Andréka and Németi (2014), however, introduce a collection of tools from definability theory that allows one to demonstrate a precise equivalence.
- The proof that Morita equivalence coincides with weak intertranslatability is due to Washington (2018).

8 From Metatheory to Philosophy

Much of twentieth-century analytic philosophy was concerned – when not explicitly, then implicitly – with theories and with the relationships between them. For example, is every spacetime theory equivalent to one with Euclidean background geometry? Or is folk psychology reducible to neuroscience? Or can there be a good reason to choose a theory over an empirically equivalent rival theory?

But what is a theory? And what does it mean to say that two theories are equivalent or that one theory is reducible to another? Carnap had the audacious idea that philosophy can follow mathematics' method of explication: to take an intuitive notion and to find a nearby neighbor in the realm of precisely defined mathematical concepts. In this book, we've tried to follow Carnap's lead; and indeed, we hope that we've done a bit better than Carnap, because mathematics has come a long way in the past hundred years. We now have mathematical concepts – such as categories, functors, and natural transformations – the likes of which Carnap never dreamed about.

In this book, we've attempted to explicate the concept of a theory, as well as some of the relations between theories that scientists and philosophers find it useful to discuss. With these explications in the background, we can now return to some of the big questions of philosophy of science, such as, "what is the proper attitude to take toward a successful scientific theory?"

8.1 Ramsey Sentences

No analytic philosopher's education is complete until she learns the magic of the Ramsey sentence. The idea was proposed by Frank Ramsey (1929) and was reinvented by Carnap in the 1950s – or, more accurately, Carnap forgot that he learned about it from Herbert Bohnert (see Psillos, 2000). Most contemporary philosophers know of the idea because David Lewis (1970) argued that it solves the problem of theoretical terms. In the years since Lewis' seminal paper, Ramsey sentences have become a sort of deus ex machina of analytic philosophy.

Let's start with a simple example. Suppose that P is a theoretical predicate and that O is an observational predicate. (Or, in Lewis' preferred terminology, O is antecedently understood vocabulary, and P is new vocabulary.) Now suppose that our theory T consists of a single sentence $P(c) \rightarrow O(c)$, which might be paraphrased as saying that $O(c)$ is an empirical sign that $P(c)$. (Here c is a constant symbol. We omit first-order quantifiers to keep things simple.) To form the Ramsey sentence of T, we simply perform an instance of second-order existential generalization:

$$\frac{P(c) \to O(c)}{\exists X(X(c) \to O(c))}.$$

The sentence below the line is called the Ramsey sentence T^R of the theory T. Thus, while the original theoretical statement T mentions some particular property P, the Ramsey sentence T^R simply says that there is some or other property that plays the appropriate role. It may feel – and has felt to many philosophers – that the truth of T^R somehow magically endows the term P with meaning. In particular, philosophers are wont to say things like, "P is whatever it is that plays the role described by T^R."

Since Ramsey sentences draw upon the resources of second-order logic, the neophyte is left to wonder: does the philosophical magic here depend on something special that happens in second-order logic, something that only the most technically sophisticated philosophers can understand? We think that the answer to this question is no. In fact, Ramsifying a theory simply weakens that theory in the same way that existentially quantifying a first-order sentence weakens that sentence. Consider the following pedestrian example.

Example 8.1.1 Let $\Sigma = \{m\}$, where the name m is a theoretical term. Let T be the theory $\exists x(x = m)$ in Σ. Then the Ramsey sentence T^R of T is the sentence $\exists x(x = x)$, which is just a tautology. That is, T^R is the empty theory in the empty signature. It is easy to see that the inclusion $I : T^R \to T$ is conservative but not essentially surjective. In particular, there is no formula ϕ of Σ such that $(I\phi)(x) \equiv (x = m)$. The fact that I is not essentially surjective corresponds to the fact that $I^* : \text{Mod}(T) \to \text{Mod}(T^R)$ is not full. Here I^* is the functor that takes a model of T and forgets the extension of m. In general, then, I^*M has more symmetries than M.

We can be yet more precise about the differences between $\text{Mod}(T)$ and $\text{Mod}(T^R)$. In short, a model of T^R is simply a nonempty set X (and two such models are isomorphic if they have the same cardinality). For each $p \in X$, there is a corresponding model X_p of T where $X_p(m) = p$. For a fixed X, and $p, q \in X$, there is an isomorphism $h : X_p \to X_q$ that maps p to q. However, the automorphism group of X_p is smaller than the automorphism group of X. Indeed, $\text{Aut}(X_p)$ consists of all permutations of X that fix p, hence is isomorphic to $\text{Aut}(X\backslash\{p\})$.

We can see then that T and T^R are not intertranslatable (or definitionally equivalent). Nonetheless, there is a sense in which mathematicians would have no qualms about passing from T^R to the more structured theory T. Indeed, once we've established that the domain X is nonempty (which, of course, is a presupposition of first-order logic), we could say, "let m be one of the elements of X." This latter statement does not involve any further theoretical commitment over what T^R asserts. ⌟

Our advice then to the neophyte is not to allow herself to be intimidated by second-order quantification. In fact, we will argue that passage from a theory T to its

Ramsified version T^R either forgets too much of what the original theory said or says *more* than what the original theory said – depending on which notion of second-order logical equivalence one adopts. Before we do this, let's pause to recall just how much philosophical work Ramsey sentences have been asked to do. We will look at three applications. First, Carnap claims that Ramsey sentences solve the problem of dividing the analytic and synthetic parts of a scientific theory. Second, Lewis claims that Ramsey sentences solve the problem of theoretical terms and, in particular, the problem of giving meaning to "mentalese" in a physical world. Third, contemporary structural realists claim that Ramsey sentences give a way of isolating the structural claims of a scientific theory.

Carnap's Irenic Realism

One theme running throughout Carnap's work is a rejection of what he sees as false dilemmas. In one sense, Carnap is one of the most pragmatic philosophers ever in the Western tradition, as he places extreme emphasis on questions such as: what questions are worth asking, and what problems are worth working on? Now, one can imagine a philosophy graduate student asking herself: what question should I try to answer in my dissertation? If she's a particularly ambitious (or perhaps overconfident) student, she might decide to determine whether materialism or dualism is true. Or she might decide to determine whether scientific realism or instrumentalism is true. Carnap's advice to her would be to work on such questions is not a good use of your time.

In the early twentieth century, the debate between scientific realism and instrumentalism centered around the question: do theoretical entities – i.e., the things named by scientific theories, but which are not evident in our everyday experience – exist? Or, shifting to a more explicitly normative manner of speech: are we entitled to believe in the existence of these entities, and perhaps even obliged to do so? The realist says yes to these questions, and the instrumentalist says no. Carnap attempts to steer a middle way. He says that the questions are ill-posed.

Toward the end of his career, Carnap hoped that Ramsey sentences could help show why there is no real argument between realism and instrumentalism. In particular, if T is a scientific theory containing some theoretical terms $r_1, \ldots, r_n$, then Carnap parses T into two parts: the Ramsey sentence T^R and the sentence $T^R \to T$ that has since been dubbed the "Carnap sentence." Carnap claims that the Ramsey sentence T^R gives the empirical (synthetic) content of T, whereas $T^R \to T$ gives the definitional (analytic) part of T. The latter claim can be made plausible by realizing that $T^R \to T$ is trivially satisfiable, simply by stipulating appropriate extensions for $r_1, \ldots, r_n$.

Psillos (2000) argues that Carnap's equation of synthetic content with the Ramsey sentence makes him a structural realist – in which case he is subject to Newman's objection, which impales him on the horns of the realism–instrumentalism dilemma. Friedman (2011) disagrees, arguing that Carnap's invocation of the Ramsey sentence successfully implements his neutralist stance. Debate on this issue continues in the literature – see, e.g., Uebel (2011); Beni (2015).

Ramsey Sentence Functionalism

In the philosophy of mind, Ramsey sentences came to play a central role through the work of Lewis (1966, 1972, 1994) and, more generally, in a point of view known as **functionalism**. To be sure, Lewis claims not to know whether or not he is a functionalist, and most functionalists don't talk explicitly about Ramsey sentences. However, by the 1980s, the connection between functionalism and Ramsey had been firmly established (see Shoemaker, 1981).

Around 1970, materialist reductionism had gone out of style. Philosophers concluded that folk psychology cannot, and should not, be reduced – neither to descriptions of behavior nor to physiological descriptions. However, philosophers weren't ready to give up the physicalist project, and, in particular, they didn't want to entertain the possibility that there is an autonomous realm of mental objects or properties. The goal then is to explain how mental properties are anchored in physical properties, even if the former cannot be explicitly defined in terms of the latter.

Functionalism, and functional definitions, are supposed to provide a solution to this problem. According to functionalism, mental properties are defined by the role that they play in our total theory T, which involves both mental concepts (such as "belief" and "desire") and physical concepts (such as "smiling" or "synapse firing"). How then are we supposed to cash out this notion of being "defined by role"? It's here that Ramsey sentences are invoked as providing the best formal explication of functional definitions.

Contemporary analytic philosophers routinely mention Ramsey sentences in this connection. Nonetheless, long ago, Bealer (1978) argued that this attempt to define mental properties – call it "Ramsey sentence functionalism" – is inconsistent. According to Bealer, functionalism has both a negative and a positive theses. On the negative side, functionalism is committed to the non-reductionist thesis: mental properties (m-properties) cannot be explicitly defined in terms of physical properties (p-properties). On the positive side, m-properties *are* defined in terms of the role they play vis-à-vis each other and the p-properties.

Let T be a theory in signature $\Sigma \cup \{r_1, \dots, r_n\}$, where we think of Σ as p-vocabulary, and of $r_1, \dots, r_n$ as m-vocabulary. We then adopt the following proposal (which defenders of functionalism are welcome to reject or modify):

> T provides functional definitions of $r_1, \dots, r_n$ in terms of Σ just in case, in each model M of the Ramsey sentence T^R, there are unique realizing properties $M(r_1), \dots, M(r_n)$.

It's easy to see then that T provides functional definitions of $r_1, \dots, r_n$ in terms of Σ only if T implicitly defines $r_1, \dots, r_n$ in terms of Σ. Indeed, if M and N are models of T, then $M|_\Sigma$ and $N|_\Sigma$ are models of T^R, and it follows from the uniqueness clause that $M(r_i) = N(r_i)$. It then follows from Beth's theorem that T explicitly defines $r_1, \dots, r_n$ in terms of Σ.

Bealer's argument, if successful, shows that functionalism is inconsistent: the positive thesis of functionalism entails the negation of the negative thesis. Surprisingly, however, functionalism lives on, apparently oblivious of this little problem of inconsistency. In fact, functionalism hasn't just survived; it is flourishing and spreading its tendrils – indeed, it has become an overarching philosophical ideology: **the Canberra plan**. The goal of the Canberra plan is to find a place in the causal nexus of physical properties for all the stuff that makes up our daily lives – things like moral and aesthetic values, laws, society, love, etc. (For further discussion, see Menzies and Price [2009].)

Structural Realism

In more recent times, Ramsey sentences have been invoked in support of a trendy view in philosophy of science: structural realism. In the early 1990s, structural realism was the new kid on the block in discussions of scientific realism and antirealism. As forcefully recounted by Worrall (1989), there are good arguments against both scientific realism and scientific antirealism. Against scientific realism, there is the *pessimistic metainduction*, which points to the long history of failed scientific theories as evidence that our current favorite scientific theories will probably also fail. Against scientific antirealism, there is the *no miracles argument*, which points to the success of scientific theories as something crying out for an explanation. In good Hegelian fashion, Worrall seeks a synthesis of the extremes of realism and antirealism – a position that offers the best of both worlds. His proposal is structural realism, according to which the part of a theory to take seriously is its pronouncements on issues of *structure*.

Worrall illustrates the idea of "preserved structure" with a specific example. In particular, before Einstein's special theory of relativity, it was thought that there was a substance, the "aether," in which electric and magentic waves propagated. After the Michelson–Morley experiment and the success of special relativity, there was no longer any use for the aether. Thus, the transition to special relativity might be taken to be a particularly clear example of failed reference – showing, in particular, that pre-Einsteinian physicists ought not to have taken their theory so seriously.

Nonetheless, says Worrall, it would have been a mistake for pre-Einsteinian physicists to treat their theory instrumentally, i.e., merely as a tool for making predictions. For the form of the equations of motion was preserved through the transition to special relativity – hence, they would have done well to trust their equations. The general lesson, says Worrall, is to trust your theory's structure but not the underlying stuff it purports to be talking about.

Worrall's example is highly suggestive, and we might like to apply it in a forward-looking direction. In particular, take one of our current-day successful scientific theories T, such as quantum mechanics. The pessimistic metainduction suggests that T will be wrong about something. But can we already make an educated guess about which parts of T will be preserved and which part will go on the scrap heap with other rejected theories?

Worrall and Zahar (2001), Cruse and Papineau (2002), and Zahar (2004) provide a specific proposal for picking out the structural commitments of a theory T: they are given by its Ramsey sentence T^R. This idea certainly has some intuitive appeal – trading on an analogy to coordinate-free descriptions of space. For a naive or straightforward description of physical space, we might use triples of real numbers, i.e., the mathematical space $\mathbb{R}^3$. But now our description of space has superfluous structure. In particular, we assigned the origin $0 \in \mathbb{R}^3$ to some particular point in space – but we didn't mean to indicate that the denoted point is any different than any other point in space. Thus, our description breaks the natural symmetry of space, and it would be natural to look for another description that respects these symmetries. Indeed, that's precisely the idea behind the move from using vector spaces to using affine spaces to describe space.

Now, just as a vector-space description of space breaks its symmetry, so our theoretical descriptions in general might fail to respect the symmetry between properties. For example, we didn't need to use the word "electron" to describe those things that are found in the energy shells around an atom's nucleus – we could simply say that something or other plays the relevant role. And that's exactly what the Ramsified theory says. Thus, it might seem that T^R provides a more intrinsic description than the original theory T.

Nonetheless, the intuitive appeal of Ramsey sentences fades quickly in the light of critical scrutiny. Most famously, already in 1928, Newman argued that Bertrand Russell's structuralism trivializes, for these structural claims are true whenever their observational consequences are true (see Newman, 1928). The so-called Newman objection to structural realism has been the centerpiece of recent debates about Ramsey-sentence structuralism. But even before we get to that level of scrutiny, there is something quite strange in the idea of passing to the Ramsified theory T^R to get rid of redundancy. Let's recall that a formal theory T doesn't actually refer to things like electrons or protons – it's formulated in an uninterpreted calculus. Hence, T doesn't actually have any referring terms.

It seems that the impulse to Ramsify is no other than the original impulse to use uninterpreted mathematical symbols to represent physical reality. You'll recall that one of the key maneuvers in the development of non-Euclidean geometries was de-interpreting words like "line," thereby liberating mathematicians to focus attention solely on the relation that "line" plays relative to other (uninterpreted) terms in their formal calculus.

In any case, what's really at stake here is the question of what attitude we should take toward the best scientific theories of our day and age. At one extreme, radical scientific realists assert that we should give nothing less than *full* assent to these theories, interpreted literally. To draw an analogy (that scientific realists will surely eschew), the extreme scientific realist is akin to the radical religious fundamentalist, and in particular to those fundamentalists who say that one must interpret scriptures literally. The point of that injunction, we all know, is to enable religious leaders to foist their opinions on others. At the opposite extreme, an extreme scientific antirealist sees science as having no epistemic authority whatsoever – i.e., a successful scientific theory doesn't call for any more epistemic attention on our part than, say, Zoroastrianism.

In the light of this somewhat hyperbolic characterization of the anti/realism debate, we can see various alternative positions as granting a selective epistemic authority to successful scientific theories. Consider an analogy: suppose that you know a highly skilled car mechanic, Jacob. You completely trust Jacob when it comes to his opinions on automobile-related issues. For example, if he says that you need a new alternator, then you won't doubt him, even if it costs you a lot of money. Nonetheless, if Jacob tells you that you need a new kidney, or that you should vote for a certain candidate, you might well ignore his opinion – since he's speaking on a topic that lies outside his proper expertise.

Now, selective scientific realists consider successful scientific theories to be epistemically authoritative, but only when they speak on topics within their expertise. The different brands of selective realism are distinguished by how they understand the expertise of science. For example, a constructive empiricist (such as van Fraassen) trusts a successful scientific theory T when it makes predictions about empirical phenomena (presupposing, as he does, that it makes sense to speak of predictions and empirical phenomena – precisely the point to which Boyd and Putnam object). Similarly, a structural realist (such as Worrall) trusts a successful scientific theory T on its structural pronouncements. But if T says something about things in themselves (or whatever is *not* structure), then the structural realist treats it as no more of an authority than your auto mechanic is on politics.

The previous considerations suggest that varieties of selective scientific realism can be classified by means of different notions of theoretical equivalence. For example, the strict empiricist thinks that the important part of a theory is its empirical content; and hence, if two theories T_1 and T_2 agree on empirical content, then there is no epistemically relevant difference between them. Similarly, a structural realist thinks that the important part of a theory is its pronouncements about structure; and hence, if two theories T_1 and T_2 agree on structure, then there is no epistemically relevant difference between them. In the particular case of Ramsey-sentence structuralism, the structural pronouncements of a theory T_i are captured by its Ramsey sentence T_i^R. Hence, if $T_1^R \equiv T_2^R$, then there is no epistemically relevant difference between T_1 and T_2.

Unfortunately, the statement "$T_1^R \equiv T_2^R$" doesn't have an obvious meaning, since there is no single, obviously correct notion of second-order logical consequence. What this means is that we get different notions of "same structure" depending on which notion of second-order consequence we adopt. Let's review, then, some salient notions of second-order logical consequence.

Second-order logic is a complicated subject in its own right, and has been the source of much dispute among analytic philosophers. We refer the reader to studies such as Shapiro (1991) and Bueno (2010) for more details. For present purposes, it will suffice to make some minor modifications of first-order logical grammar: first, we add a list of second-order variables $X, Y, \ldots$ Each second-order variable has a specific arity $n \in \mathbb{N}$, which means that it can stand in the place of an n-ary relation symbol. We then permit formulas such as $X(x_1, \ldots, x_n)$, with a second-order variable of arity n applied to n first-order variables. We also add an existential quantifier $\exists X$ that can be applied to quantify over second-order variables.

Now there are two important facts to keep in mind about second-order logic. The first fact to keep in mind is that second-order logic has is intrinsically incomplete – hence there is no tractable syntactic relation "$\vdash$" of second-order provability. The second fact to keep in mind is that there are several candidates for the semantic relation "$\vDash$" of entailment. Depending on which choice we make for this relation, we will get a different notion of logical equivalence.

DEFINITION 8.1.2 A second-order Σ-frame $\mathscr{F} = (M, (\mathscr{E})_{n \in \mathbb{N}})$ consists of a first order Σ-structure M and, for each $n \in \mathbb{N}$, a subset $\mathscr{E}_n$ of $\mathscr{P}(M^n)$. We let $\mathscr{E}^{\mathscr{F}} = \bigcup_{n \in \mathbb{N}} \mathscr{E}_n$. Here the sets in $\mathscr{E}^{\mathscr{F}}$ will give the domain of the second-order quantifiers in frame $\mathscr{F}$.

In order to define the relation $\vDash$, we will also make use of the notion of a variable assignment. Given a Σ-frame $\mathscr{F}$, a first-order variable assignment g assigns each variable x to an element $g(x) \in M$. A second-order variable assignment G assigns each variable X of arity n an element $G(X) \in \mathscr{E}_n$. We then define
$M[G, g] \vDash \exists X \phi$ iff for some $E \in \mathscr{E}_n$, $M[G_X^E, g] \vDash \phi$, where G_X^E is the second-order variable assignment that agrees with G on everything besides X, which it assigns to E.

Now to define the relation $\vDash$ between sentences, we have to decide which second-order Σ-frames to quantify over. We get three different notions, depending on the family we choose:

1. For **full semantics**, we permit only those Σ-frames in which $\mathscr{E}_n = \mathscr{P}(M^n)$.
2. For **Henkin semantics**, we permit all Σ-frames in which $\mathscr{E}_n$ is closed under first-order definability.
3. For **frame semantics**, we permit all Σ-frames.

Recall that the more structures there are, the more counterexamples and, hence, the fewer implications. Accordingly, full semantics has more entailments than Henkin semantics, and Henkin semantics has more entailments than frame semantics. Hence, full semantics yields a more liberal notion of equivalence than Henkin semantics, which yields a more liberal notion of equivalence than frame semantics.

In the following discussion, we will take T_i, for $i = 1, 2$, as a theory in signature $\Sigma \cup \Sigma_i$, where Σ_i is disjoint from Σ. We let T_i^* be the result of replacing terms in Σ_i with (possibly second-order) variables, and we let T_i^R be the corresponding Ramsey sentence of T_i. We now give a general schema for Ramsey equivalence of theories.

DEFINITION 8.1.3 Two theories T_1 and T_2 are **Ramsey equivalent** if T_1^R is logically equivalent to T_2^R.

The three choices of frames discussed earlier give rise to three notions of Ramsey equivalence.

- RE_1 = loose Ramsey equivalence = Ramsey sentences are equivalent relative to full semantics.
- RE_2 = moderate Ramsey equivalence = Ramsey sentences are equivalent relative to Henkin semantics.
- RE_3 = strict Ramsey equivalence = Ramsey sentences are equivalent relative to frame semantics.

Obviously, then, we have $RE_3 \Rightarrow RE_2 \Rightarrow RE_1$.

We can now give a sharpened formulation of the Newman problem – in the spirit of Ketland (2004) and Dewar (2019). Recall that on the old-fashioned syntactic view of theories, two theories T_1 and T_2 are considered to be empirically equivalent if they have the same consequences in the observation language. If we now think of Σ as the observation vocabulary, then we could formulate this criterion as saying that $\mathrm{Cn}(T_1)|_\Sigma = \mathrm{Cn}(T_2)|_\Sigma$, where $\mathrm{Cn}(T_i)|_\Sigma$ indicates the restriction of the set of consequences to those that contain only observation terms.

One might also wish to formulate a more semantically oriented notion of empirical equivalence. For example, we might say that two theories T_1 and T_2 are empirically equivalent if their models agree on Σ-structure.

DEFINITION 8.1.4 We say that T_1 and T_2 are Σ-equivalent just in case, for each model M of T_1, there is a model N of T_2 and an isomorphism $h : M|_\Sigma \to N|_\Sigma$, and vice versa.

The following result shows that this semantic notion of empirical equivalence implies the syntactic notion.

PROPOSITION 8.1.5 *If T_1 and T_2 are Σ-equivalent, then* $\mathrm{Cn}(T_1)|_\Sigma = \mathrm{Cn}(T_2)|_\Sigma$.

Proof Suppose that T_1 and T_2 are Σ-equivalent. Let ϕ be a Σ-sentence such that $\phi \notin \mathrm{Cn}(T_2)$. By completeness, there is a model M of T_2 such that $M \not\models \phi$. Since T_1 and T_2 are Σ-equivalent, there is a model N of T_1 and an isomorphism $h : M|_\Sigma \to N|_\Sigma$. But then $N \not\models \phi$, hence $\phi \notin \mathrm{Cn}(T_1)$. It follows that $\mathrm{Cn}(T_1)|_\Sigma \subseteq \mathrm{Cn}(T_2)|_\Sigma$. The result follows by symmetry. □

However, this implication cannot be reversed – i.e., the syntactic notion of empirical equivalence doesn't imply the semantic notion.

Example 8.1.6 Let Σ be the empty signature (with equality). Let $\Sigma_1 = \{c_r \mid r \in \mathbb{R}\}$, and let T_1 be the theory in $\Sigma \cup \Sigma_1$ with axioms $c_r \neq c_s$, for all $r \neq s$. Let T_2 be the theory in Σ that says there are infinitely many things. Then $\mathrm{Cn}(T_1)|_\Sigma = \mathrm{Cn}(T_2)|_\Sigma$. However, T_2 has a countable model M, and T_1 has no countable model. Therefore, T_1 and T_2 are not Σ-equivalent. ⌟

The Newman problem for structural realism is usually phrased as saying that it's too easy for a theory's Ramsey sentence to be true – that the Ramsey sentence is "trivially realizable." We can make precise what is meant here by "too easy" in terms of the notion of theoretical equivalence. In short, Ramsey equivalence – i.e., having logically equivalent Ramsey sentences – is too liberal a notion of equivalence. In particular, empirically equivalent theories are Ramsey equivalent.

PROPOSITION 8.1.7 (Dewar) *If T_1 and T_2 are Σ-equivalent, then T_1^R and T_2^R are logically equivalent relative to full semantics.*

Proof Suppose that T_1 and T_2 are Σ-equivalent. Now let $\mathscr{F}$ be a full Σ-frame such that $\mathscr{F} \models T_1^R$. Thus, there is a second-order variable assignment G such that $\mathscr{F}[G] \models T_1^*$. Let M be the $\Sigma \cup \Sigma_1$ structure obtained by assigning $M(R) = G(X)$, where X is the variable in T_1^* that replaces R in T_1. Clearly M is a model of T_1. Since T_1 and T_2 are Σ-equivalent, M is Σ-isomorphic to a model N of T_2. This model N of T_2 defines a second-order variable assignment G' such that $\mathscr{F}[G'] \models T_2^*$, and hence $\mathscr{F} \models T_2^R$. □

The notion of empirical equivalence imposes no constraints whatsoever on what the theories T_1 and T_2 say in their theoretical vocabulary – and for this reason, nobody but the most extreme empiricist should adopt weak Ramsey equivalence as their standard.

Moving back toward the right-wing side of the spectrum of theoretical equivalence, one might hope that moderate Ramsey equivalence would provide a more reasonable standard. But the following result shows that any two mutually interpretable theories satisfy RE_2.

PROPOSITION 8.1.8 (Dewar) *If T_1 and T_2 are Σ-equivalent and mutually interpretable, then T_1^R and T_2^R are logically equivalent relative to Henkin semantics.*

Proof Suppose that T_i is a theory in $\Sigma \cup \Sigma_i$. We will show that if $F : T_1 \to T_2$ is a translation (which is the identity on Σ), then $T_2^R \vDash T_1^R$, where the $\vDash$ symbol is entailment relative to Henkin semantics, and T_i^R is the result of Ramsefying out Σ_i. Suppose then that $F : T_1 \to T_2$ is a translation and that $\mathscr{H}$ is a Henkin structure (of signature Σ) such that $\mathscr{H} \vDash T_2^R$. Thus, $\mathscr{H}[G] \vDash T_2^*$ relative to some second-order variable assignment G. Consider then the first-order structure M for signature $\Sigma \cup \Sigma_1$ that agrees with $\mathscr{H}$ on Σ, and such that $M(P) = G(X_P)$, for each $P \in \Sigma_2$, where X_P is the second-order variable that replaces P in T_1^*. It is clear then that $M \vDash T_2$. Now we will use the fact that the translation $F : T_1 \to T_2$ gives rise to a functor $F^* : \text{Mod}(T_2) \to \text{Mod}(T_1)$ (6.6.5). In particular, $(F^*M)(Q) = M(F(Q))$ for each relation symbol $Q \in \Sigma \cup \Sigma_1$. Now define a second-order variable assignment G' by setting

$$G'(X_Q) = (F^*M)(Q) = M(F(Q)),$$

for each variable X_Q that occurs in T_1^*. (Again, we use X_Q to denote the variable that replaces a relation symbol Q that occurs in T_1.) To see that G' is a Henkin-admissible assignment, note that $F(Q)$ is a Σ_2-formula, and so $M(F(Q))$ is a first-order definable subset of M. By construction, each first-order definable subset of M is an element of $\mathscr{E}^{\mathscr{H}}$. Now, it's clear that $\mathscr{H}[G'] \vDash T_1^*$, and hence that $\mathscr{H} \vDash T_1^R$. Since $\mathscr{H}$ was an arbitrary Henkin frame, it follows that $T_2^R \vDash T_1^R$. By symmetry, if there is a translation $G : T_2 \to T_1$, then $T_1^R \vDash T_2^R$. Therefore, if T_1 and T_2 are mutually interpretable, then T_1^R and T_2^R are Henkin equivalent. □

There is one last hope for the Ramsefier: that strict Ramsey equivalence (RE_3) will provide the right notion of structural equivalence. Unfortunately, RE_3 proves to be the worst candidate for structuralism, since intertranslatable theories need not satisfy RE_3.

Example 8.1.9 Let $\Sigma_1 = \{r\}$, and let $\Sigma_2 = \{r'\}$, where both r and r' are unary predicates. Let $T_1 = \{\exists x r(x)\}$, and let $T_2 = \{\exists x \neg r'(x)\}$. The reconstrual $F(r) = \neg r'(x)$ induces a homotopy equivalence between T_1 and T_2 – i.e., T_1 and T_2 are intertranslatable. However, the Ramsey sentences of T_1 and T_2 are not frame equivalent. In particular, consider any frame $\mathscr{F}$ with first-order domain M, and $\mathscr{E}_1^{\mathscr{F}} = \{M\}$ – i.e., M is the only admissible subset of M. Then $\mathscr{F} \vDash T_1^R$ but $\mathscr{F} \nvDash T_2^R$. ⌟

Since strict Ramsey equivalence (RE_3) is more conservative ("right wing") than definitional equivalence, we don't expect structural realists to find it congenial. But what about those hard-core realists – like David Lewis or Ted Sider – who pin their theoretical hopes on natural properties and reference magnetism? Might they actually want a criterion of equivalence that is even more conservative than definitional equivalence? In fact, it seems that frame semantics might be a good way to capture the idea that to describe a possible world, you need to say not only what things exist, but also what the natural properties are. We should note, however, that adopting a first-order signature Σ already goes some way to picking out natural properties. When we specify a Σ-structure M, we get a natural property $M(\phi)$ for each formula ϕ of Σ. It's not clear then why a theorist who has adopted a first-order signature Σ would need to additionally specify a notion of natural properties.

The previous results can be summarized in the following diagram:

$$\begin{array}{ccccc} EE & \Longleftarrow & MI & \Longleftarrow & IT \\ \Downarrow & & \Downarrow & & \not\searrow \\ RE_1 & \Longleftarrow & RE_2 & \Longleftarrow & RE_3 \end{array}$$

Here "EE" is empirical equivalence (explicated semantically), "MI" is mutual interpretability over Σ, and "IT" is intertranslatability over Σ, which is equivalent to definitional equivalence. It appears then that none of the notions of Ramsey equivalence gets us near the promising area in the neighborhood of intertranslatability. Most philosophers, we think, would agree that intertranslatability is a reasonable – if somewhat strict – explication of the idea that two theories have the same logical structure.

8.2 Counting Possibilities

If you page through an analytic philosophy journal, it won't be long before you see the phrase "possible world." Many philosophical discussions focus on this concept, and it is frequently used as a basis from which to explicate other concepts – Humean supervenience, counterfactuals, laws of nature, determinism, physicalism, content, knowledge, etc. When the logically cautious philosopher encounters this concept, she will want to know what rules govern its use. Where things get really tricky is when philosophers start invoking facts about the structure of the space of possible worlds – e.g., how many worlds there are, which worlds are similar, and which worlds are identical. These sorts of assumptions play a significant role in discussions of fundamental ontology. To take a paradigm example, Baker (2010) argues that if two models are isomorphic, then they represent the same possible world.

Analytic philosophers might be the primary users of the phrase "possible world," but they aren't the only ones using the concept. Scientists talk about possible worlds all the time. However, at least in the exact sciences, there are explicit rules governing the use of possible-worlds talk. Indeed, these rules are built into the structure of their theories and, more particularly, in the structure of those theories' spaces of models. Following Belot

(2017), we think that philosophers ought to try to understand the way that scientists' theories guide their use of modal concepts.

Nonetheless, it's not hard to find philosophers scratching their heads and asking themselves questions like the following:

($\star$) Consider two general relativistic spacetimes, M and N, and suppose that $h : M \to N$ is an isomorphism (e.g., a metric preserving diffeomorphism). Do M and N represent the same possible world?

($*$) Consider two Newtonian spacetimes, M and N, and suppose that $h : M \to N$ is an isomorphism (e.g., a shift). Do M and N represent the same possible world?

Belot (2017) helpfully classifies philosophers into two groups according to how they answer these questions: the *shiftless* claim that isomorphisms do not generate new possibilities, and the *shifty* claim that isomorphisms do generate new possibilities. In particular, the shiftless philosopher says that if $h : M \to N$ is an isomorphism, then M and N represent the same possibility. In contrast, the shifty philosopher allows that M and N might represent different possibilities, even though they are isomorphic. While the majority of philosophers of physics and metaphysicians have become shiftless, Belot champions the heterodox, shifty point of view. As we will now argue, all parties to the dispute have adopted a questionable presupposition, viz. that it makes sense to count possibilia.

But first, what hangs on this dispute between the shifty and the shiftless? In the first place, shiftless philosophers believe that they are on the right side of history, ontologically speaking. In particular, they believe that it would be wrong to countenance the existence of two possibilities, represented by M and N, when a single one will do the job. This way of thinking trades on vague associations with Leibniz's principle of the identity of indiscernibles: since M and N are indiscernible, there is no reason to regard them as different. Belot points out, however, that shiftless philosophers have trouble making sense of how theories can guide the use of modal concepts. In particular, he argues that the shiftless view is in danger of collapsing the distinction between deterministic and indeterministic theories.

One is tempted immediately to dismiss the shiftless position, because it patently conflicts with the standard reading of physical theories. Take, for example, a Galilean spacetime M, and let $\gamma : \mathbb{R} \to M$ be an inertial world line in M. Now, a boost $x \mapsto x + vt$ for some fixed $v > 0$ is represented by an isomorphism $h : M \to M$. Does this boost generate a new possibility? The question might seem confusing because the model on the right side of $h : M \to M$ is the same as the model on the left side. It might seem to be trivially true, then, that $h : M \to M$ does *not* generate a new possibility. But let's see what happens if we adopt the shiftless view. If h does *not* generate a new possibility, then we ought to say, of a particle in inertial motion that it could *not* be in some other state of inertial motion (because there is no other such state of inertial motion). But that claim is contrary to the way that physicists use this theory to guide their modal reasoning. When a physicist adopts Galilean relativity, she commits to the claim that there are many distinct possible states of inertial motion, and that a thing that is in one state of inertial motion *could be* in some other state of inertial motion. In other words, it matters to physicists that the isomorphism $h : M \to M$ is not the identity isomorphism and, in particular, that the world line $h \circ \gamma$ is not the same as the world line γ. Nonetheless, shiftless philosophers can't make sense of these modal claims, because they insist that isomorphisms don't generate new possibilities.

Despite the implausibility of the shiftless view, there are some very serious and smart philosophers who defend it. What is it, then, that really drives their insistence on saying that isomorphism (at the level of representations) implies identity (at the level of the represented)? We suspect that the shiftless are fumbling their way toward an insight – but an insight that is difficult to articulate when one is operating with mistaken views about mathematical objects and, in particular, about the relation between abstract and concrete objects. We blame a lot of this confusion on Quine, who decided that we have no need for the abstract–concrete distinction – in particular, that belief in the existence of abstracta is no different in principle from belief in the existence of concreta.

At risk of oversimplifying, we will first give a simple formulation of the basic insight toward which we think the shiftless philosophers are fumbling:

(†) A theory T is indifferent to the question of the identity of its models. In other words, if M and N are models of T, then T neither says that $M = N$ nor that $M \neq N$. The only question T understands is: are these models isomorphic or not?

Now, please don't get us wrong: (†) does not say that isomorphic models are identical, nor does it say that the theory T treats isomorphic models as if they were identical. No, from the point of view of T, the question, "are they identical?" simply does not make sense. According to this thesis, claims of identity, or nonidentity of models, play *no* explanatory role in the theory.

We realize that this thesis is controversial and that it might take some time for philosophers to become comfortable with it. The problem is that we learned a little bit of set theory in our young years, and we seem to assume that everything lives in a world of sets – where questions of the form "is M equal to N" always have a definite answer. Indeed, the rigid grip of set theory makes philosophers profoundly uncomfortable with contemporary mathematics, which likes to play a fast and loose game with identity conditions. Consider a simple example (due to John Burgess): suppose that we ask two different mathematicians two different questions:

(Q1) How many groups are there with two elements?

(Q2) Inside the group $\mathbb{Z}_2 \oplus \mathbb{Z}_2$, how many subgroups are there with two elements?

What we are likely to find is that mathematicians will give apparently conflicting answers. On the one hand, they will tell us that there is only one group with two elements. On the other hand, they will tell us that $\mathbb{Z}_2 \oplus \mathbb{Z}_2$ has two distinct subgroups with two elements. Obviously, if taken literally, these two answers contradict each other. But there is no genuine conflict, and mathematicians are not in crisis about the number of groups with two elements. No, the fact is mathematicians use words and symbols in a different way than we use them in everyday life – e.g., when we count the number of apples in a basket.

To reinforce this point, recall that categorical equivalence doesn't respect the number of objects in a category. Consider, for example, the following two categories: let **C** be the category with one object and one identity morphism. Let **D** be the category with two objects a, b, one identity morphism from each object to itself, and a pair of morphisms $f : a \to b$ and $g : b \to a$ that are inverse to each other. Then **C** and **D** are equivalent categories – which entails that "this category" doesn't really have a definite number of objects. It is not correct to say that it has one object, and it's not correct to say that it has two. Or, perhaps better: it is just as correct to say that it has one object as it is to say that it has two.

Here, then, is our positive proposal:

> For the purposes of interpreting a theory T, the collection $\text{Mod}(T)$ of its set-theoretic models should be treated as nothing more nor less than a *category*. In particular, the philosopher of science shouldn't say things about $\text{Mod}(T)$ that are not invariant under categorical equivalence, nor should they argue over questions – such as "how many models does T have?" – whose answer is not invariant under categorical equivalence.

If this proposal is adopted, then there is no debate to be had between the shifty and the shiftless. The question they are asking – do isomorphisms generate new possibilities? – depends on a notion (the number of isomorphic possibilities) that is not invariant under categorical equivalence.

The rationale for this proposal is our belief that models of a theory T in **Sets** are *representations* of that theory; the set-theoretic description of these models is *not* itself a further theory that attempts to describe the world at an even finer-grained level of detail than was done by T. We can further clarify these points by means of a simple example.

Example 8.2.1 Suppose that Berit is a scientist with a very simple theory. Her language Σ has a single predicate symbol P, and her theory T says that there are exactly two things, one of which is a P:

$$\exists x \exists y (P(x) \wedge \neg P(y) \wedge \forall z ((z = x) \vee (z = y))).$$

Now we metatheorists know that a set-theoretic model M of T consists of a two-element set, say $X = \{a, b\}$, with a singleton set $M(P)$. Let M be the model such that $M(P) = \{a\}$, and let N be the model such that $N(P) = \{b\}$. Then the permutation $h(a) = b, h(b) = a$ gives a Σ-isomorphism $h : M \to N$. (But the permutation h is not an automorphism of M.)

Let's consider the shifty–shiftless dilemma with regard to the models M and N, with the isomorphism $h : M \to N$. The shifty philosopher (e.g., Belot) says that M and N represent distinct possibilities. The shiftless philosopher (e.g., Baker) says that M and N represent the same possibility. Who is on the side of truth?

In our opinion, both the shifty and the shiftless say misleading things about this example. On the one hand, the shifty claim is misleading, because the user of T doesn't have the language to say what would be different between M and N. She cannot say, "in

M, a is P, and in N, a is not P," because she herself doesn't have the name "a." The shiftless wants us to start counting how many models there are, but the theory T doesn't answer that question.

On the other hand, the shiftless would insist that there is only one possibility, represented redundantly by M and N. But that claim is misleading for the following reason. Berit's theory T is an extension of the theory T_0, in empty signature, that says there are exactly two things. Let $I : T_0 \to T$ be the translation of T_0 into T, and let $I^* : \text{Mod}(T) \to \text{Mod}(T_0)$ be the functor that forgets the assignment of P. Here I^*M and I^*M are both the bare two-point set X, and the isomorphism $I^*h = h : X \to X$ is the nontrivial permutation. Recall, though, that functors map identity morphisms to identity morphisms. Hence, if the isomorphism $h : M \to N$ is considered to be an identity (as the shiftless seem to do), then it would follow that I^*h is the identity morphism. Thus, contra the shiftless, we cannot identify M and N and forget that there was a nonidentity isomorphism $h : M \to N$. If we do that, then we won't be able to see how the theory T is related to the theory T_0.

The confusion here is somewhat similar to Skølem's paradox (about the existence of uncountable sets in models of ZF set theory), where we run into trouble if we don't distinguish between claims made in the object language and claims made in the metalanguage. In the present case, one might be tempted to think of the theory T as saying things such as

In model M, a is a P.

Of course, T says no such thing, since it doesn't have names for models or for elements in models.

The other problem here is in the way that we've set up the problem – by speaking as if the representation relation holds between M (or N) and the world. To the contrary, the representation relation holds between Berit's language and the world, and we (the metatheorists) are representing Berit's theorizing using our own little toy theory (which presumably includes some fragment of set theory, because that's a convenient way to talk about collections of formulas, etc.). Berit herself doesn't claim that M (or N) represents the world – rather, the metatheorist claims that M and N represent ways that Berit's language could represent the world. Accordingly, Berit doesn't claim that $M = N$, or that $M \neq N$; those are metatheoretical assertions – and do not add to the stock of knowledge about the world. ⌟

Before proceeding, we should deal with an obvious objection to the view we've put forward. Some philosophers will point out that it is simply false to say that physicists don't count the number of possibilities. Indeed, it's precisely by counting the number of possibilities that physicists derive notions such as entropy.

We do not disagree with this point, but it doesn't conflict at all with our positive proposal (to talk about models of a theory as a category). Category theory is a framework that is almost infinitely flexible: what we can talk about in a categorically invariant way depends on how we – or physicists – define the relevant category. For the case at hand, if X is a classical phase space, then it is assumed that X is a discrete category – i.e., that there are no nontrivial isomorphisms between elements of X. Thus, in this case, there is no question about whether to count two isomorphic possibilities as the same, because we (or better, the physicists) have chosen not to admit isomorphic possibilities.

To be clear, we explicitly reject the idea that there is a single relation "being isomorphic" that either holds or does not hold between concrete objects. On the contrary, the notion of isomorphism applies to abstractions, and different notions of isomorphism are valid for different levels of abstraction. It's up to us to decide which level of abstraction serves our purposes in reasoning about concrete, physical reality. (In particular, models of a theory are not concrete realities, and that's why they cannot either be identical or nonidentical.)

For all of its other virtues, one of the defects of the semantic view of theories is that it obscures the object language-metalanguage distinction, a distinction that is absolutely necessary to make sense of the notion of symmetry of representations. To be more accurate, the targets of this criticism are advocates of the "language-free" or "semantic-L" view (see Halvorson, 2013). The picture we get from the language-free semantic view is that mathematical structures are out there in the world, and that they are either isomorphic to each other or they are not. Of course, that picture completely ignores the fact that isomorphisms are defined in terms of language or, to put it more accurately, that isomorphisms relate mappings $M : \Sigma \to \mathbf{Sets}$ and $N : \Sigma \to \mathbf{Sets}$, which have a common domain Σ. Thus, in particular, arbitrary mathematical structures are neither isomorphic nor non-isomorphic.

The object language Σ serves as the reference point in defining a notion of symmetry. The object language tells us what must be held fixed, and the metalanguage tells us what can be varied. In particular, a model M of a theory T can have a nontrivial automorphism group because of two features of the formal setup:

1. The metalanguage describes the world in finer-grained language than the object language.
2. Distinctions that are not made by the object language are not significant for the kinds of explanations that the theory T gives.

If we drop either one of those components, then we will most likely make a hash of the notion of symmetry. Without the metalanguage, there is no way to see any difference between a and b, and so no way to express the change the occurs in the permutation $a \mapsto b$. But if we think of the metalanguage as a better object language, then we shouldn't count $a \mapsto b$ as a symmetry, since these two things are distinguishable in the metalanguage. Thus, it's precisely the mismatch between object language and metalanguage that provides us with a rich notion of symmetry; and, conversely, the importance of the notion of symmetry gives us reason to maintain a distinction between object language and metalanguage.

The distinction between object language and metalanguage is one of the most interesting ideas in twentieth century logic and philosophy – and it remains one of the least well understood. Obviously, Carnap made a lot of this distinction, and, in fact, he seems to use it as his primary analogy in formulating the distinction between internal and external questions and, more generally, in understanding the relationship between theories in the exact sciences and our other, nonscientific beliefs and attitudes. In contrast, Quine seems to reject the idea that there is an important difference of status between object and metalanguage. He seems to propose, instead, that the ascent to metalanguage should be seen as an extension of one's object language – and so assertions in the metalanguage have exactly the same force as assertions in the object language.

8.3 Putnam's Paradox

Perhaps the most notorious argument from logical metatheory to philosophy is Hilary Putnam's model-theoretic argument against realism (Putnam, 1977, 1980). Here is how the argument goes.

> Suppose that theory T is consistent, i.e., T does not imply $\perp$, or equivalently, T has a model. Now let W represent the collection of all actually existing objects, i.e., W represents "the world." Besides consistency, we will make two other minimal mathematical assumptions about T: First, the cardinality of the language is not so large as to force belief in the existence of too many objects. In short, we require that $|\Sigma| \leq |W|$. Second, the theory T doesn't entail that there are at most n things, for $n \in \mathbb{N}$.
>
> We then proceed as follows: by the Löwenheim–Skølem theorem, there is a model M of T such that $|M| = |W|$. This means, of course, that there is a bijection $f : M \to W$. Now we define another model of T, still called W, by setting $W(p) = f(M(p))$ for each relation symbol p in the theory T. But then the the world is a model of T. That is, T is true.

This argument is intended to show that if T is consistent, then T is true – actually true, in the real world. There is one obvious way to try to block this argument, and that's to say that the model W may not be the "intended" assignment of relation, function, and constant symbols to things in the real world. However, Putnam tries to block that response essentially by calling upon your charity. Imagine that T is the theory held by some other person, and that you're going to try your best to believe that what that person says is true. In other words, you are going to give her the benefit of the doubt whenever possible. Then what Putnam has shown is essentially that there is a way of giving her the benefit of the doubt.

This simple-looking argument is so subtle, and there are many ways we might respond to it. But let me be completely clear about my view of this argument: it is absurd. This version of Putnam's argument is not merely an argument for antirealism, or internal realism, or something like that. This version of the argument would prove that all consistent theories should be treated as equivalent: there is no reason to choose one over the other. Thus, Putnam's paradox is essentially an argument for one of the most radically liberal views of theoretical equivalence imaginable. The only more radical view is the Zenonian view, according to which all theories are equivalent.

To keep things concrete, let's suppose that T is Mette's theory. The goal of Putnam's argument is to show that Mette's theory is true. In my view, the problematic assumption in the argument is the following:

(S) The world can be described as an object W in the universe of **Sets**.

The question to be raised here is: *who* is using the theory of sets to describe the world? Putnam's presentation makes it seem that either: (1) it's unproblematic and theory-neutral to describe the world as a set, or (2) a realist must describe the world as a set. We don't agree with either claim.

Let's remember that nobody here – including Putnam – is free from language and theory. When Putnam describes the world as a set, it might seem that he is making minimal assumptions about it. But the opposite is true. When you have a set, you have all of its subsets; and when you have two sets, you have all of the functions between them. To even say these things, we need the rich and expressive language of set theory.

Thus, Putnam has set things up in a misleading way by (1) describing the world as a set but (2) failing to note who is responsible for this description of the world as a set. Suddenly it becomes clear why Putnam's argument goes through, and why it's trivial. Putnam assumes that Mette's theory T is set-theoretically consistent, which simply means that Mette's theory can be translated into the background theory T_0 that was used to describe the world. That is, there is a translation $F : T \to T_0$. Putnam rightly concludes that the T_0-theorist could take Mette's theory T to be true. What Putnam does not show is that *anybody*, regardless of their background theory, could take Mette's theory to be true.

Putnam's argument should actually not make any assumption about W – i.e., it should be like a black box. However, Putnam begins by assuming that there is already a fixed interpretation of ZF into the world – i.e., we know what objects are, and collections of objects, and functions between objects, etc. He then asks whether T has one (or perhaps even many) interpretations into this already understood domain. And of course, the answer is yes.

Thus, Putnam assumes that he is permitted a trans-theoretical language to speak of the domains M and W. By "trans-theoretical" here, I mean simply that the language of T_0 (in this case, ZF) is not the same as the language of the theory T. In particular, for Putnam's argument to go through, he needs to be able to make distinctions in W that simply cannot be made by users of the theory T.

To make these ideas more concrete, let's consider an example: Let $\Sigma = \{c, d\}$, where c and d are constant symbols. Let T be the theory in Σ that says $c \neq d$, and $\forall x((x = c) \vee (x = d))$. (This example violates the strictures of Putnam's Löwenheim–Skølem based argument, but the point will not depend on those details.) Of course, there is only one model of T up to isomorphism. And yet, a skeptical worry arises! Imagine two people, Mette and Niels, both of whom accept T, and both of whom think that the world is the set $\{a, b\}$. And yet, Mette says that c denotes a, whereas Niels says that c denotes b. Do Mette and Niels disagree? The answer is yes and no.

We have already misdescribed the situation. Mette cannot say that "c denotes a," because a is not a name in her language. Similarly, Niels cannot say that "c denotes b." It is the metatheorist who can say: "Mette uses c to denote a," and "Niels uses d to denote b." But how does the metatheorist's language get a grip on the world? How can he tell what Mette and Niels are denoting, and that they are different things? Now, Putnam might claim that it is not he, but the realist, who thinks that the world is made of things, and that when our language use is successful, our names denote these things.

So far I agree. The realist does think that. But the realist can freely admit that even he has just another theory, and that his theory cannot be used to detect differences in how other people's theories connect up with the world. All of us – Mette, Niels, Hilary, you, and I – are on the same level when it comes to language use. None of us has the metalinguistic point of view that would permit us to see a mismatch between language and world.

Now, I suspect that some people might think that I've simply affirmed Putnam's conclusion – i.e., that I have embraced internal realism. I can neither affirm nor deny that claim (largely because of unclarity in the meaning of "internal realism"). But I insist that *if* Putnam's argument works, then we have no reason to discriminate between (ideal) consistent theories, and we should adopt an absolutely radical left-wing account of theoretical equivalence. I, for one, am loath to think that good theories are so easy to find.

Consider another scenario, where now I, rather than Putnam, get to choose the rules of the game. In other words, I have my own theory T_0 of which I believe the world W is a model. Then along comes Putnam and says that any consistent theory can be interpreted into the world W. But if my background theory T_0 is not ZF, then I don't see W as a set, and Putnam's argument cannot even get started. In particular, I don't necessarily grant that there is an isomorphism $f : M \to W$ between a model M of T and this model W of my theory T_0. For one, what would I even mean by the word "isomorphism"? I, the user of the theory T_0, know about isomorphisms between models of my theory. However, M is a model of a different theory T, written in a different signature Σ, and so there may be no standard of comparison between models of T and models of T_0.

There is still another, more severe problem for Putnam's argument. For a scientific theory to be "ideal," it's really not enough for it to correctly report every actual fact. In must do more! There are a few ways to get a handle on what more a good scientific theory must do. David Lewis recognized that the "best theory" is not simply one that gets every fact correct. Instead, the "best" achieves an ideal balance of strength and simplicity. Here "strength" means reporting the facts, and simplicity means . . . well, we all know it when we see it, right? Whether or not we philosophers have a good account of simplicity, the fact is that Lewis was right that there is (at least) a second component to theory evaluation, and it has something to do with systematicity, or choosing the right language, or cutting nature at the joints.

Thus, when I'm looking at a scientific theory, I'm not just interested in whether it's true. You could write down every truth in a massive encyclopedia, and I wouldn't consider it to be the best scientific theory. There are better and worse ways to say the truth. And what this means for our considerations here is that not all true theories are created equal; thus, certainly not all ideal consistent theories are equal.

We might want to go to the trouble of explaining when I, user of theory T_0, would grant that T can be interpreted into a model M of my theory. In the simplest sort of case, I would require that for each relation symbol p of T, there is a formula Fp of the appropriate arity of my language Σ_0 such that p can be interpreted as $M(Fp)$. As a user of theory T_0, I only recognize those subsets of M that can be described via the predicates of my theory. In particular, I don't necessarily have the resources to name the elements of the domain M, and I don't necessarily have the resources to collect arbitrary elements of M and form subsets out of them. I can only talk about "things that satisfy ϕ", where ϕ is one of the predicates of my language.

So, suppose then that T is consistent relative to my theory T_0: for each model M of my theory, there is a model M^* of T with the same domain as M, and such that for each relation symbol p of Σ, $M^*p = M(Fp)$ for some formula Fp of my language Σ_0. However, even in this scenario, I wouldn't necessarily consider the theory T to be adequate, for it may fail to pick up the relationships between various models of my theory. I'd want to know that the user of T recognizes the same connections between models that I do. In particular, where I see an elementary embedding $h : M \to N$, I would require the user of T to see a corresponding elementary embedding $h^* : M^* \to N^*$ between models of his theory T. And that just means that $h \mapsto h^*$ completes the definition of a functor from Mod(T_0) to Mod(T), where the object part is given by $M \mapsto M^*$. We then have the following result.

PROPOSITION 8.3.1 *Let F be a map of Σ-formulas to Σ_0-formulas such that the map $M(\phi) \mapsto M(F\phi)$ defines a functor from* Mod(T_0) *to* Mod(T). *Then $F : T \to T_0$ is a translation.*

Proof Define a reconstrual $G : \Sigma \to \Sigma_0$ by setting $Gp = Fp$. We claim that $T_0 \vdash G\phi \leftrightarrow F\phi$ for all formulas ϕ of Σ. For this, it suffices to run through the clauses in the definition of F. For example, we need to check that $F(\phi_1 \wedge \phi_2) \equiv F(\phi_1) \wedge F(\phi_2)$, where $\equiv$ means provable equivalence modulo T_0. But this is easy to check: let M be an arbitrary model of T_0. Then

$$\begin{aligned} M(F(\phi_1 \wedge \phi_2)) &\equiv M^*(\phi_1 \wedge \phi_2) \\ &\equiv M^*(\phi_1) \cap M^*(\phi_2) \\ &\equiv M(F(\phi_1)) \cap M(F(\phi_2)) \\ &\equiv M(F(\phi_1) \wedge F(\phi_2)). \end{aligned}$$

(Here I've ignored for simplicity the fact that ϕ_1 and ϕ_2 might have different free variables.) The clauses for the other connectives and for the quantifiers are similar. □

The upshot of this result for Putnam's argument is as follows: a user of a theory T_0 should only grant that T can be true if there is a translation of T into T_0. This result is not surprising at all. In real life, this is the sort of criterion we do actually employ. If I hear someone else speaking, I judge that what they are saying "could be true" if I can reconstrue what they are saying in my language. If there is no way that I can interpret their utterances into *my* language, then I am forced to regard those utterances as false or meaningless.

As Otto Neurath pointed out, and as Quine liked to repeat, we cannot start the search for knowledge from scratch. Each of us already has a theory, or theories. And we have a notion of permissible translations between theories that regulates (or describes) our attitude about which other theories could potentially be correct. If a theory T can be conservatively translated into my theory T_0, then I will think that T might possibly say something true (perhaps if its terms are charitably interpreted). But even then, I would not necessarily judge T to be true. Indeed, if my standard of theoretical equivalence is weak intertranslatability, then I will judge T to be potentially true (even under the most charitable interpretation) only if T and T_0 are weakly intertranslatable. (And do recall that weak intertranslatability is a fairly conservative criterion of equivalence.)

What Putnam has shown, at best, is that *relative* to a background theory T_0 of bare sets, a theory T that has a model in **Sets** could be charitably interpreted as true by a user of T_0. The result is really not very interesting – except insofar as it reminds us of the dangers of uncritically accepting set theory as our background metatheory. Indeed, set theory makes nontrivial existence claims – e.g., the claim that any two points in a model are related by a permutation.

The things I've just said might sound quite similar to Lewis' (1984) response to Putnam's argument. Lewis attempts to block the argument precisely by denying the permissibility of the relevant permutation – or, what's the same, of denying that each subset of WORLD picks out a genuine property. But Lewis' response is not, by itself, sufficient to block Putnam's argument. Suppose indeed that we've identified a privileged subclass $\mathcal{N}$ of natural properties among the subsets of WORLD. We can also require, as Lewis does, that a predicate symbol p of the signature Σ must be assigned to a set $M(p) \in \mathcal{N}$. In other words, M cannot assign p to any old subset of WORLD.

What Lewis has done here, in effect, is to propose an extension of Putnam's background theory T_0, by means of adding predicate symbols to the signature Σ_0 in order to designate the subsets in $\mathcal{N}$. Let T_1 be Lewis' strengthened background theory – the theory that describes the world as a set WORLD, with a privileged family $\mathcal{N}$ of subsets of WORLD to represent the natural properties. Then Lewis' requirement that the predicates of T be interpreted as elements of $\mathcal{N}$ is tantamount to the requirement that there is an interpretation of T into Lewis' background theory T_1. Since T_1 is expressively weaker than Putnam's background theory T_0, it is more demanding to ask for an interpretation of T into T_1 than it is to ask for an interpretation of T into T_0.

Lewis' requirement can block Putnam's trivializing maneuver: for some choices of $\mathcal{N}$, there are theories T that are set-theoretically consistent but that cannot be translated into T_1. To take one trivial example, suppose that T_1 has three natural properties: the empty set, the entire world, and some proper subset of the world. Suppose also that T includes the axiom

$$\exists x Px \wedge \exists y Qy \wedge \neg \exists z(Pz \wedge Qz).$$

Then T is set-theoretically consistent, but T cannot be translated into T_1.

Nonetheless, Lewis' demands here are not strong enough. In general, for any sufficiently rich family of natural properties $\mathcal{N}$, too many theories T will be interpretable into Lewis' background theory T_1. And hence, if Lewis grants Putnam's call for charitable interpretation, then Lewis must grant that those theories are true. That concedes too much. It is easy to think of examples that would make a realist choke. For example, suppose that Gargamel has a theory that says there are many gods, and there are no electrons. If Lewis countenances just a single natural property with instances, then Gargamel's theory can be translated into Lewis' background theory – and, by the principle of charity, should be counted as true.

We will not engage now in further formal investigation of these matters, e.g., to ask how many natural properties there need to be in order for a given theory T to be interpretable into Lewis' background theory. We don't think that question is very interesting – because we've already gone off on a bad track. There are two interrelated problems here. The first problem is that Lewis' background theory T_1 has little to recommend it, even if we are inclined to accept that there are "natural properties." (And anyone who uses first-order logic implicitly does accept the existence of natural properties – they are precisely the properties that are definable in her language.) The second, and deeper, problem is that Lewis, like Putnam, seems to be supposing that all parties – or at least all metaphysical realists – can agree on some particular fixed background theory T_*. We reject that assumption, and as a result, Putnam's paradox simply dissolves.

8.4 Realism and Equivalence

According to the standard stereotype, the logical positivists were *antirealists* or *instrumentalists* about scientific theories. Moreover, this antirealist stance was facilitated by means of the syntactic analysis of scientific theories, according to which a theory T's language has some purely observational terms O, and its empirical content can be identified with $T|_O$. With this formal analysis, the positivists could then articulate their particular versions of epistemic and semantic empiricism:

- Epistemic empiricism: the reasons we have to believe T derive from reasons we have for believing $T|_O$.
- Semantic empiricism: the meaning of terms in $\Sigma \backslash O$ derives exclusively from the meaning of terms in O.

The extreme instrumentalist would say that the terms in $\Sigma \backslash O$ have no meaning: there are merely instruments to facilitate making predictions. The attenuated instrumentalist tries to find a way for terms in Σ to inherit meaning from terms in O.

In the 1960s and 1970s, the syntactic view of theories was discredited, and the tide seemed to have turned decisively against antirealism – or at least against this stereotyped antirealism. Without a clear delineation of the empirical part of a theory, it was no longer possible to think that warrant or meaning could flow upward from the observationally relevant parts of a theory.

Van Fraassen characterized the state of play in 1976: "After the demise of logical positivism, scientific realism has once more returned as a major philosophical position" (van Fraassen, 1976, 623). He goes on to characterize scientific realism as commitment to the following thesis:

> The aim of science is to give us *a literally true story of what the world is like*; and the proper form of acceptance of a theory is to believe that it is true. (van Fraassen, 1976, 623)

As is well known, van Fraassen then gave several strong arguments against scientific realism, before going on to develop his positive alternative: *constructive empiricism*.

In the years that followed, there was much back-and-forth debate: van Fraassen on the side of constructive empiricism – and dozens of other philosophers on the side of scientific realism. The terms of this debate had been set by van Fraassen, and these terms were rarely (if ever) questioned. In particular, the scientific realists seem to have been happy enough with van Fraassen's characterization of their position; their job was merely to bring out its merits.

However, if we look more closely, it becomes apparent that the debate wasn't so clear-cut. During the 1960s and 1970s, scientific realists were fond of saying that the philosophical position of scientific realism is itself a scientific hypothesis, and that the reasons for believing it are of the same nature as the reasons for believing any other scientific theory. In particular, they claimed that the hypothesis of scientific realism is the *best explanation* for the success of the scientific enterprise.

Now, van Fraassen certainly questioned the latter claim. But more interestingly, he chose not to play by the same game as the scientific realists. For van Fraassen, the reasons for being a constructive empiricist are different in kind from the reasons for accepting a scientific theory. For those who were following the debate closely, it became clear that the choice between realism and antirealism about science was not a simple disagreement about which hypothesis better explains a common domain of phenomena. There was a deeper and more elemental disagreement about the goals of philosophical reflection.

For many philosophers of the next generation, the question of scientific realism versus scientific antirealism had receded too far into the upper reaches of metaphilosophy. The simple "pro and con" arguments of the 1970s and 1980s were not going to get us anywhere, seeing that the opposed parties were using different standards to evaluate these arguments. Thus, the next generation of philosophers of science moved downward – back to the analysis of specific scientific theories. Although they may not openly use these words, I suspect that many philosophers of science now feel that "realism or antirealism?" is a pseudoquestion, or at least not a particularly interesting question.

Speaking of pseudoquestions, what makes a question pseudo? Here is one criterion: a question is pseudo if getting an answer to it wouldn't change anything you do. By that standard, it's easy to see why the realism–antirealism debate might seem like a pseudodebate. Would a scientist do anything differently tomorrow if he converted to constructive empiricism? Wouldn't he go on looking for the elegant and powerful theories, and using them to make predictions and give explanations?

This last thought suggests a better way to understand what's really at stake in the realism–antirealism debate. I suggest that the debate can be fruitfully reconceived as a battle over standards of theoretical equivalence. In particular, a realist is somebody who adopts – or recommends that people adopt – stricter standards of theoretical equivalence. Conversely, an antirealist is somebody who adopts – or recommends that people adopt – looser standards of theoretical equivalence. In short, realists are conservatives about theoretical equivalence, and antirealists are liberals about theoretical equivalence.

This construal of the realism–antirealism debate matches well with various well-known cases. Consider, for example, the case of the logical positivists. We tend to think that they were antirealists because they said, "the content of the theory T resides in its observable part $T|_O$." But there are a lot of unclear words here, such as "content" and "residing" and "observable," and so this doesn't make for a very sharp statement of a philosophical thesis. However, one concrete implication of these positivist words is that if $T|_O = T'|_O$, then we should treat T and T' as equivalent. For example, suppose that two scientists, say Werner and Erwin, have apparently conflicting theories T and T' with the same empirical content, i.e., $T|_O = T'|_O$. Then the positivist would recommend that Werner and Erwin reconcile, for there can be no reason to prefer T over T' or vice versa. The difference between their theories is no more important than the difference between theories written in German and French. In contrast, if T says *anything* that conflicts with T', then the scientific realist thinks that one of the two must be better than the other, and that we should actively pursue inquiries to determine which it is.

This picture of the realism debate also makes sense of what structural realists were trying to achieve. In short, structural realists urge that not every single detail of a successful scientific theory should be taken with equal seriousness. In particular, they argue that if two theories T and T' differ only with respect to content, and not with respect to structure, then one can have no reason to prefer T over T', or vice versa. The normative core of structural realism, then, is to propose a notion of theoretical equivalence that lies somewhere to the left of the extreme right realist view and somewhere to the right of the extreme left views of the logical positivists, Nelson Goodman, and Putnam in the later stages of his career.

Scientists and philosophers – and, in fact, everyone – have implicit standards of equivalence that they employ to judge between truth claims, especially when those claims seem prima facie to conflict. If you believe "God doesn't exist," and your French colleague believes "*Dieu n'existe pas*," then you know that there is no dispute to be settled. Not only are those two sentences compatible with each other; they are equivalent. Even within a single language, we can say the same thing in different ways. Imagine that your friends Anne and Bent disagree about the number of roses in the vase on the counter. Anne says, "there are six roses," and Bent says, "there are a half dozen roses." In such a case, you would surely advise Anne and Bent to kiss and make up, since their dispute is merely verbal.

Those cases are easy. But there are more difficult cases in life – especially as we move into the more abstruse regions of the sciences. (And that's not even to speak of cases such as differences in matters of politics or religion.) For example, there is a debate among evolutionary biologists about the units of selection: is it the individual or the species? Many a friendship between scientists has been broken because of disagreement on issues like this one. But what if there really was no dispute between them? What if they were saying the same thing in different terms?

You might think such a scenario is unimaginable. But if the history of science can be trusted, then there have been numerous cases where prima facie disagreement has later been judged to be spurious. For example, in the mid-1920s, Werner Heisenberg developed a theory that made use of non-commutative algebra in order to predict the outcomes of measurements. This theory, called *matrix mechanics*, was hailed by many as a breakthrough, for it unified the ad hoc recipes that plagued the old quantum theory of Bohr and Sommerfeld. However, others abjured matrix mechanics, on the grounds that it was incomprehensible and unvisualizable and entailed bizarre claims, most notably the existence of "quantum jumps." Thus, a competing theory was developed by Erwin Schrödinger, a theory based on completely different ideas and mathematical techniques. According to Schrödinger's theory, there are waves moving through physical space (or a higher-dimensional configuration space), and particles such as electrons are simply harmonic resonances in these waves.

Thus, Heisenberg presented one theory, T_1, to account for the quantum phenomena, and Schrödinger presented another theory, T_2, to account for the same phenomena. While both these were empirically adequate, the battle between Heisenberg and Schrödinger was fierce, including name-calling, a fight for prominence at professional meetings, and competition for funding and university positions. The behavior of Heisenberg and Schrödinger clearly indicated that they saw this debate as *genuine* and in need of resolution.

The conclusion of this story is typically told as follows. Based on some suggestions that Schrödinger himself made, a young mathematician, John von Neumann, formulated a conjecture: Heisenberg's matrix mechanics T_1 is equivalent to Schrödinger's wave mechanics T_2. Von Neumann then went on to prove this theorem, to the great satisfaction of most participants involved – especially those like Niels Bohr, who didn't want to choose between Heisenberg and Schrödinger. As a result, *the debate came to an end.* Since T_1 and T_2 are equivalent theories, there is no question about which one is better, at least not in any epistemically or ontologically relevant sense. There is no decision to be made about whether to accept T_1 or T_2.

Such is the nature of judgments of theoretical equivalence. When one judges that theories T_1 and T_2 are equivalent, one judges that accepting T_1 is tantamount to accepting T_2. Conversely, if one feels that T_1 might be favored over T_2, or vice versa, then one judges that these theories are *not* equivalent.

Are there equivalent theories? Setting aside the Heisenberg-Schrödinger theory as controversial, still every sane person will admit that at least some theories are equivalent. For example, say that T_1 is the theory written down in the textbook *General Relativity* by Robert Wald that is sitting on the shelf in my office, and that T_2 is the theory written down in the textbook *General Relativity* by Robert Wald that is sitting on the shelf in Carlo Rovelli's office. Of course, we all know that T_1 and T_2 are equivalent theories. In fact, most of us just say that these are the *same* theory, and that's why we use a definite description for it: "*the* general theory of relativity." But if we boil everything down to fundamental physics, then we can only say that there are two distinct collections of ink splotches, one in an office in Princeton and another in an office in Marseilles.

In my experience, philosophers tend to react to this silly sort of example by flying to the realm of abstract entities. They say something like this: the two books contain sentences that pick out the same *propositions*, and that's why we say that the sentences represent the same theory. Now, I don't disagree with this claim; I only doubt its utility. If you give me two languages I don't understand, and theories in the respective languages, then I have no way of knowing whether those theories pick out the same propositions. And that's precisely the sort of case we face with something like matrix and wave mechanics. Employing new formalism that is not yet very well understood, it is unclear whether these theories say the same thing. Thus, we need some criterion for equivalence that is *checkable*, at least in principle. In other words, we need to know when two sentences pick out the same proposition.

There are essentially two ways to proceed from here. On the one hand, we can ask: what features must two theories have in common in order to be equivalent? In philosophical jargon: what are the necessary conditions for theoretical equivalence? This question can also be given a mathematical gloss: what are the *invariants* of theoretical equivalence? For example, some people would say that for two (single-sorted) first-order theories T_1 and T_2 to be equivalent, they must agree on the number of existing objects. There are other conditions we might try to impose, but which are a bit more difficult to cash out in terms of formal logic. For example, many contemporary philosophers would say that two equivalent theories must have the same *primitive notions* – i.e., those objects, properties, etc., that ground the other things that the theory mentions.

The second question we could ask has a more top-down flavor: could we simply define an equivalence relation on the collection **Th** of all theories? Disregarding the fact that **Th** is a proper class and not a set, there are many such equivalence relations, all of which yield some notion of theoretical equivalence. Among these untold number of equivalence relations, some have relatively simple or elegant definitions. Indeed, each one of the notions of equivalence we have canvassed in this book – e.g., definitional equivalence, Morita equivalence, and categorical equivalence – defines an equivalence relation on the class of all first-order theories.

An ideal method, I think, is to take both procedures into account. On the one hand, we need not accept a definition of equivalence if it violates necessary conditions to which we are committed. On the other hand, some of us might feel compelled to abandon an intuitive necessary condition of equivalence – i.e., some intuitive invariant of theoretical equivalence – if it conflicts with what otherwise seems the most reasonable formulation of an equivalence relation on **Th**.

We can see these sorts of choices and trade-offs being made all the time in philosophy. On the more conservative side, philosophers such as David Lewis and Ted Sider lay heavy stress on choosing the right primitives. At times it seems as if they would go so far as to say that there is a *privileged language* for metaphysics so that no theory in this language could be equivalent to a theory that is not in this language. (One wonders, however, how they individuate languages.)

One could imagine an even more conservative stance on theoretical equivalence. For example, suppose that Σ is a fixed signature (say, the preferred signature for metaphysics), and T_1 and T_2 are theories in Σ that have the same consequences (equivalently, have the same models). Should we then consider T_1 and T_2 to be equivalent? I suspect that Sider would say yes. But I also suspect that some philosophers would have said no, for they might have thought that there are preferred ways of axiomatizing a theory. Indeed, if you really believe that some facts are more basic than all the others, then shouldn't those facts be the ones enunciated in the axioms, so that all other facts are seen as flowing from them? Thus, we get an even finer-grained equivalence relation on **Th** if we demand that equivalent theories are in the same signature *and* have the same axioms.

Even that requirement – having the same axioms – is not the most conservative imaginable. We might even require that the theories literally have the same notation. For example, in formulating group theory, we could use the symbol $\circ$ for the binary relation, or we could use the symbol $\bullet$. Who knows, perhaps one of these two symbols more perspicuously represents the structure of the binary function in the world that we are trying to represent. At the farthest end of this spectrum, one could adopt a pure "Heraclitean" account of theoretical equivalence, according to which no two theories are the same. In other words, the criterion of theory identity could be made out to be *literal identity* – of symbols, axioms, etc.

Conservative views of theoretical equivalence tend to align with "realist" views about science or metaphysics. Roughly speaking, if you think that the world has real structure, then you'll think that a good theory has to represent the structure that is out there. If two theories disagree about that structure, then they cannot be equivalent. Going in the opposite direction, liberal views of theoretical equivalent tend to align with "antirealist" views about science and metaphysics. We see this tendency with Nelson Goodman in the 1960s and with Hilary Putnam in the 1970s. Putnam's move toward antirealism was augured by his giving many examples of theories that he says are equivalent, but which realists regard as being inequivalent. For example, Putnam claims that Euclidean geometry based on points is equivalent to Euclidean geometry based on lines – even though the models of these two theories can have different cardinalities.

Long before Putnam turned in this direction, the connection between antirealism and liberal views of theoretical equivalence had already been established. I'm thinking here of the logical positivists and their infamous notion of *empirical equivalence*. The idea here is that two theories T_1 and T_2 are empirically equivalent just in case they share the same observable consequences – and regardless of what else these might say. So, to take an extreme example, if T_2 is T_1 plus the sentence, "there is a new unobservable particle," then T_1 and T_2 are empirically equivalent.

Now, for the logical positivists – or at least, for some of them – empirical equivalence is equivalence enough. For they identified the content of a proposition with that proposition's empirical consequences; and it follows from this that if two propositions ϕ and ψ have the same empirical consequences, then they have the same content – i.e., they are the same proposition. Stepping back up to theories, as collections of propositions, the positivist view of content entails that two theories are equivalent *tout court* if they are empirically equivalent.

The positivist view of theoretical equivalence is quite liberal, and certainly unacceptable to scientific and metaphysical realists. Most of us have the intuition that theories can say different things about unobservable things, even if those theories agree in all their observational consequences. In this case, we have to reject empirical equivalence as a sufficient condition for theoretical equivalence.

A case can be made that Putnam's view of theoretical equivalence eventually became – at least tacitly and in practice – even more liberal than empirical equivalence. In putting forward the model-theoretic argument, Putnam essentially makes an argument for the following claim:

If T is consistent (and has other virtues such as completeness), then T ought to be taken as true.

Now, in application to *two* consistent theories T and T', we have the following result:

If T and T' are consistent (and have other virtues such as completeness), then T and T' ought both to be taken as true.

In other words, consistent, ideal theories are true in all conditions, hence in all the same conditions, and so they are equivalent. That is a radically liberal view, almost Zeno-like in its implications. For in this case, there is only one equivalence class of consistent theories.

What I've left out from this story so far are all the intermediate (and more plausible) views of theoretical equivalence – views that we have been discussing throughout this book, such as definitional equivalence or Morita equivalence. To put everything together, consider the diagram that follows, which places the different views of theoretical equivalence on a one-dimensional spectrum from maximally liberal (Zenonian) to maximally conservative (Heraclitean).

Zeno $\leftarrow$ categorical $\leftarrow$ w-intertranslatable $\leftrightarrow$ Morita $\leftarrow$ s-intertranslatable $\leftrightarrow$ CDE $\leftarrow$ logical $\leftarrow$ Heraclitus

So, given this wide range of different notions of equivalence, how are we to choose among them? And do we need to choose among them? I would say that we don't have to explicitly choose among them – but that our attitudes toward them mirror our attitudes toward real life cases, or at least to cases that come up in other philosophical discussions. Consider, for example, North's (2009) argument for the inequivalence of Hamiltonian and Lagrangian mechanics. She says, "Hamiltonian and Lagrangian mechanics are not equivalent in terms of statespace structure. This means that they are not equivalent, period." In other words, she's putting a model of Hamiltonian mechanics next to a model of Lagrangian mechanics and comparing structure. Seeing that these structures are not "equivalent," she declares that the theories are not equivalent. We see then that, at the very least, North adopts a criterion that is more conservative than categorical equivalence, which is blind to the internal structure of individual models. (In fact, Barrett [2018a] shows that Hamiltonian and Lagrangian mechanics are categorically equivalent.) Most likely, North's criterion is further to the right than even Hudetz's definable categorical equivalence (see Hudetz, 2018a), for she doesn't consider questions as to whether Lagrangian structure can be defined in terms of Hamiltonian structure, and vice versa.

We can see a similar thought process going on with critics of quantifier variance. Indeed, we can think of debates about quantifier variance as debates about which notion of theoretical equivalence to adopt. The opponents of quantifier variance insist that equinumerosity of models is a necessary condition for theoretical equivalence. Thus, they draw the line short of Morita equivalence, which allows that equivalent theories can have models of different cardinalities. In contrast, defenders of quantifier variance claim that theories can be intertranslatable even if they violate that cardinality constraint. The question boils down to which criterion of theoretical equivalence is the better one to adopt.

I believe that this is one of the most interesting questions that philosophers can ask, precisely because it's a non-factual question. Or, to put it more accurately, the answer that one gives to such a question determines what one thinks is a factual question – and so it's not the kind of question that two parties can easily resolve by appeal to a shared stock of facts. Nonetheless, we've made a lot of progress on the technical side, so we now have a much more clear sense of what's at stake and the price we must pay for adopting some particular formal notion of equivalence as an intuitive guide to our practice of judging between theories.

Consider, for example, the distinction between definitional equivalence and Morita equivalence – or what is the same, between strong and weak intertranslatability. The line between these two notions of equivalence seems to correspond pretty well to the distinction between metaphysical realists and, well, those who aren't quite metaphysical realists. (The metaphysical realist might insist that if theories are equivalent, then their models have the same number of objects.) However, we shouldn't forget that Morita equivalence isn't all that liberal. It's certainly far more conservative than what Putnam was suggesting in the model-theoretic argument.

We can also see that the ontology of Morita equivalent (i.e., weakly intertranslatable) theories can never be radically different from each other. If $F : T \to T'$ is a homotopy equivalence (between single-sorted theories), then for each model M of T', there is a model F^*M of T, whose domain is explicitly constructed by the recipe:

$$(F^*M)(\sigma) = M(\sigma') \times \cdots \times M(\sigma')/\sim,$$

where $\sim$ is an equivalence relation defined by the theory T'. There are a couple of important points here. First, the ontology of F^*M results from simple logical constructions of the ontology of M. Borrowing terminology from Bertrand Russell, we could say that the elements of F^*M are *logical constructs* of elements of M. Second, the recipe for constructing F^*M from M is *uniform* – i.e., it doesn't depend on M. In other words, it's not just that each model of T consists of logical constructs of elements of a model of T'; it's that the type of construction is uniform. It's in this extended and, nonetheless, quite strong sense that T has the same ontology as T'.

Moreover, since F is assumed to be a homotopy equivalence, we can say the same thing in reverse order: each model of T' consists of logical construct of elements of a model of T, and this construction is uniform on models. One bonus insight here is seeing how the relation "being a logical construct of" differs from the mereological parthood relation. Consider the specific example of the point and line formulations of affine geometry (see Section 7.4). Here the points are logical constructs of lines, and the lines are logical constructs of points. It's tempting to think then that points are logical constructs of points – but that would be incorrect. The reason that inference doesn't go through is that "being a logical construct of" is not like the mereological notion of parthood. To get a line, we don't simply take two points; we take an equivalence class of two points. Thus, there is no sense here in which a line results from taking a composite of points. The opposite direction is even more clear. We can construct points from lines, but certainly a point is not made out of lines.

The upshot of these considerations is that moving from definitional equivalence to Morita equivalence is not as radical a generalization (or liberalization) as it might seem at first. Even for the ontological purists, a case could be made that Morita equivalence involves only the slightest relaxation of the constraint that equivalent theories should have equinumerous domains.

In contrast, categorical equivalence is extremely liberal from an ontological point of view. It's possible, indeed, to have categorically equivalent theories where there is no reasonable sense in which the ontology of the first's models can be constructed from the ontology of the second's models.

There are, however, some intermediate cases that are worth considering. Some of these are discussed by Hudetz (2018a). Here we just look at one example that will be familiar from Chapter 3. Consider the categories **Bool**, of Boolean algebras, and **Stone**, of Stone spaces. As we proved, **Bool** is equivalent to the opposite of **Stone**, where the arrows have been flipped. Moreover, the functors relating these two categories do have a strongly constructive flavor. The functor $F : \mathbf{Bool} \to \mathbf{Stone}^{op}$ is the representable functor $\hom(-, 2)$, where 2 is the two-element Boolean algebra. The functor $G : \mathbf{Stone}^{op} \to \mathbf{Bool}$ takes the clopen subsets. In both cases, the functor involves construction of an object of one category out of an object of the second category, and possibly some reference object, such as 2.

Could these latter sorts of functors be taken as representing genuine theoretical equivalences? There are two clarifications we need to raise for that question. First, the question doesn't even make sense until we say something more about how a category, which may not be of the form $\mathrm{Mod}(T)$ for a first-order theory, can represent a theory. Second, for many physical theories – and *pace* Quine – the elements of a mathematical domain X are not necessary meant to represent objects in the physical world. Consider the following example, which – besides being extremely interesting in its own right – illustrates several of these points.

General relativity (GTR), qua mathematical object, can roughly be taken to be the category **Lor** of Lorentzian manifolds, equipped with an appropriate collection of smooth mappings between them. There has been a longstanding debate – stimulated, no doubt, by Quine's criterion of ontological commitment – about whether accepting GTR demands that one accept the existence of spacetime points. Perhaps partially in response to that claim, Earman noted that GTR could also be formulated in terms of mathematical objects called "Einstein algebras." The relationship between Lorentzian manifolds and Einstein algebras is suggestively parallel to the relationship between Stone spaces and Boolean algebras. This parallel was confirmed by Rosenstock et al. (2015), who showed that **Lor** is dual to the category **EAlg** of Einstein algebras.

If one takes categorical equivalence as the criterion for theoretical equivalence, then the Einstein algebra formulation of GTR is no better nor worse than the Lorentzian manifold formulation. However, one might also wish to draw a stronger conclusion: one might wish to say that Rosenstock et al.'s proof shows that accepting GTR does *not* involve ontological commitment to spacetime points.

However, that conclusion would be hasty. The implicit argument pattern here would run as follows:

Let T be a theory with a sort σ. If T is equivalent to T', and T' doesn't quantify over σ, then to accept T cannot involve ontological commitment to things of type σ.

To see that this inference pattern proves too much, we can consider some simple examples. First, consider the example of the theory T in sort $\Sigma = \{\sigma\}$ that says there are exactly two things, and consider the theory T' in sort $\Sigma' = \{\sigma'\}$ that says there are exactly two things. By the preceding inference rule we would have to conclude that accepting T does not demand ontological commitment to things of type σ, merely because there is another sort symbol σ'. This is silly. The difference between σ and σ' could be simply notational.

Perhaps then the argument pattern is meant to be a bit more nuanced.

Let T be a theory with sort σ. If T is equivalent to a theory T', and T' has no sort σ' that is "isomorphic" to σ, then accepting T does not involve ontological commitment to things of type σ.

The word "isomorphic" was put into quotes because we would still need to explicate what we mean by it. But that could be done; e.g., we might say that an equivalence $F : T \to T'$ shows that σ and σ' isomorphic if $F(\sigma) = \sigma'$ and $E_{x,y} \equiv (x =_\sigma y)$ for variables x, y of sort σ. But in this case, the proposed criterion simply begs the question against the idea that Morita equivalent theories can have the "same ontology." To take Morita equivalence seriously as a criterion of theoretical equivalence means simply that there is no cross-theoretical reference point for counting objects or quantifying over them.

8.5 Flat versus Structured Views of Theories

For the past fifty years, philosophers' discussions of the nature of scientific theories has been dominated by the dilemma: are theories sets of sentences, or are theories collections of models? But the point of this debate has become less and less clear. Most of us these days are non-essentialists about mathematical explications. For example, most of us don't think that scientific theories really are sets of axioms or collections of models. Instead, we think that different explications are good for different purposes. There is, nonetheless, a big question lurking in the background – viz. the question of whether we should conceive of theories as "flat," or whether we should conceive of them as "structured." And this question comes up whether one thinks that theories are made of sentences or whether one thinks that they are made of models.

The syntactic view of theories is usually formulated as follows:

A theory is a *set* of sentences.

This formulation provides a *flat* view: a theory consists of a collection of things, and not in any relations between those things or structure on those things. In contrast, a *structured* view of theories says that scientific theories are best represented by structured mathematical objects. For example, a structured syntactic view of theories might say that a theory consists of both sentences and inferential relations between those sentences.

A flat version of the semantic view might be formulated as

A theory is a *set* (or *class*) of models.

In contrast, a structured version of the semantic view will say that a theory consists of a structured collection of models. For example, a theory might consist of models with certain mappings between these models (such as elementary embeddings), or a theory might consist of models and certain "nearness" relations between those models.

Both the syntactic and the semantic views of theories are typically presented as flat views. In the latter case, I suspect that the flat point of view is accidental. That is, most proponents of the semantic view are not ideologically committed to the claim that a theory is a bare set (or class) of models. They may not have realized the implications of that claim or that there is an alternative to it.

In contrast, in the case of syntactically oriented views, some twentieth-century philosophers were ideologically committed to a flat view – perhaps due to their worries about intensional and/or normative concepts. The main culprit here is Quine, whose criticism of the analytic–synthetic distinction is directed precisely against a structured view of theories. On a structured syntactic view of theories, the essential structure of a theory includes not just some number of sentences, but also the logical relations between those sentences. In this case, commitment to a theory would involve claims about inferential relations – in particular, claims about which sentences are logical consequences of the empty set. In other words, a structured syntactic view of theories presupposes an analytic–synthetic distinction.

Quine's powerful criticisms of the analytic–synthetic distinction raise worries for a structured syntactic picture of theories. But is all well with the unstructured, or flat, syntactic view? I maintain that the unstructured view has *severe* problems that have never been addressed. First of all, if theories are sets of sentences, then what is the criterion of equivalence between theories? A mathematically minded person will be tempted to say that between two *sets*, there is only one relevant condition of equivalence, namely equinumerosity. But certainly we don't want to say that two theories are equivalent if they have the same number of sentences! Rather, if two theories are equivalent, then they should have some further structure in common. What structure should they have in common? I would suggest that, at the very least, equivalent theories ought to share the same inferential relations. But if that's the case, then the content of a theory includes its inferential relations.

8.6 Believing a Scientific Theory

The difference between scientific realists and antirealists is supposed to be that the former believe scientific theories, and the latter do not – or at least they don't believe everything that these theories say. For example, constructive empiricists like van Fraassen don't necessarily believe what scientific theories say about unobservable things. This classification is based on a presupposition, viz. that we understand what it means to "believe everything a scientific theory says." But there is something wrong with these presupposition. On none of the reasonable analyses is a scientific theory nothing more than some claims about the world. If that's right, then the appropriate attitude to a successful scientific theory cannot be exactly the same thing as simple belief.

To see what's at issue here, it will be helpful to revisit an old objection to the semantic view of theories. According to the semantic view of theories, a scientific theory is a class of models. Now, the objector to the semantic view points out that there is a grammatical problem: in the phrase "*S* believes that *X*," the second argument *X* needs to be filled by something toward which a person can bear a propositional attitude. The argument *X* cannot be replaced by a name such as "Thor," or predicate such as "purple," much less by a name for a class of things, such as "the set of . . ." In particular, it makes no grammatical sense to say that "*S* believes that $\mathscr{M}$," where $\mathscr{M}$ is a class of models.

The semanticists have a ready reply to this objection:

Semantic Analysis of Belief (SAB): When a theory *T* is given by means of a class $\mathscr{M}$ of models, then belief in *T* means belief that the world is isomorphic to one of the models in $\mathscr{M}$.

There are many problems with SAB, most notably the opacity of the notion of a model being isomorphic with the concrete world (see, e.g., Van Fraassen, 2008). However, there is another problem with SAB that we find even more serious, because it bears directly on questions of a normative nature, e.g., to what one commits oneself when one accepts a scientific theory. In particular, believing a theory involves further commitments beyond those that are expressed by SAB.

Consider a specific example. Let *T* be Einstein's general theory of relativity (GTR). According to SAB, a person believes GTR iff she believes that the world is isomorphic to one of the models of GTR. But that analysis is inaccurate in both directions: it captures both more and less than physicists actually believe when they accept GTR. First, it captures more, because it seems to commit physicists to the belief that there is some privileged model of GTR that gives the best overall picture of the physical world. If you know how GTR works, then you might laugh at that thought. Just imagine two relativists – say, a cosmologist and a black hole theorist – sitting down to argue over whose model gives a more perspicuous representation of reality. They won't do that, because they are well aware that these models are accurate representations for certain purposes and not for others. And what's more, it's we – the users of physics – who choose the intended application of the theory. Thus, SAB says more than physicists will actually want to say about their theories.

Second, the semantic analysis of belief (SAB) also omits some of the content that physicists pack into their theories. Indeed, SAB locates the content of a theory in one or other particular model, ignoring the fact that physicists routinely invoke the existence of other models, not to speak of a rich system of relations between models. Indeed, if a model *M* is removed from its context in Mod(*T*), then it can no longer do the representational and explanatory work that it's expected to do. Consider again the case of GTR. As we noted before, GTR is a powerful theory not because it is overly specific, but because it is widely applicable – offering different, but related, models for a wide variety of situations. GTR finds the unity between these situations, including counterfactual situations. (David Lewis said that a good theory balances informativeness and simplicity. However, there are different ways of being informative: saying what is unique about your situation or saying what is common among many different situations.)

Furthermore, some of the most powerful explanations in GTR draw on facts about how a model sits inside the space of all models, some of which we know not to represent the actual world. For example, what explains the fact that our universe began in a singularity? According to GTR, singular spacetimes are generic, i.e., they densely pack the space of cosmological solutions to Einstein's field equations; hence, the reason our universe begins in a singularity is because most nomologically possible universes begin this way.

The fact that GTR uses all of its models, and the relations between them, is only reinforced by looking at simple examples from first-order logic. If we take a first-order theory T, then typically a single model M of T does not contain enough information to reconstruct T. In other words, if you give me a model M of T, I couldn't reliably reconstruct the theory T of which it was a model. What that means is that M contains less information than the theory T itself. The content of the syntactic object T is not contained in a single model M, but in the structured collection $\text{Mod}(T)$ of all its models. What this means in turn is that accepting T cannot be reduced to a claim about one of the models in $\text{Mod}(T)$; instead, accepting T must involve some sort of attitude toward the entire collection $\text{Mod}(T)$.

The point we are making here ties all the way back to the preface of the book, where we tried to justify our omission of modal logic. There we claimed that accepting a first-order theory – with no explicit modal operators – involves modal commitments. We're making the same point here. To accept a theory T isn't just to take a stand on how the world *is*; it is also to take a stand on how the world *could be*. More is true. To accept a theory T involves choosing a language Σ, and this language determines how we parse the space of possibilities – e.g., which possibilities we consider to be isomorphic, and which we consider not to be isomorphic. (If you've read the previous chapters carefully, you're also aware that the language Σ determines the topological structure of $\text{Mod}(T)$.) In short, the syntactic approach to theories had the advantage (largely unnoticed by its proponents) that the syntactic object T packs in a lot of information about what is possible and about how to classify possibilities. One of the dangers of the semantic view is forgetting how much scientific theories say.

The fault here doesn't lie completely with the semantic view of theories. In fact, there's an analogous problem for those, such as Quine, who accept a flat syntactic view of theories (see Section 8.5). According to the flat syntactic view, a theory T is a *set* of sentences. Indeed, Quine – among other flat syntacticists – sometimes equates belief in T with belief in a set of sentences. But that cannot be quite right, as we can see again from actual scientific theories, as well as from simple examples from first-order logic.

As for examples from first-order logic, let Σ_1 be the empty signature, and let T_1 be the theory in Σ_1 that says there are exactly two things. Let $\Sigma_2 = \{c\}$, where c is a constant symbol, and let T_2 be the theory in Σ_2 that says there are exactly two things. Here T_1 and T_2 share the same axiom, but they aren't equivalent theories by any reasonable standard – not even by categorical equivalence. The first theory's model has automorphism group $\mathbb{Z}_2$, whereas the second theory's model has trivial automorphism group (since the denotation of c is fixed). Nonetheless, T_1 and T_2 agree on the statements that they make about any particular model: they both say that there are two things. The user of T_2 has an extra name c, but her using this name does not amount to any claim about how things are. Thus, we have a puzzle: on a world-by-world basis, T_1 and T_2 say the same thing; and yet, it's not reasonable to think that T_1 and T_2 say the same thing.

The solution to this little puzzle is to recognize that believing a theory cannot be reduced to believing that a certain collection of sentences is true. At the very least, believing a theory also requires that we adopt a language – or an "ideology," as Quine liked to call it. However, Quine wasn't completely clear on what the reasons might be for accepting an ideology. The issues became slightly clearer when Lewis suggests that our choice of ideology corresponds to our beliefs about which properties are "natural," and when Sider (2013) suggests that choice of ideology is tantamount to assertion that the world has a certain structure. While we don't necessarily agree with this way of describing the situation, we agree that ideology plays a theoretical role.

If we claim that a theory is a collection of sentences, then we ought also to accept the claim that theories are equivalent only if they contain the same sets of sentences. Or, to be more accurate, two theories are equivalent just in case each sentence in the first is equivalent to a sentence in the second, and vice versa. But now, what standard of equivalence should we use for the sentences? The only reasonable standard – two sentences are equivalent if they express the same proposition – is of no use in comparing actual scientific theories. Thus, the only reasonable account of the identity of scientific theories treats theories as a *structured* objects, in which case equivalence means having the same structure. And then we have a challenge question: what does it mean to believe or accept a structured object?

It might be illuminating to compare a scientific theory with the kinds of beliefs for which people live and die – e.g., religious beliefs. As you know, many western religions have creeds that are supposed to capture the key tenets of the system of belief. Now, suppose that you were to try to write down the central tenets of a scientific theory as a creed. For example, you might take a copy of Robert Wald's *General Relativity* and start searching through it for the basic "truth claims" of the theory. However, you'll quickly grow frustrated, as it doesn't seem to make any specific claims. GTR doesn't say what happened on December 7, 1941, nor does it say how many planets are in our solar system, nor does it say (before one selects a particular model for application) how old the universe is. Instead, GTR consists of some mathematics and some recommendations about how to apply this mathematics to various situations. And yet, there is never any hint that GTR is a bad theory because it's not specific enough. Quite to the contrary, GTR is a good theory precisely because it is so general.

One might be tempted to think that the creed of GTR is summed up in its basic equation, Einstein's field equation (EFE). In this case, to accept GTR would be to say:

(†) I believe that $R_{ab} - \frac{1}{2} g_{ab} R = T_{ab}$.

This is an interesting possibility to consider, and there are two attitudes we could take to it. I will call these two attitudes the physicist's attitude, and the metaphysician's attitude (almost certainly caricaturing both). In my experience, physicists don't say things like (†). Certainly, they write down EFE, and they use it to generate descriptions of situations that they take to be accurate. But I've never heard a physicist say, "I believe that Einstein's field equation is true." These physicists seem to have a positive attitude toward EFE – perhaps we should call it "acceptance," but I don't think we could call it "belief."

In contrast, the naive scientific realist might say something like: "The success of GTR gives us reason to think that EFE is true." In order to make sense of EFE being true, these realists will then cast about for referents for the terms that occur in it. For example, in the spirit of David Armstrong, they might say something like, "The symbols R_{ab} and T_{ab} refer to natural properties, and EFE is the statement that a second-order relation holds between these properties." This kind of realist seems to think that there aren't enough mundane physical objects to account for the meaning of the abstract statements of science. Accordingly, he makes up names for new things that can serve as the referents for the symbols in scientific statements such as EFE. After some subtle wordplay, we're supposed to be able to feel what it means to really believe that EFE is true.

However, this naive realist way of looking at EFE doesn't capture the way that these kinds of equations function in physics. As with any other differential equation in physics, EFE is used as a guide for differentiating between what is nomologically possible and what is not. A differential equation doesn't say how things are; it says how things could be.

It might sound like I'm simply endorsing instrumentalism, i.e., saying that the theoretical statements of science are mere instruments from which to derive predictions. But that accusation depends on a false dilemma between naive realism and instrumentalism – a dilemma that is sadly reinforced by formal semantics. In formal semantics we have a simple, black-and-white distinction between interpreted and uninterpreted terms. Accordingly, we're tempted to think that the terms of EFE are either interpreted (hence EFE is either true or false) or uninterpreted (hence EFE is just an instrument). But this is the wrong way to think about things. The symbols in EFE in themselves are neither interpreted nor uninterpreted. It is we, users of the theory, who endow these symbols with an interpretation. What's more, we might well want to interpret the symbols differently for different applications.

The existence of more than one model – or, to speak more accurately, of more than one application – is not a bug of scientific theories; it is a feature. What is lost in informativeness is gained in applicability. But the more flexible a theory is in its applications, the less sense it makes to think of our attitude toward that theory as simple "belief." Perhaps this is one reason why we need another word, such as "acceptance." As van Fraassen pointed out long ago, to accept a theory cannot be reduced to an attitude that the theory somehow mirrors the world. Acceptance of a theory involves a sort of appropriation, where the theory serves as a guide to future action.

I've been considering the question, "what does it mean to accept a scientific theory?" and have found ample reason to reject the idea that it's nothing more than a special case of belief. Accepting a scientific theory may involve believing that some things are true, but it also involves a more complex set of attitudes – such as adopting certain standards for explanation, certain rules for reasoning about counterfactual scenarios, etc.

8.7 Notes

- For more technical details on second-order logic, see Shapiro (1991); Manzano (1996). Philosophers have argued quite a bit about the advantages and disadvantages of second-order logic. For example, Quine argued that second-order logic is "set theory in sheep's clothing." See, e.g., Bueno (2010).
- Carnap gives his mature view of Ramsey sentences in Carnap (1966). For more on the role of Ramsey sentences in Carnap's philosophy of science, see Psillos (2000, 2006); Friedman (2011); Demopoulos (2013).
- For more on the Ramsey sentence functionalism, see Shoemaker (1981).
- For a detailed, but older, discussion of the technical issues surrounding Ramsey sentences, see Tuomela (1973, chapter 3). For a recent discussion of the prospects of Ramsey sentence structuralism, see Ketland (2004); Melia and Saatsi (2006); Ainsworth (2009); Dewar (2019).
- For general surveys of structural realism, see Frigg and Votsis (2011); Ladyman (2014). The idea behind structural realism goes much further back than the 1980s. Something similar had been proposed by Poincaré and Russell in the early 1900s, and then again by Grover Maxwell in the 1960s. What's new about the 1990s reincarnation of structural realism is (1) the explicit claim that it can solve the pessimistic metainduction and (2) the explication of structure in terms of Ramsey sentences. Needless to say, the idea behind structural realism could survive, even if – as we've argued – Ramsey sentences don't provide a useful explication of the structure of a theory.
- My view on counting possibilities was influenced by Weatherall (2016b).
- Putnam's model-theoretic argument first appeared in Putnam (1977, 1980), with antecedents in Quine's permutation arguments for ontological relativity. The most influential response to Putnam is Lewis' (1984), which is the *locus classicus* of his version of metaphysical realism which emphasizes the notion of *natural properties*. That torch has been taken up by Sider (2013). The response we gave to Putnam's argument follows the spirit of Van Fraassen (1997). For an excellent overview of Putnam's arguments, see Button (2013).

Bibliography

Adámek, J., Sobral, M., and Sousa, L. (2006). Morita equivalence of many-sorted algebraic theories. *Journal of Algebra*, 297(2):361–371.

Ahlbrandt, G. and Ziegler, M. (1986). Quasi finitely axiomatizable totally categorical theories. *Annals of Pure and Applied Logic*, 30(1):63–82.

Ainsworth, P. M. (2009). Newman's objection. *The British Journal for the Philosophy of Science*, 60(1):135–171.

Andreas, H. (2007). *Carnap's Wissenschaftslogik: Eine Untersuching zur Zweistufenkonzeption*. Mentis.

Andréka, H., Madaraśz, J., and Németi, I. (2008). Defining new universes in many-sorted logic. *Mathematical Institute of the Hungarian Academy of Sciences, Budapest*, 93.

Andréka, H., Madaraśz, J., and Németi, I. (2005). Mutual definability does not imply definitional equivalence, a simple example. *Mathematical Logic Quarterly*, 51(6):591–597.

Andréka, H. and Németi, I. (2014). Comparing theories: The dynamics of changing vocabulary. A case-study in relativity theory. In *Johan van Benthem on Logic and Information Dynamics*. Springer.

Awodey, S. (2010). *Category Theory*. Oxford University Press.

Awodey, S. and Forssell, H. (2013). First-order logical duality. *Annals of Pure and Applied Logic*, 164(3):319–348.

Awodey, S. and Klein, C. (2004). *Carnap Brought Home: The View from Jena*. Open Court Publishing.

Baker, D. J. (2010). Symmetry and the metaphysics of physics. *Philosophy Compass*, 5(12):1157–1166.

Barnes, D. W. and Mack, J. M. (1975). *An Algebraic Introduction to Mathematical Logic*. Springer-Verlag.

Barrett, T. W. (2015). On the structure of classical mechanics. *The British Journal for the Philosophy of Science*, 66(4):801–828.

Barrett, T. W. (2018a). Equivalent and inequivalent formulations of classical mechanics, *The British Journal for the Philosophy of Science*, https://doi.org/10.1093/bjps/axy017.

Barrett, T. W. (2018b). What do symmetries tell us about structure? *Philosophy of Science*, 85(4), 617–639.

Barrett, T. W. and Halvorson, H. (2016a). Glymour and Quine on theoretical equivalence. *Journal of Philosophical Logic*, 45(5):467–483.

Barrett, T. W. and Halvorson, H. (2016b). Morita equivalence. *The Review of Symbolic Logic*, 9(3):556–582.

Barrett, T. W. and Halvorson, H. (2017a). From geometry to conceptual relativity. *Erkenntnis*, 82(5):1043–1063.

Barrett, T. W. and Halvorson, H. (2017b). Quine's conjecture on many-sorted logic. *Synthese*, 194(9):3563–3582.

Bealer, G. (1978). An inconsistency in functionalism. *Synthese*, 38(3):333–372.

Bell, J. and Machover, M. (1977). *A Course in Mathematical Logic*. North-Holland.

Belot, G. (1998). Understanding electromagnetism. *The British Journal for the Philosophy of Science*, 49(4):531–555.

Belot, G. (2017). Fifty million Elvis fans can't be wrong. *Noûs*, 52(4):946–981.

Ben-Menahem, Y. (2006). *Conventionalism: From Poincaré to Quine*. Cambridge University Press.

Beni, M. D. (2015). Structural realism without metaphysics: Notes on Carnap's measured pragmatic structural realism. *Organon F*, 22(3):302–324.

Beth, E. and Tarski, A. (1956). Equilaterality as the only primitive notion of Euclidean geometry. *Indagationes Mathematicae*, 18:462–467.

Beth, E. W. (1956). On Padoa's method in the theory of definition. *Journal of Symbolic Logic*, 21(2):194–195.

Bickle, J. (1998). *Psychoneural Reduction: The New Wave*. MIT Press.

Bickle, J. (2013). Multiple realizability. *Stanford Encyclopedia of Philosophy*. https://plato.stanford.edu/entries/multiple-realizability/.

Blanchette, P. (2012). The Frege-Hilbert controversy. *The Stanford Online Encyclopedia of Philosophy*. https://plato.stanford.edu/entries/frege-hilbert/.

Blatti, S. and Lapointe, S. (2016). *Ontology after Carnap*. Oxford University Press.

Boolos, G. S., Burgess, J. P., and Jeffrey, R. C. (2002). *Computability and Logic*. Cambridge University Press.

Borceux, F. (1994). *Handbook of Categorical Algebra*. Cambridge University Press.

Bourbaki, N. (1970). *Théorie des Ensembles*. Hermann.

Breiner, S. (2014). *Scheme Representation for First-Order Logic*. PhD thesis, Carnegie Mellon University.

Bueno, O. (2010). A defense of second-order logic. *Axiomathes*, 20(2-3):365–383.

Burgess, J. P. (1984). Synthetic mechanics. *Journal of Philosophical Logic*, 13(4):379–395.

Burgess, J. P. (2005). *Fixing Frege*. Princeton University Press.

Button, T. (2013). *The Limits of Realism*. Oxford University Press.

Butz, C. and Moerdijk, I. (1998). Representing topoi by topological groupoids. *Journal of Pure and Applied Algebra*, 130:223–235.

Carnap, R. (1928). *Der Logische Aufbau der Welt*. Springer Verlag.

Carnap, R. (1934). *Logische Syntax der Sprache*. Springer.

Carnap, R. (1935). *Philosophy and Logical Syntax*. Kegan Paul.

Carnap, R. (1950). Empiricism, semantics, and ontology. *Revue Internationale de Philosophie*, pages 20–40.

Carnap, R. (1956). The methodological character of theoretical concepts. In *The Foundations of Science and the Concepts of Psychology and Psychoanalysis*, pages 38–76. University of Minnesota Press.

Carnap, R. (1966). *Philosophical Foundations of Physics*. Basic Books.

Carnap, R. and Schilpp, P. A. (1963). *The Philosophy of Rudolf Carnap*. Cambridge University Press.

Chalmers, D., Manley, D., and Wasserman, R. (2009). *Metametaphysics: New Essays on the Foundations of Ontology*. Oxford University Press.

Coffa, A. (1986). From geometry to tolerance: Sources of conventionalism in nineteenth-century geometry. In Colodny, R., editor, *From Quarks to Quasars: Philosophical Problems of Modern Physics*, pages 3–70. University of Pittsburgh Press.

Coffa, J. A. (1993). *The Semantic Tradition from Kant to Carnap: To the Vienna Station*. Cambridge University Press.

Cori, R. and Lascar, D. (2000). *Mathematical Logic*. Oxford University Press.

Coxeter, H. S. M. (1955). The affine plane. *Scripta Mathematica*, 21:5–14.

Creath, R. and Friedman, M. (2007). *The Cambridge Companion to Carnap*. Cambridge University Press.

Cruse, P. and Papineau, D. (2002). Scientific realism without reference. In Marsonet, M., editor, *The Problem of Realism*, pages 174–189. Ashgate.

Curiel, E. (2014). Classical mechanics is Lagrangian; it is not Hamiltonian. *The British Journal for the Philosophy of Science*, 65(2):269–321.

Davidson, D. (1970). Mental events. In Foster, L. and Swanson, J. W., editors, *Essays on Actions and Events*, pages 107–119. Clarendon Press.

de Bouvére, K. L. (1965). Synonymous theories. In *Symposium on the Theory of Models*, pages 402–406. North-Holland Publishing Company.

Demopoulos, W. (2013). *Logicism and Its Philosophical Legacy*. Cambridge University Press.

Dewar, N. (2017b). Sophistication about symmetries. *The British Journal for the Philosophy of Science*.

Dewar, N. (2018a). On translating between two logics. *Analysis*, 78:622–630.

Dewar, N. (2018b). Supervenience, reduction, and translation. Preprint.

Dewar, N. (2019). Ramsey equivalence. *Erkenntnis*, 84(1):77–99.

Dicken, P. and Lipton, P. (2006). What can Bas believe? Musgrave and van Fraassen on observability. *Analysis*, 66(291):226–233.

Dizadji-Bahmani, F., Frigg, R., and Hartmann, S. (2010). Who's afraid of Nagelian reduction? *Erkenntnis*, 73(3):393–412.

Dorr, C. (2014). Quantifier variance and the collapse theorems. *The Monist*, 97(4):503–570.

Dukarm, J. J. (1988). Morita equivalence of algebraic theories. *Colloquium Mathematicae*, 55(1):11–17.

Dwinger, P. (1971). *Introduction to Boolean Algebras*. Physica-Verlag.

Eilenberg, S. and Mac Lane, S. (1942). Group extensions and homology. *Annals of Mathematics*, 43(4):757–831.

Eilenberg, S. and Mac Lane, S. (1945). General theory of natural equivalences. *Transactions of the American Mathematical Society*, 58:231–294.

Engelking, R. (1989). *General Topology*. Heldermann Verlag.

Feferman, S. (1974). Applications of many-sorted interpolation theorems. In *Proceedings of the Tarski Symposium*, volume 25, pages 205–223.

Fewster, C. J. (2015). Locally covariant quantum field theory and the problem of formulating the same physics in all spacetimes. *Philosophical Transactions of the Royal Society A*, 373(2047):20140238.

Field, H. (1980). *Science without Numbers*. Princeton University Press.

Fletcher, S. (2016). Similarity, topology, and physical significance in relativity theory. *British Journal for the Philosophy of Science*, 67(2):365–389.

Freyd, P. (1964). *Abelian Categories*. Harper and Row.

Friedman, H. M. and Visser, A. (2014). When bi-interpretability implies synonymy. *Logic Group Preprint Series*, 320:1–19.

Friedman, M. (1982). Review of *The Scientific Image*. *Journal of Philosophy*, 79(5):274–283.

Friedman, M. (1999). *Reconsidering Logical Positivism*. Cambridge University Press.

Friedman, M. (2011). Carnap on theoretical terms: Structuralism without metaphysics. *Synthese*, 180(2):249–263.

Frigg, R. and Votsis, I. (2011). Everything you always wanted to know about structural realism but were afraid to ask. *European Journal for Philosophy of Science*, 1(2):227–276.

Gajda, A., Krynicki, M., and Szczerba, L. (1987). A note on syntactical and semantical functions. *Studia Logica*, 46(2):177–185.

Givant, S. and Halmos, P. (2008). *Introduction to Boolean Algebras*. Springer.

Glymour, C. (1971). Theoretical realism and theoretical equivalence. In Buck, R. C. and Cohen, R. S., editors, *PSA: Proceedings of the Biennial Meeting of the Philosophy of Science Association*, pages 275–288. Springer.

Glymour, C. (1977). The epistemology of geometry. *Noûs*, 11:227–251.

Glymour, C. (1980). *Theory and Evidence*. Princeton University Press.

Gödel, K. (1929). *Über die Vollständigkeit des Logikkalküls*. PhD thesis, University of Vienna.

Goldblatt, R. (1987). *Orthogonality and Spacetime Geometry*. Springer.

Halmos, P. and Givant, S. (1998). *Logic as Algebra*. Cambridge University Press.

Halvorson, H. (2011). Natural structures on state space. *Manuscript*.

Halvorson, H. (2012). What scientific theories could not be. *Philosophy of Science*, 79(2): 183–206.

Halvorson, H. (2013). The semantic view, if plausible, is syntactic. *Philosophy of Science*, 80(3):475–478.

Halvorson, H. (2016). Scientific theories. In Humphreys, P., editor, *The Oxford Handbook of the Philosophy of Science*. Oxford University Press.

Harnik, V. (2011). Model theory vs. categorical logic: Two approaches to pretopos completion (aka T^{eq}). In Hart, B., editor, *Models, Logics, and Higher-Dimensional Categories*, page 79. American Mathematical Society.

Hawthorne, J. P. (2006). Plenitude, convention, and ontology. In *Metaphysical Essays*, pages 53–70. Oxford University Press.

Healey, R. (2007). *Gauging What's Real: The Conceptual Foundations of Contemporary Gauge Theories*. Oxford University Press.

Hellman, G. (1985). Determination and logical truth. *The Journal of Philosophy*, 82(11):607–616.

Hellman, G. P. and Thompson, F. W. (1975). Physicalism: Ontology, determination, and reduction. *The Journal of Philosophy*, 72(17):551–564.

Herrlich, H. (2006). *Axiom of Choice*. Springer.

Herrlich, H. and Keremedis, K. (2000). The Baire category theorem and choice. *Topology and Its Applications*, 108(2):157–167.

Hilbert, D. (1930). *Grundlagen der Geometrie*. Teubner.

Hirsch, E. (2011). *Quantifier Variance and Realism: Essays in Metaontology*. Oxford University Press.

Hirsch, E. and Warren, J. (2017). Quantifier variance and the demand for a semantics. *Philosophy and Phenomenological Research*.

Hodges, W. (1993). *Model Theory*. Cambridge University Press.

Hudetz, L. (2018a). Definable categorical equivalence. *Philosophy of Science*, 2019 86(1): 47–75.

Hudetz, L. (2018b). *The Logic of Scientific Theories*. PhD thesis, University of Salzburg.

Hudson, R. (2010). Carnap, the principle of tolerance, and empiricism. *Philosophy of Science*, 77(3):341–358.

Hylton, P. (2007). *Quine*. Routledge.

Johnstone, P. T. (1986). *Stone Spaces*. Cambridge University Press.

Johnstone, P. T. (2003). *Sketches of an Elephant: A Topos Theory Compendium*. Oxford University Press.

Kanger, S. (1968). Equivalent theories. *Theoria*, 34(1):1–6.

Keisler, H. J. (2010). The ultraproduct construction. www.math.wisc.edu/keisler/ultraproducts-web-final.pdf.

Ketland, J. (2004). Empirical adequacy and Ramsification. *The British Journal for the Philosophy of Science*, 55(2):287–300.

Kleene, S. C. (1952). *Introduction to Metamathematics*. van Nostrand.

Knox, E. (2014). Newtonian spacetime structure in light of the equivalence principle. *The British Journal for the Philosophy of Science*, 65(4):863–880.

Koppelberg, S. (1989). General theory of Boolean algebras. In Monk, J. and Bonnet, R., editors, *Handbook of Boolean Algebras*, volume 3. North-Holland.

Kuratowski, K. (1966). *Topology*. Academic Press.

Ladyman, J. (2014). Structural realism. *Stanford Online Encyclopedia of Philosophy*. https://plato.stanford.edu/entries/structural-realism/.

Lawvere, F. W. (1964). An elementary theory of the category of sets. *Proceedings of the National Academy of Sciences*, 52(6):1506–1511.

Lawvere, F. W. and Rosebrugh, R. (2003). *Sets for Mathematics*. Cambridge University Press.

Leinster, T. (2014). Rethinking set theory. *American Mathematical Monthly*, 121(5):403–415.

Leitgeb, H. (2011). Logic in general philosophy of science: Old things and new things. *Synthese*, 179(2):339–350.

Lewis, D. (1966). An argument for the identity theory. *The Journal of Philosophy*, 63(1):17–25.

Lewis, D. (1970). How to define theoretical terms. *The Journal of Philosophy*, 67(13):427–446.

Lewis, D. (1972). Psychophysical and theoretical identifications. *Australasian Journal of Philosophy*, 50(3):249–258.

Lewis, D. (1984). Putnam's paradox. *Australasian Journal of Philosophy*, 62(3):221–236.

Lewis, D. (1994). Reduction of mind. In Guttenplan, S., editor, *Companion to the Philosophy of Mind*, 412–431. Blackwell.

Lloyd, E. (1984). *A Semantic Approach to the Structure of Evolutionary Theory*. PhD thesis, Princeton University.

Love, A. C. and Hüttemann, A. (2016). Reduction. In Humphreys, P., editor, *The Oxford Handbook of Philosophy of Science*. Oxford University Press.

Mac Lane, S. (1948). Groups, categories and duality. *Proceedings of the National Academy of Sciences*, 34(6):263–267.

Mac Lane, S. (1971). *Categories for the Working Mathematician*. Springer.

Makkai, M. (1985). Ultraproducts and categorical logic. In *Methods in Mathematical Logic*, pages 222–309. Springer.

Makkai, M. (1987). Stone duality for first order logic. *Advances in Mathematics*, 65(2):97–170.

Makkai, M. (1991). *Duality and Definability in First Order Logic*. American Mathematical Society.

Makkai, M. (1995). First order logic with dependent sorts with applications to category theory. www.math.mcgill.ca/makkai/folds/foldsinpdf/FOLDS.pdf.

Makkai, M. and Reyes, G. E. (1977). *First Order Categorical Logic*. Springer.

Manes, E. G. (1976). *Algebraic Theories*. Springer.

Manzano, M. (1993). Introduction to many-sorted logic. In Meinke, K. and Tucker, J., editors, *Many-Sorted Logic and Its Applications*, pages 3–86. Wiley.

Manzano, M. (1996). *Extensions of First-Order Logic*. Cambridge University Press.

Marker, D. (2006). *Model Theory: An Introduction*. Springer.

Maxwell, G. (1962). The ontological status of theoretical entities. In Feigl, H. and Maxwell, G., editors, *Scientific Explanation, Space, and Time*, pages 3–27. University of Minnesota Press.

McLaughlin, B. and Bennett, K. (2018). Supervenience. *Stanford Online Encyclopedia of Philosophy*. https://plato.stanford.edu/entries/supervenience/.

McSweeney, M. (2016a). An epistemic account of metaphysical equivalence. *Philosophical Perspectives*, 30(1):270–293.

McSweeney, M. (2016b). *The Metaphysical Basis of Logic*. PhD thesis, Princeton University.

Melia, J. and Saatsi, J. (2006). Ramseyfication and theoretical content. *The British Journal for the Philosophy of Science*, 57(3):561–585.

Menzies, P. and Price, H. (2009). Is semantics in the plan? In Braddon-Mitchell, D. and Nola, R., editors, *Conceptual Analysis and Philosophical Naturalism*, pages 159–182. MIT Press.

Mere, M. C. and Veloso, P. (1992). On extensions by sorts. *Monografias em Ciências da Computaçao, DI, PUC-Rio*, 38:92.

Moerdijk, I. and Vermeulen, J. (1999). Proof of a conjecture of A. Pitts. *Journal of Pure and Applied Algebra*, 143(1-3):329–338.

Monk, J. D. (2014). The mathematics of Boolean algebras. https://plato.stanford.edu/entries/boolalg-math/.

Munkres, J. R. (2000). *Topology*. Prentice Hall.

Myers, D. (1997). An interpretive isomorphism between binary and ternary relations. In *Structures in Logic and Computer Science*, pages 84–105. Springer.

Nagel, E. (1935). The logic of reduction in the sciences. *Erkenntnis*, 5(1):46–52.

Nagel, E. (1961). *The Structure of Science*. Harcourt, Brace, and World, Inc.

Nestruev, J. (2002). *Smooth Manifolds and Observables*. Springer.

Newman, M. H. (1928). Mr. Russell's "causal theory of perception". *Mind*, 37(146):137–148.

North, J. (2009). The "structure" of physics: A case study. *The Journal of Philosophy*, 106:57–88.

Park, W. (2012). Friedman on implicit definition: In search of the Hilbertian heritage in philosophy of science. *Erkenntnis*, 76(3):427–442.

Pearce, D. (1985). *Translation, reduction and equivalence*. Peter Lang, Frankfurt.

Pelletier, F. J. and Urquhart, A. (2003). Synonymous logics. *Journal of Philosophical Logic*, 32(3):259–285.

Petrie, B. (1987). Global supervenience and reduction. *Philosophy and Phenomenological Research*, 48(1):119–130.

Pinter, C. C. (1978). Properties preserved under definitional equivalence and interpretations. *Mathematical Logic Quarterly*, 24(31-36):481–488.

Poizat, B. (2012). *A Course in Model Theory*. Springer.

Price, H. (2009). Metaphysics after Carnap: The ghost who walks. In Chalmers, D., Manley, D., and Wasserman, R., editors, *Metametaphysics: New Essays on the Foundations of Ontology*, pages 320–346. Oxford University Press.

Psillos, S. (2000). Carnap, the Ramsey-sentence and realistic empiricism. *Erkenntnis*, 52(2): 253–279.

Psillos, S. (2006). Ramsey's Ramsey-sentences. In Galavotti, M., editor, *Cambridge and Vienna: Vienna Circle Institute Yearbook*, pages 67–90. Springer.

Putnam, H. (1962). What theories are not. In Nagel, E., Suppes, P., and Tarski, A., editors, *Logic, Methodology and Philosophy of Science: Proceedings of the 1960 International Congress*, pages 240–251. Stanford University Press.

Putnam, H. (1977). Realism and reason. In *Proceedings and Addresses of the American Philosophical Association*, volume 50, pages 483–498.

Putnam, H. (1980). Models and reality. *The Journal of Symbolic Logic*, 45(3):464–482.

Putnam, H. (1992). *Renewing Philosophy*. Harvard University Press.

Putnam, H. (2001). Reply to Jennifer Case. *Revue Internationale de Philosophie*, 4(218).

Quine, W. V. (1937). New foundations for mathematical logic. *American Mathematical Monthly*, pages 70–80.

Quine, W. V. (1938). On the theory of types. *The Journal of Symbolic Logic*, 3(04):125–139.

Quine, W. V. (1951a). On Carnap's views on ontology. *Philosophical Studies*, 2(5):65–72.

Quine, W. V. (1951b). Two dogmas of empiricism. *The Philosophical Review*, 60:20–43.

Quine, W. V. (1956). Unification of universes in set theory. *The Journal of Symbolic Logic*, 21(03):267–279.

Quine, W. V. (1960). *Word and Object*. MIT.

Quine, W. V. (1963). *Set Theory and Its Logic*. Harvard University Press.

Quine, W. V. (1964). Implicit definition sustained. *The Journal of Philosophy*, 61(2):71–74.

Quine, W. V. (1975). On empirically equivalent systems of the world. *Erkenntnis*, 9(3):313–328.

Quine, W. V. (1976). *The Ways of Paradox, and Other Essays*. Harvard University Press.

Quine, W. V. and Goodman, N. (1940). Elimination of extra-logical postulates. *The Journal of Symbolic Logic*, 5(3):104–109.

Ramsey, F. P. (1929). Theories. In *F.P. Ramsey Philosophical Papers*. Cambridge University Press.

Rasiowa, H. and Sikorski, R. (1950). A proof of the completeness theorem of Gödel. *Fundamenta Mathematicae*, 37(1):193–200.

Rasiowa, H. and Sikorski, R. (1963). *The Mathematics of Metamathematics*. Pańistwow Wydaawnictwo Naukowe.

Restall, G. (2002). *An Introduction to Substructural Logics*. Routledge.

Ribes, L. and Zalesskii, P. (2000). *Profinite Groups*. Springer.

Rieffel, M. A. (1974). Morita equivalence for C^*-algebras and W^*-algebras. *Journal of Pure and Applied Algebra*, 5:51–96.

Robinson, R. (1959). Binary relations as primitive notions in elementary geometry. In Henkin, L., Suppes, P., and Tarski, A., editors, *The Axiomatic Method with Special Reference to Geometry and Physics*, pages 68–85. North-Holland.

Rooduijn, J. (2015). Translating theories. Bachelor's Thesis, Universiteit Utrecht.

Rosenstock, S., Barrett, T. W., and Weatherall, J. O. (2015). On Einstein algebras and relativistic spacetimes. *Studies in History and Philosophy of Modern Physics*, 52:309–316.

Rosenstock, S. and Weatherall, J. O. (2016). A categorical equivalence between generalized holonomy maps on a connected manifold and principal connections on bundles over that manifold. *Journal of Mathematical Physics*, 57(10):102902.

Royden, H. L. (1959). Remarks on primitive notions for elementary Euclidean and non-Euclidean plane geometry. In Henkin, L., Suppes, P., and Tarski, A., editors, *The Axiomatic Method with Special Reference to Geometry and Physics*, pages 86–96. North-Holland.

Russell, B. (1901). Mathematics and the metaphysicians. In *Mysticism and Logic*, pages 57–74. Dover.

Russell, B. (1914a). Logic as the essence of philosophy. In *Our Knowledge of the External World*, pages 26–48. Routledge.

Russell, B. (1914b). On the scientific method in philosophy. In *Mysticism and Logic*, pages 75–96. Dover.

Sarkar, S. (2015). Nagel on reduction. *Studies in History and Philosophy of Science*, 53:43–56.

Scheibe, E. (2013). *Die Reduktion physikalischer Theorien: Ein Beitrag zur Einheit der Physik*. Springer-Verlag.

Schlick, M. (1918). *Allgemeine Erkenntnislehre*. Springer.

Schmidt, A. (1951). Die Zulässigkeit der Behandlung mehrsortiger Theorien mittels der üblichen einsortigen Prädikatenlogik. *Mathematische Annalen*, 123(1):187–200.

Schwabhäuser, W. and Szczerba, L. (1975). Relations on lines as primitive notions for Euclidean geometry. *Fundamenta Mathematicae*, 82(4):347–355.

Schwabhäuser, W., Szmielew, W., and Tarski, A. (1983). *Metamathematische Methoden in der Geometrie*. Springer.

Scott, D. (1956). A symmetric primitive notion for Euclidean geometry. *Indagationes Mathematicae*, 18:457–461.

Shapiro, S. (1991). *Foundations without foundationalism: A case for second-order logic*. Clarendon Press.

Shoemaker, S. (1981). Some varieties of functionalism. *Philosophical Topics*, 12(1):93–119.

Sider, T. (2009). Ontological realism. In Chalmers, D., Manley, D., and Wasserman, R., editors, *Metametaphysics*, pages 384–423. Oxford University Press.

Sider, T. (2013). *Writing the Book of the World*. Oxford University Press.

Sikorski, R. (1969). *Boolean Algebras*. Springer.

Soames, S. (2014). *The Analytic Tradition in Philosophy*. Princeton University Press.

Suppe, F. (1974). *The Structure of Scientific Theories*. University of Illinois Press, Urbana, Illinois.

Suppe, F. (1989). *The Semantic Conception of Theories and Scientific Realism*. University of Illinois Press.

Suppe, F. (2000). Understanding scientific theories: An assessment of developments, 1969-1998. *Philosophy of Science*, pages S102–S115.

Svenonius, L. (1959). A theorem on permutations in models. *Theoria*, 25(3):173–178.

Swanson, N. and Halvorson, H. (2012). On North's "The structure of physics". *Manuscript*.

Szczerba, L. (1977). Interpretability of elementary theories. In *Logic, Foundations of Mathematics, and Computability Theory*, pages 129–145. Springer.

Szczerba, L. (1986). Tarski and geometry. *The Journal of Symbolic Logic*, 51(4).

Szczerba, L. and Tarski, A. (1979). Metamathematical discussion of some affine geometries. *Fundamenta Mathematicae*, 104(3):155–192.

Tarski, A. (1929). Les fondements de la géométrie des corps. *Ksiega Pamiatkowa Pierwszego Polskiego Zjazdu Matematycznego*, pages 29–33.

Tarski, A. (1956). A general theorem concerning primitive notions of Euclidean geometry. *Indagationes Mathematicae*, 18:468–474.

Tarski, A. (1959). What is elementary geometry? In Henkin, L., Suppes, P., and Tarski, A., editors, *The Axiomatic Method with Special Reference to Geometry and Physics*, pages 16–29. North-Holland.

Tennant, N. (1985). Beth's theorem and reductionism. *Pacific Philosophical Quarterly*, 66(3-4):342–354.

Tennant, N. (2015). *Introducing Philosophy: God, Mind, World, and Logic*. Routledge.

Tsementzis, D. (2017a). First-order logic with isomorphism. https://arxiv.org/abs/1603.03092.

Tsementzis, D (2017b). A syntactic characterization of Morita equivalence. *Journal of Symbolic Logic*, 82(4):1181–1198.

Tuomela, R. (1973). *Theoretical Concepts*. Springer.

Turner, J. (2010). Ontological pluralism. *The Journal of Philosophy*, 107(1):5–34.

Turner, J. (2012). Logic and ontological pluralism. *Journal of Philosophical Logic*, 41(2): 419–448.

Uebel, T. (2011). Carnap's ramseyfications defended. *European Journal for Philosophy of Science*, 1(1):71–87.

van Benthem, J. (1982). The logical study of science. *Synthese*, 51(3):431–472.

van Benthem, J. and Pearce, D. (1984). A mathematical characterization of interpretation between theories. *Studia Logica*, 43(3):295–303.

van Fraassen, B. (1976). To save the phenomena. *The Journal of Philosophy*, 73(18):623–632.

van Fraassen, B. (1980). *The Scientific Image*. Oxford University Press.

van Fraassen, B. (1989). *Laws and Symmetry*. Oxford University Press.

van Fraassen, B. (1997). Putnam's paradox: Metaphysical realism revamped and evaded. *Noûs*, 31(s11):17–42.

van Fraassen, B. (2008). *Scientific Representation: Paradoxes of Perspective*. Oxford University Press.

van Fraassen, B. (2011). Logic and the philosophy of science. *Journal of the Indian Council of Philosophical Research*, 27:45–66.

van Inwagen, P. (2009). Being, existence, and ontological commitment. In Chalmers, D., Manley, D., and Wasserman, R., editors, *Metametaphysics*, pages 472–506. Oxford University Press.

van Oosten, J. (2002). Basic category theory. www.staff.science.uu.nl/~ooste110/syllabi/catsmoeder.pdf.

van Riel, R. and van Gulick, R. (2014). Scientific reduction. *Stanford Encyclopedia of Philosophy*. https://plato.stanford.edu/entries/scientific-reduction/.

Veblen, O. and Young, J. W. (1918). *Projective Geometry*, volume 2. Ginn and Company.

Visser, A. (2006). Categories of theories and interpretations. In *Logic in Tehran. Proceedings of the Workshop and Conference on Logic, Algebra and Arithmetic, Held October 18–22, 2003*. ASL.

Warren, J. (2014). Quantifier variance and the collapse argument. *The Philosophical Quarterly*, 65(259):241–253.

Washington, E. (2018). On the equivalence of logical theories. Bachelor's Thesis, Princeton University.

Weatherall, J. O. (2016a). Are Newtonian gravitation and geometrized Newtonian gravitation theoretically equivalent? *Erkenntnis*, 81(5):1073–1091.

Weatherall, J. O. (2016b). Regarding the "hole argument". *The British Journal for the Philosophy of Science*, pages 1–22.

Weatherall, J. O. (2016c). Understanding gauge. *Philosophy of Science*, 83(5):1039–1049.

Weatherall, J. O. (2018). Categories and the foundations of classical field theories. In Landry, E., editor, *Categories for the Working Philosopher*. Oxford University Press.

Willard, S. (1970). *General Topology*. Dover.

Winnie, J. A. (1967). The implicit definition of theoretical terms. *The British Journal for the Philosophy of Science*, 18(3):223–229.

Winnie, J. A. (1986). Invariants and objectivity: A theory with applications to relativity and geometry. In Colodny, R., editor, *From Quarks to Quasars*. University of Pittsburgh Press.

Worrall, J. (1989). Structural realism: The best of both worlds? *Dialectica*, 43(1-2):99–124.

Worrall, J. and Zahar, E. (2001). Ramseyfication and structural realism. In *Poincaré's Philosophy*, pages 236–251. Open Court.

Zahar, E. (2004). Ramseyfication and structural realism. *Theoria. Revista de Teoría, Historia y Fundamentos de la Ciencia*, 19(1):5–30.

Zawadowski, M. W. (1995). Descent and duality. *Annals of Pure and Applied Logic*, 71(2): 131–188.

Index